Nutrition at the Interface of Sleep and Circadian Rhythms: Implications for Health

Nutrition at the Interface of Sleep and Circadian Rhythms: Implications for Health

Editors

Egeria Scoditti
Sergio Garbarino

MDPI • Basel • Beijing • Wuhan • Barcelona • Belgrade • Manchester • Tokyo • Cluj • Tianjin

MDPI

Editors
Egeria Scoditti
National Research Council—Institute of Clinical Physiology
Lecce Section
Laboratory of Vascular Biology and Nutrigenomics
Italy

Sergio Garbarino
University of Genoa
Italy

Editorial Office
MDPI
St. Alban-Anlage 66
4052 Basel, Switzerland

This is a reprint of articles from the Special Issue published online in the open access journal *Nutrients* (ISSN 2072-6643) (available at: https://www.mdpi.com/journal/nutrients/special_issues/Nutrition_Circadian).

For citation purposes, cite each article independently as indicated on the article page online and as indicated below:

LastName, A.A.; LastName, B.B.; LastName, C.C. Article Title. *Journal Name* **Year**, *Volume Number*, Page Range.

ISBN 978-3-0365-6708-2 (Hbk)
ISBN 978-3-0365-6709-9 (PDF)

Contents

About the Editors

Egeria Scoditti

Egeria Scoditti is a permanent researcher at the Institute of Clinical Physiology (IFC) of the National Research Council (CNR), Lecce and Pisa, Italy. She holds a degree in Biology and a PhD in Innovative Strategies in Biomedical Research. She also has a specialization in Clinical Biochemistry. Her research focuses on the molecular basis of atherosclerosis, obesity and type 2 diabetes, lipid and glucose metabolism, immunity, circadian rhythms and sleep–wake regulation. The role of natural compounds and nutrients from the Mediterranean diet in the prevention and treatment of chronic degenerative diseases with a focus on gene–diet interactions is also studied. She is author of more than 78 research papers and several book chapters, with more than 3300 citations.

Sergio Garbarino

Sergio Garbarino is Adjunct Professor at the Department of Neurosciences (DINOGMI), University of Genoa, and at the Department of Woman, Child and Public Health, University Cattolica del Sacro Cuore, Rome (Italy). He holds a degree in Medicine and Surgery, a specialization in Neurology, in Neurophysiopathology, and in Occupational Medicine, and a PhD in Neuroscience and in Neurophysiopathology. He is a Board Member of the Italian Association of Sleep Medicine. He was member of Ministerial Commissions in Italy and the European Union on issues related to sleep disorders and public health for the drafting of decree laws. His interest is on sleep, sleep–wake rhythm disorders and neuropsychological adaptation, with a focus on the related safety and health implications in the general population and at-risk workers. He is the author of more than 160 research papers and book chapters and the editor of several books. His work has been cited over 3830 times.

Preface to "Nutrition at the Interface of Sleep and Circadian Rhythms: Implications for Health"

The bidirectional interaction of nutrition with central and peripheral endogenous circadian clocks and with other rhythmic behaviors, especially the sleep–wake cycle, has been often overlooked in research and preventive medicine; however, it represents a novel frontier for health maintenance and disease prevention. Nutrient composition, meal timing and its entrainment in the circadian sleep–wake cycle are integral parts of physiology. Indeed, an unbalanced diet and irregular meal timing significantly impact on sleep quantity, quality, and timing; contrarily, sleep patterns interact with the quality and timings of someone's dietary intake. These aspects are crucially important in today's society, which is increasingly characterized by night shifts and social jetlag and, as a consequence, by unhealthy nutrition, sleep deprivation and circadian misalignment, increasing the risk of major chronic diseases including cardiometabolic, neurodegenerative, cognitive and oncologic diseases. This Special Issue, "Nutrition at the Interface of Sleep and Circadian Rhythms: Implications for Health", comprises 12 manuscripts (reviews and research articles) highlighting the latest research on the interaction of nutrition with sleep and circadian rhythms as well as the related health implications. It aims at providing novel knowledge, contributing to the maintenance of public and individual wellbeing, performance and health, as well as the prevention of major chronic diseases.

Egeria Scoditti and Sergio Garbarino
Editors

nutrients

MDPI

Editorial

Nutrition, Sleep, Circadian Rhythms, and Health Implications: "Come Together"

Egeria Scoditti [1,*] and Sergio Garbarino [2]

1 National Research Council (CNR), Institute of Clinical Physiology (IFC), 73100 Lecce, Italy
2 Department of Neuroscience, Rehabilitation, Ophthalmology, Genetics and Maternal/Child Sciences (DINOGMI), University of Genoa, 16132 Genoa, Italy
* Correspondence: egeria.scoditti@ifc.cnr.it

Over the last few years, novel and important aspects of nutrition that are often overlooked in nutritional epidemiology, experimental research, and recommendations for health maintenance and disease prevention concerning the circadian rhythmicity of feeding, as well as the bidirectional interaction of nutrition with central and peripheral endogenous circadian clocks, and with other rhythmic behaviors including the sleep–wake cycle have received increasing attention from the research community. Besides total energy intake and diet composition, the daily rhythms of eating patterns are an integral part of homeostasis [1]. Feeding is a resetting and entraining cue for the circadian clock [2]. Unbalanced nutrition and irregular meal timing patterns, which occur in today's modern 24/7 society, especially due to night shifts or social jet lag, are risk factors for circadian disruptions in physiological processes, including metabolism, neurologic, immune, cardiovascular, and endocrine functions, among others, thus enhancing the risk for numerous chronic diseases [1]. Furthermore, dietary intake and timing have a significant impact on sleep quality and duration and can therefore increase the risk of sleep disturbance and related health outcomes [3]. Conversely, sleep patterns influence eating behavior, diet quality, and total energy intake [4].

The importance of these aspects become more evident when considering the significant implications for public health, disease prevention, and disease management that arise from the altered crosstalk between diet, sleep, and the circadian system due to genetics, unhealthy behaviors, (patho)physiological states, and/or environmental factors. Notwithstanding this, the integrated role of diet, sleep, and circadian rhythms is not accounted for in public health policy or in health promotion plans, thus representing a current gap that should be filled.

This Special Issue, entitled "Nutrition at the Interface of Sleep and Circadian Rhythms: Implications for Health" focused on the interaction as well as the related health implications of nutrition with sleep and circadian rhythms, gathering 12 papers, including eight original research articles and four reviews.

The prospective study performed by Al-Musharaf et al. [5] investigated sleep patterns among a cohort of 140 Saudi women aged 18–39 years old throughout all three trimesters of pregnancy. They found worsening sleep quality and a shortened sleep duration from early to late pregnancy, with low socio-economic status, low serum vitamin D levels, greater energy intake, and sitting time as significant and independent predictors of worsening changes in sleep patterns during pregnancy.

The study by Fenton et al. [6] was a three-armed randomized controlled trial in 116 adults (70% female, 44.5 years) with overweight and obesity exploring the efficacy of a multi-component m-Health weight-loss intervention (enhanced) targeting sleep, diet, and physical activity to improve dietary intake over 6 months, and after a 12-month follow-up, compared to a waitlisted control group and the traditional dietary and physical activity intervention, the multi-component weight-loss intervention resulted in reduced total energy

Citation: Scoditti, E.; Garbarino, S. Nutrition, Sleep, Circadian Rhythms, and Health Implications: "Come Together". *Nutrients* **2022**, *14*, 5105. https://doi.org/10.3390/nu14235105

Received: 14 November 2022
Accepted: 15 November 2022
Published: 1 December 2022

and sodium intake as well as increased fruit intake in adults at six months compared to the control group. Importantly, the enhanced intervention group improved their dietary intake relative to the traditional group at 12 months, with a higher intake of nutrient-dense foods and protein and a lower intake of energy-dense nutrient-poor foods.

Another prospective study conducted in 607 individuals aged 60 years of age or older found that a higher energy intake at dinner compared to during other eating occasions was a risk factor for the development of metabolic syndrome, mostly due to abdominal obesity and hypertriglyceridemia [7]. This study confirms the impact of meal timing on cardiometabolic health.

The short-term longitudinal study by Mathew et al. [8] assessed the bidirectional relationship between caffeinated beverage consumption and objective and subjective sleep in adolescents. Adolescents with a more variable sleep duration and midpoint were found to have higher average odds of consuming caffeinated beverages on a given day. The consumption of one or more caffeinated beverages predicted later sleep onset that night and a later wake time the next morning versus no consumption. These data could be important when considering that adolescents commonly report poor sleep patterns and increasingly adopt the habit of consuming caffeine, which may have a significant influence on sleep health.

Another paper [9] pertains the effect of energy balance on sleep in adolescents. In that study, 28 male adolescents, 14 with obesity and 14 normal-weight age-matched controls, underwent an experimental protocol that included a controlled balanced diet adjusted to energy requirements (eucaloric) and a diet offered ad libitum for three days in random order, after which sleep was measured by polysomnography. The results indicate that the eucaloric diet was able to improve sleep features, including sleep latency and N1 stage, in adolescents with obesity compared to the ad libitum diet. Interestingly, sleep improvements occurred in the absence of any substantial modification in macronutrient proportions and were correlated to reduced energy intake, especially during the evening meal.

The study by Strojny et al. [10] pointed to the effect of chronotype, a measure of the interindividual variability in circadian rhythmicity as reflected in the preferred timing of the sleep–wake cycle, on the effectiveness of a 3-week weight loss intervention (caloric restriction) in 131 adults with obesity. At the end of the intervention, both the morning and evening chronotypes reported a similar body weight and body mass index (BMI) reduction, though the evening chronotype, who had earlier meal and sleep times than their usual habits imposed on them, experienced a tendency towards greater losses in body fat compared to the morning chronotype.

Another randomized cross-over trial in female adolescents with the habit of skipping breakfast tested the effects of breakfast consumption compared to breakfast omission for seven days on free-living physical activity energy expenditure and dietary intake [11]. Physical activity and energy expenditure during physical activity did not differ between breakfast consumption and breakfast omission, and the total daily energy intake was almost identical between the conditions. However, breakfast omission was associated with an increased perceived morning appetite and a tendency for increased carbohydrate intake as an energy-intake compensation behavior. Consuming breakfast, on the contrary, contributed to higher fiber intake in breakfast-skipping adolescent girls.

Pacifico et al. [12] analyzed the effects of dietary patterns naturally rich in photosensitizers, such as vegan and vegetarian diets, on skin sensitivity to narrow-band ultraviolet B (NB-UVB) phototherapy in 119 adults suffering from psoriasis, a chronic inflammatory skin disease that is also characterized by circadian rhythmicity disorders. They found that vegan and vegetarian diets and, in particular, the intake of furocumarins are associated with greater skin sensitivity to NB-UVB phototherapy in psoriatic patients compared to omnivores, thus limiting the total number of phototherapy sessions needed. The study points to an important diet-related effect on psoriasis treatment, with implications for patient management and clinical outcomes.

Accordingly, among the published reviews, Controne et al. [13] analyzed the bidirectional associations of psoriasis with diet as well as sleep. Scientific evidence was discussed regarding the effect of unhealthy diets, mainly the Western diet, and of sleep disorders on psoriasis risk, development, and clinical outcomes, with the analysis of potential underlying common mechanisms involving an alteration of immune-mediated responses and chronic inflammation. The review draws attention to the importance of patient lifestyles and sleep patterns in the prevention and/or treatment of psoriasis.

In the review by Sejbuk et al. [14], modifiable factors affecting sleep quality were discussed, addressing, in particular, the role of nutrition, stimulants, and physical activity. The gathered data suggest that proper nutrition, in terms of caloric intake, micro- and macronutrient composition and balance as well as eating timing, may benefit sleep quality, while stimulants, including alcohol, nicotine, caffeine, and cannabis, negatively influence sleep quality. Physical activity and, in particular, a sufficient amount of moderate- to high-intensity exercise, preferably not performed in the late evening, emerges as an important player in improving sleep quality and preventing insomnia. Pending future research addressing individual lifestyle factors could be a strategy to significantly modulate sleep and, through this, to impact wellbeing and disease prevention.

In line with this notion, Scoditti et al. [15] synthetized evidence from the literature on the effect of the Mediterranean diet, one of the most studied and healthful dietary patterns, on sleep features in healthy individuals. Evidence, though limited to epidemiological studies, converge upon a benefit of the Mediterranean diet on sleep, promoting adequate sleep duration and quality and preventing sleep disturbances. The plausible mediating role of the antioxidant, anti-inflammatory, immunomodulatory, and neuroprotective properties of the foods and nutrients in the Mediterranean diet as well as the modulation of the gut microbiota in sleep improvement was described. The evidence further strengthens the notion that diet may influence sleep and could contribute to sleep hygiene.

Garbarino et al. [16] reviewed evidence from the literature regarding the role of nutrition (composition and timing) as a (de)synchronizer of the internal circadian clock in peripheral tissues, with the possible role of chrononutrition in contrasting the effects of circadian misalignment and sleep deprivation, which prevails in modern society, mainly due to shift work schedules and social jet lag. As an excellent example, chocolate, an energy-dense nutrient-rich food containing high levels of flavonoids, was discussed regarding its beneficial effect on mental and cognitive functions, the cardiovascular system, and metabolism as well as its attenuation of the stress- and sleep-deprivation-induced alteration of the circadian sleep–wake rhythm, especially when consumed at breakfast and generally during the active phase.

This Special Issue provides novel knowledge on the complex interplay of diet, sleep, and circadian rhythms and contributes to spurring further research as well as elaborate novel directions for improving public and individual wellbeing and health and strategies for the prevention of major diseases.

Author Contributions: Conceptualization, E.S. and S.G.; methodology, E.S.; investigation, E.S.; writing—original draft preparation, E.S.; writing—review and editing, E.S. and S.G. All authors have read and agreed to the published version of the manuscript.

Funding: This research received no external funding.

Informed Consent Statement: Not applicable.

Data Availability Statement: Not applicable.

Conflicts of Interest: The authors declare no conflict of interest.

References

1. Pickel, L.; Sung, H.K. Feeding Rhythms and the Circadian Regulation of Metabolism. *Front. Nutr.* **2020**, *7*, 39. [CrossRef] [PubMed]
2. Hatori, M.; Vollmers, C.; Zarrinpar, A.; DiTacchio, L.; Bushong, E.A.; Gill, S.; Leblanc, M.; Chaix, A.; Joens, M.; Fitzpatrick, J.A.; et al. Time-restricted feeding without reducing caloric intake prevents metabolic diseases in mice fed a high-fat diet. *Cell Metab.* **2012**, *15*, 848–860. [CrossRef] [PubMed]
3. St-Onge, M.P.; Zuraikat, F.M. Reciprocal Roles of Sleep and Diet in Cardiovascular Health: A Review of Recent Evidence and a Potential Mechanism. *Curr. Atheroscler Rep.* **2019**, *21*, 11. [CrossRef]
4. Chaput, J.P. Sleep patterns, diet quality and energy balance. *Physiol. Behav.* **2014**, *134*, 86–91. [CrossRef] [PubMed]
5. Al-Musharaf, S. Changes in Sleep Patterns during Pregnancy and Predictive Factors: A Longitudinal Study in Saudi Women. *Nutrients* **2022**, *14*, 2633. [CrossRef] [PubMed]
6. Fenton, S.; Burrows, T.L.; Collins, C.E.; Rayward, A.T.; Murawski, B.; Duncan, M.J. Efficacy of a Multi-Component m-Health Diet, Physical Activity, and Sleep Intervention on Dietary Intake in Adults with Overweight and Obesity: A Randomised Controlled Trial. *Nutrients* **2021**, *13*, 2468. [CrossRef] [PubMed]
7. Hermenegildo-Lopez, Y.; Donat-Vargas, C.; Sandoval-Insausti, H.; Moreno-Franco, B.; Rodriguez-Ayala, M.; Rey-Garcia, J.; Banegas, J.R.; Rodriguez-Artalejo, F.; Guallar-Castillon, P. A Higher Intake of Energy at Dinner Is Associated with Incident Metabolic Syndrome: A Prospective Cohort Study in Older Adults. *Nutrients* **2021**, *13*, 3035. [CrossRef] [PubMed]
8. Mathew, G.M.; Reichenberger, D.A.; Master, L.; Buxton, O.M.; Chang, A.M.; Hale, L. Too Jittery to Sleep? Temporal Associations of Actigraphic Sleep and Caffeine in Adolescents. *Nutrients* **2021**, *14*, 31. [CrossRef] [PubMed]
9. Saidi, O.; Rochette, E.; Del Sordo, G.; Dore, E.; Merlin, E.; Walrand, S.; Duche, P. Eucaloric Balanced Diet Improved Objective Sleep in Adolescents with Obesity. *Nutrients* **2021**, *13*, 3550. [CrossRef] [PubMed]
10. Strojny, Z.; Rutkowski, R.; Kanikowska, A.; Zawada, A.; Juchacz, A.; Grzymislawski, M.; Sato, M.; Litwinowicz, M.; Korybalska, K.; Breborowicz, A.; et al. No Significant Effect of the Individual Chronotype on the Result of Moderate Calorie Restriction for Obesity-A Pilot Study. *Nutrients* **2021**, *13*, 4089. [CrossRef] [PubMed]
11. Zakrzewski-Fruer, J.K.; Seall, C.; Tolfrey, K. Breakfast Consumption Suppresses Appetite but Does Not Increase Daily Energy Intake or Physical Activity Energy Expenditure When Compared with Breakfast Omission in Adolescent Girls Who Habitually Skip Breakfast: A 7-Day Randomised Crossover Trial. *Nutrients* **2021**, *13*, 4261. [CrossRef] [PubMed]
12. Pacifico, A.; Conic, R.R.Z.; Cristaudo, A.; Garbarino, S.; Ardigo, M.; Morrone, A.; Iacovelli, P.; di Gregorio, S.; Pigatto, P.D.M. Grada, A.; et al. Diet-Related Phototoxic Reactions in Psoriatic Patients Undergoing Phototherapy: Results from a Multicenter Prospective Study. *Nutrients* **2021**, *13*, 2934. [CrossRef] [PubMed]
13. Controne, I.; Scoditti, E.; Buja, A.; Pacifico, A.; Kridin, K.; Fabbro, M.D.; Garbarino, S.; Damiani, G. Do Sleep Disorders and Western Diet Influence Psoriasis? A Scoping Review. *Nutrients* **2022**, *14*, 4324. [CrossRef]
14. Sejbuk, M.; Mironczuk-Chodakowska, I.; Witkowska, A.M. Sleep Quality: A Narrative Review on Nutrition, Stimulants, and Physical Activity as Important Factors. *Nutrients* **2022**, *14*, 1912. [CrossRef] [PubMed]
15. Scoditti, E.; Tumolo, M.R.; Garbarino, S. Mediterranean Diet on Sleep: A Health Alliance. *Nutrients* **2022**, *14*, 2998. [CrossRef]
16. Garbarino, S.; Garbarino, E.; Lanteri, P. Cyrcadian Rhythm, Mood, and Temporal Patterns of Eating Chocolate: A Scoping Review of Physiology, Findings, and Future Directions. *Nutrients* **2022**, *14*, 3113. [CrossRef] [PubMed]

nutrients

Article

Efficacy of a Multi-Component m-Health Diet, Physical Activity, and Sleep Intervention on Dietary Intake in Adults with Overweight and Obesity: A Randomised Controlled Trial

Sasha Fenton [1,2], Tracy L. Burrows [1,3], Clare E. Collins [1,3], Anna T. Rayward [1,4], Beatrice Murawski [1,2] and Mitch J. Duncan [1,2,*]

1 Priority Research Centre for Physical Activity and Nutrition, The University of Newcastle, University Drive, Callaghan, NSW 2308, Australia; sasha.fenton@uon.edu.au (S.F.); tracy.burrows@newcastle.edu.au (T.L.B.); clare.collins@newcastle.edu.au (C.E.C.); anna.rayward@newcastle.edu.au (A.T.R.); beatrice.murawski@uon.edu.au (B.M.)
2 School of Medicine and Public Health, College of Health, Medicine and Wellbeing, The University of Newcastle, University Drive, Callaghan, NSW 2308, Australia
3 School of Health Sciences, College of Health, Medicine and Wellbeing, The University of Newcastle, University Drive, Callaghan, NSW 2308, Australia
4 School of Education, College of Human and Social Futures, The University of Newcastle, University Drive, Callaghan, NSW 2308, Australia
* Correspondence: mitch.duncan@newcastle.edu.au

Citation: Fenton, S.; Burrows, T.L.; Collins, C.E.; Rayward, A.T.; Murawski, B.; Duncan, M.J. Efficacy of a Multi-Component m-Health Diet, Physical Activity, and Sleep Intervention on Dietary Intake in Adults with Overweight and Obesity: A Randomised Controlled Trial. *Nutrients* **2021**, *13*, 2468. https://doi.org/10.3390/nu13072468

Academic Editors: Sergio Garbarino and Egeria Scoditti

Received: 23 June 2021
Accepted: 17 July 2021
Published: 19 July 2021

Abstract: This three-arm randomised controlled trial evaluated whether (1) a multi-component weight loss intervention targeting diet, physical activity (PA), and sleep was effective at improving dietary intake over six months and 12 months, compared with a control, and (2) the enhanced diet, PA, and sleep intervention was more effective at improving dietary intake than the traditional diet and PA intervention. A total of 116 adults (70% female, 44.5 years, BMI 31.7 kg/m^2) were randomised to either traditional diet and PA intervention; enhanced diet, PA, and sleep intervention; or wait-list control. To examine between-group differences, intervention groups were pooled and compared with the control. Then, the two intervention groups were compared. At six months, the pooled intervention group consumed 1011 fewer kilojoules/day (95% CI −1922, −101), less sodium (−313.2 mg/day; 95% CI −591.3, −35.0), and higher %EI from fruit (+2.1%EI; 95% CI 0.1, 4.1) than the controls. There were no differences in intake between the enhanced and traditional groups at six months. At 12 months, the pooled intervention and control groups reported no significant differences. However, compared to the traditional group, the enhanced reported higher %EI from nutrient-dense foods (+7.4%EI; 95% CI 1.3, 13.5) and protein (+2.4%EI; 95% CI 0.1, 4.6), and reduced %EI from fried/takeaway foods (−3.6%EI; 95% CI −6.5, −0.7), baked sweet products (−2.0%EI; 95% CI −3.6, −0.4), and packaged snacks (−1.1%EI; 95% CI −2.2, −0.3). This weight loss intervention reduced total energy and sodium intakes as well as increased fruit intake in adults at six months. The enhanced intervention group reported improved dietary intake relative to the traditional group at 12 months.

Keywords: diet; nutrition; physical activity; sleep; overweight; obesity; weight loss; RCT; m-health

1. Introduction

In 2015, 39% of the global population was overweight or obese [1]. Excess adiposity increases the risk of chronic diseases, such as type two diabetes and cardiovascular disease, decreases life expectancy, and has negative personal, social, and economic consequences [2]. Contributing to the prevalence of overweight and obesity is poor diet quality [3], which is characterised by high consumption of energy-dense, nutrient-poor foods and drinks containing saturated fat and/or added salt and sugar, and alcohol; with inadequate consumption of fruits, vegetables, and whole grains [4,5]. Given the association between poor diet quality and weight gain [3], aligning dietary patterns with national guidelines is a recommended strategy for treatment of overweight and obesity, along with increasing physical activity [6]. Weight loss interventions most commonly target reduced total fat and increased fruit and vegetable intakes to create an energy deficit [7–9]. However, weight loss studies rarely report participants' dietary changes in addition to weight loss outcomes [10,11]. Of those studies that have reported dietary outcomes, change in total energy intake or macronutrient intake was most commonly reported, while change in overall diet quality or foods was rarely reported [12]. While changes energy and macronutrient intakes are important, identifying changes in the consumption of different types of healthy/unhealthy foods provides valuable information about food patterns that are amenable to change in the context of overweight and obesity. This information could be used to inform future weight loss interventions and may assist in improving effectiveness

Recently, sleep has been recognised as another health behaviour that is potentially modifiable for weight management [13,14]. Short sleep duration has been linked to poor dietary behaviours with several experimental studies reporting increased energy intake, higher number of meals and snacks consumed, and increased night-time eating in response to short sleep duration [14–17]. The physiological mechanisms that drive increased food intake and energy intake following short sleep are not well understood, but short sleep appears to increase activity in the brain's hedonic (reward) system. Evidence from experimental studies suggests that during periods of restricted sleep duration, the presentation of food stimuli provokes greater activity patterns in brains areas associated with pleasure and reward, compared with viewing the same food stimuli following normal sleep [18–22]. Another proposed mechanism is through the regulation of appetitive hormones. Several experimental studies have reported elevated concentration of the hunger-stimulating hormone ghrelin and reduced concentration of the satiety hormone leptin in response to short sleep duration [23–25]. In terms of the increase in energy intake that can occur when sleep is insufficient, meta-analyses have reported an increase of 1059–1612 kilojoules (kJ) (253–385 kcal) per day in adults whose sleep was restricted to $\leq$5.5 h per night, compared with controls [18,26,27]. As such, increasing sleep duration among short-sleepers may contribute to reducing energy intake and improving energy balance [28] for enhanced weight loss.

Few randomised controlled trials (RCTs) have measured the effect of a dedicated sleep intervention on dietary intake [29–32]. One RCT (n = 42) measured the effect of extended sleep duration on aspects of dietary intake including energy, macronutrient, sugar, fibre, and caffeine intake and diet quality over four weeks in adults with normal weight [29]. The study reported that extended sleep duration reduced free-sugar intake and increased adherence to dietary guidelines [29]. Another RCT (n = 22) that measured the effect of six weeks of extended sleep duration compared with control on energy, macronutrient, and sodium intake for eight days reported no significant change in dietary outcomes [30]. However, preliminary data from an RCT (n = 60) in adults with overweight and obesity has shown that two weeks of sleep extension ($\approx$1.2 h per night) resulted in significantly lower energy intake ($-$207 kcal per day), compared with control [31]. There is limited evidence investigating dietary outcomes in response to a sleep intervention over longer periods.

The primary objective of the current study was to evaluate whether a multi-component m-Health weight loss intervention in adults with overweight and obesity that targeted dietary, physical activity, and sleep behaviours was effective at improving dietary intake over six months, and longer-term at 12 months, compared to a wait-list control. The secondary objective was to evaluate whether the intervention targeting dietary, physical activity, and sleep behaviours (enhanced) was more effective at improving dietary intake than the dietary and physical activity intervention (traditional). The hypotheses were that the enhanced and the traditional interventions would achieve greater improvements in dietary intake than the wait-list control at six and 12 months, and that the enhanced intervention would achieve greater improvements in dietary intake than the traditional intervention.

2. Materials and Methods

2.1. Trial Design

This is a secondary analysis of the Move, Eat & Sleep study, which has been described in detail elsewhere [33]. Move, Eat & Sleep was a 3-arm parallel-group RCT with a six-month intervention period (primary endpoint) and follow-up at 12 months. Participants were recruited May–September 2017 from Newcastle, NSW, Australia, primarily by media stories, social media advertising, and participant registries. Sample size calculations were based on the primary outcome measure, weight (kilograms), as described elsewhere [34]. A minimum of 114 participants was required to provide adequate power, and a total of 116 were randomised. Participants were stratified by baseline body mass index (BMI) (25.0–29.9, 30.0–40.0 kg/m^2) and randomly allocated (1:1 ratio) by the project manager to one of three groups using secure, web-based, permuted block randomisation that was generated by an independent statistician. Ethical approval was granted by the Human Research Ethics Committee of the University of Newcastle (H-2017–0039), and the trial was prospectively registered with the Australian New Zealand Clinical Trials Registry (ACTRN12617000735358; U1111-1219-2050). Informed written consent was obtained from all individual participants included in the study. The reporting of this study adheres to the Consolidated Standards of Reporting Trials (CONSORT) guidelines [35].

2.2. Participants

Individuals were eligible for participation if they reported being aged 18–65 years, a BMI of 25.0–40.0 kg/m^2 (overweight/obesity classification), had access to an iOS/Android device with Internet access, and had the ability to attend four assessments over 12 months. Individuals were ineligible if they were using a tracking device for physical activity and/or sleep, were pregnant, reported the presence of a sleep disorder diagnosed by a medical practitioner, were taking medication related to sleep or weight management, had a condition that precludes participation in physical activity and/or modification of diet and/or sleep, had lost $\geq$4.5 kg of body weight in the last three months, intended to participate in another weight loss program, had previous weight loss surgery, or were employed as a shift-worker (on a rotating roster).

2.3. Intervention

The intervention components have been described in detail previously [34]. Briefly, the Move, Eat & Sleep study was a multi-component m-Health behaviour-change weight loss intervention. The traditional intervention group targeted change in dietary and physical activity behaviours, while the enhanced intervention group targeted change in dietary behaviours, physical activity, and sleep health. The wait-list control group received access to the intervention after the 12 month follow-up. Participants received intervention content specific to their group allocation and self-directed under free-living conditions. The intervention content was guided by specific behaviour change techniques (e.g., goal-setting, self-monitoring) to operationalise constructs from the self-regulatory and social cognitive theories [36–38].

The dietary intervention was delivered via intervention materials and one face-to-face dietary session with a dietitian, where participants were given personalised dietary advice based on assessment of their current dietary intake, as measured by the Australian Eating Survey® (FFQ) and personalised nutrition report (Australian Eating Survey® Version 2, The University of Newcastle, Callaghan, NSW, Australia) [39,40]. They received information about the Australian Dietary Guidelines and the Australian Guide to Healthy Eating, which promote increased intake of nutrient-dense core food groups and reduced intake of energy-dense, nutrient-poor discretionary foods and drinks [41]. Resources for planning healthy meals, controlling portion sizes, and interpreting food labels were also provided. Participants were given a personalised daily energy intake target of 2000 kJ less than their estimated energy requirement to create an energy deficit, as this is the amount required to achieve $\approx$0.5 kg weight loss per week [42]. Intervention group participants were instructed to set dietary behaviour goals and self-monitor their behaviours using the Balanced app, which contains 10 daily food goals: (1) eat two serves of fruit, (2) eat five serves of vegetables, (3) choose whole-grains, (4) choose low-fat dairy, (5) choose lean meats/alternatives, (6) have a soft drink/energy drink-free day, (7) have an alcohol-free day, (8) choose healthy snack options, (9) have a fast-food-free day, (10) drink plenty of water. Participants were instructed to choose the number of food goals they would aim to achieve daily. This approach has been used previously to improve indicators of diet quality [43]. Participants were instructed to self-monitor their daily energy intake against their recommended energy intake goal using the ControlMyWeight™ app by CalorieKing four days per week (CalorieKing Wellness Solutions Inc., La Mesa, CA, USA). Participants were also recommended to self-monitor and receive feedback about their diet quality by completing the free online Healthy Eating Quiz™ (version 3, The University of Newcastle, Callaghan, NSW, Australia) [44] monthly throughout the intervention period.

The physical activity intervention component promoted increasing daily steps, moderate-to-vigorous intensity physical activity (MVPA), and resistance training (RT) activity to align with physical activity and sedentary behaviour guidelines [45]. Participants were asked to set physical activity goals and self-monitor daily step counts, minutes of physical activity, and number of RT sessions using the Balanced app and a provided Fitbit Alta activity tracker (Fitbit Inc, San Francisco, CA, USA).

The sleep intervention component delivered to the enhanced intervention group provided information about sleep duration and quality recommendations, the importance of sleep health, and daily sleep hygiene practices for participants to implement to promote healthy sleep. Information on stress management techniques (i.e., progressive muscle relaxation, deep breathing, mindfulness), and guidance on cognitive (e.g., self-efficacy, outcome expectations, goal-setting, action planning, self-monitoring, feedback) and behavioural self-regulation strategies (e.g., sleep hygiene practices such as limited bed and wake time variability, limiting caffeine consumption) to promote sleep health was also provided [34,46,47]. Enhanced group participants were asked to set and self-monitor sleep goals for sleep time, wake time, and sleep hygiene practices to help them achieve adequate sleep duration, consistent sleep timing, and improved sleep quality. To promote compliance and reinforce the intervention components, participants received personalised weekly feedback about their progress in relation to their goals.

2.4. Measures

All outcome measures were completed at baseline, six months, and 12 months at The University of Newcastle, Australia by trained assessors using a standardised protocol. Assessors were blinded to participant group allocation at each time point, and participants were blinded to group allocation until the completion of baseline assessments.

2.4.1. Sociodemographics and Anthropometry

Sociodemographic data were collected at baseline using the Qualtrics^{XM} online survey platform (Qualtrics, Provo, UT, USA). Weight and height were measured to the nearest 0.1 kg and 0.1 cm, respectively, on a combined stadiometer and calibrated digital scale (Biospace BSM370 Portable Automatic BMI Stadiometer, Biospace Co, Ltd., Seoul, Korea) and used to calculate BMI (weight (kg)/height (m^2)). Waist circumference was measured to the nearest 0.5 cm using a measuring tape (Seca 203, Seca Gmph & Co., Hamburg, Germany). Glycated haemoglobin (HbA1c) was measured using a capillary blood sample and analysed using the validated A1C Now+ device (Polymer Technology Systems).

2.4.2. Dietary Intake

Dietary outcomes (energy intake, macronutrient intake (including alcohol), micronutrient intake, nutrient-dense food intake, energy-dense nutrient-poor food intake, diet quality, and caffeine intake) were measured using the Australian Eating Survey (AES), which is a validated 120-item semi-quantitative FFQ that measures the frequency of consumption of foods items and types over the previous three to six months [39,40]. Standard adult portion sizes were derived from the 1995 National Nutrition Survey data or from the product standard serving size. Nutrient intakes were calculated from the most current food composition database of Australian foods, the AusNut 1999 database (All Foods), and AusFoods (Brands) Revision 5. The AES also generates a diet quality score, the Australian Recommended Food Score (ARFS) calculated from a subset of AES questions [39,40]. Diet quality scores are based on core food groups recommended in the Australian Dietary Guidelines [39,48] and calculated by summing points per food item. The score range is from 0 to 73, with higher scores indicative of higher diet quality and categorised as: 'needs work' (<33), 'getting there' (33–38), 'excellent' (39–46), or 'outstanding' (47+) [44]. Daily caffeine consumption (drinks containing caffeine) was measured using a question adapted from the AUDIT-C alcohol screening test [49], with minor word changes.

2.4.3. Physical Activity and Sleep Health

Weekly minutes of MVPA were measured using The Active Australia Survey (AAS). The AAS has acceptable levels of validity and test–retest reliability [50,51], and it is sensitive to change [52]. Sleep quality was measured using the Pittsburgh Sleep Quality Index (PSQI) [53]. The PSQI is a 19-item survey that measures indicators of sleep health (sleep duration, sleep onset latency, sleep efficiency, sleep disturbances, daytime dysfunction, sleep medication use, and subjective sleep quality) over the previous month. The scores (0–3) for each sleep indicator are summed to provide an overall sleep quality score ranging from 0 to 21, where '0' indicates no sleeping difficulty and '21' indicates severe difficulty. Sleep quality scores >5 are indicative of poor sleep quality [53]. Sleep duration is calculated from PSQI questions about bed and wake times. The PSQI has demonstrated good reliability (Cronbach $\alpha = 0.83$), sensitivity to change, and strong psychometric properties [53,54].

2.5. Statistical Methods

Analyses followed intention-to-treat principles using generalised linear mixed models (GLMM). To examine between-group differences, first, the pooled intervention group (traditional and enhanced) was compared to the control group. The groups were combined to maximise statistical power. Then, the two intervention groups were compared to each other. Group differences in key variables at six months were examined using GLMM using an ANCOVA (baseline-adjusted) approach. The models included fixed effects for the baseline value of the outcome, group, and the BMI stratification variable (categorised as 25 to <30 and 30 to 40 kg/m^2). Group differences in caffeine intake at both time points were examined using logistic regression with adjustment for the baseline value of the outcome. GLMM were used to measure group differences in dietary intake, physical activity, and sleep quality and duration at 12 months. A random intercept for ID accounted for repeated measures. All models used a response distribution and link function as appropriate to the outcome, with an alpha level of 0.05. Sensitivity analyses explored the influence of participants living in the same residence allocated to the same intervention group. Analyses were conducted using Stata version 15 [55]. Effect sizes were calculated using the equation: Cohen's $d = (M_{1\ \text{change score}} - M_{2\ \text{change score}})/SD_{\text{pooled [change scores]}}$. The magnitude of effects is interpreted using the criteria, small (0.2), medium (0.5), and large (0.8), as defined by Cohen [56].

3. Results

3.1. Participants

The Move, Eat & Sleep Study enrolled 116 participants: traditional intervention group, $n = 41$; enhanced intervention group, $n = 39$; wait-list control group, n = 36 (Figure S1). The retention rate was 70% at six months with 81 participants completing dietary assessments (traditional, $n = 32$; enhanced, $n = 28$; wait-list control, $n = 21$), and 47% at 12 months (54 participants: traditional, $n = 23$; enhanced, $n = 14$; wait-list control, $n = 17$). The mean age of participants was 44.5 ± 10.4 years (range 19–65 years), 71% were female and 94%

identified as Caucasian. Eighty-seven percent of participants were employed, and 78% reported ≥14 years of education. The mean baseline BMI was 31.7 ± 3.9 kg/m^2 (obese classification). Participants reported participation in 313 min of MVPA per week. The mean PSQI sleep quality score was 7.0, and the mean self-reported sleep duration was 6.8 h (406 min) per night (Table 1).

3.2. Dietary Intake

3.2.1. Baseline

Participants (n = 116) had mean (±SD) energy intake of 9683 (±3146) kJ per day, with 43.5% of energy from carbohydrate, 33.3% from fat (14.0% from saturated fat), 18.0% from protein, and 5.6% from alcohol. Nutrient-dense foods contributed 58.3% of total energy intake, with grain foods contributing the most energy (16.7%), followed by meats (13.1%), dairy foods (9.0%), vegetables (8.5%), fruits (6.6%), and meat alternatives (4.4%). Participants reported consuming 4.5 serves of vegetables and 1.8 serves of fruit per day. Participants had a mean ARFS (diet quality score) of 35.4 (±9.0) out of a maximum 73 points, which is categorised as 'getting there' [44], and scored the highest for variety within the vegetable subscale. Energy-dense nutrient-poor foods contributed 41.7% of total energy intake, predominantly from fried/takeaway foods (9.1% of total energy), confectionery (6.8%), baked sweet products (5.6%), and alcoholic beverages (5.6%). At baseline, approximately half (52%) of participants reported consuming ≤2 drinks containing caffeine per day, while the remaining participants reported higher caffeine consumption. Overall, dietary intake was similar between all groups at baseline (Table 1).

3.2.2. Six Months and 12 Months

Total daily energy intake: At six months, there was a significant mean difference between the pooled intervention group (traditional and enhanced groups) and the control group in energy intake. The pooled intervention groups consumed 1011 fewer kilojoules (242 kcal) per day compared with the control group (95% CI −1922, −101; p = 0.029; Cohen's d = 0.55), indicating a medium-sized intervention effect (Table 2). At 12 months, the pooled intervention group maintained a non-significant reduction in total energy intake compared with the control group (−913 kJ; 95% CI −2033, 207; p = 0.110; d = 0.47) (Table 3).

Macronutrients: No significant differences were observed between the pooled intervention group and the control group at six months for macronutrient distribution (including alcohol) or fibre intake (Table 2). Effect sizes ranged from d = 0.02 to 0.29. At 12 months, the enhanced group reported a significantly higher percentage of energy from protein than the traditional group (+2.4 %EI; 95% CI 0.1, 4.6; p = 0.040; d = 0.74). No significant between-group differences were observed for energy, macronutrient, or micronutrient intake at 12 months (Table 3).

Micronutrients: The pooled intervention groups reported statistically significantly lower sodium intakes per day compared with the controls at six months (−313.2 mg; 95% CI −591.3, −35.0; p = 0.027; d = 0.56) (Table 2). At 12 months, the reduction in sodium intake was maintained by the pooled intervention group, but it was not significantly different from the controls (−326.2 mg; 95% CI −662.6, 10.2; p = 0.057; d = 0.56) (Table 3).

Energy intake from healthy core foods and discretionary non-core foods, and diet quality: There were no significant differences between the pooled intervention group and the control group at six months in nutrient-dense food intakes, energy-dense nutrient-poor food intake, or diet quality, with small–medium effect sizes estimated (d = 0.03–0.44). At six months, the pooled intervention group reported a significantly higher percentage of energy from fruits than the control group (+2.1 %EI; 95% CI 0.1, 4.1; p = 0.040; d = 0.53); however, while still higher than at baseline, the difference in fruit intake between groups was not maintained at 12 months (+1.6 %EI; 95% CI −0.3, 3.5; p = 0.093; d = 0.47). Caffeinated beverage consumption was also not significantly different at six or 12 months. At 12 months, the pooled intervention group reported significantly lower consumption of fried/takeaway foods relative to the control group (−2.4 %EI; 95% CI −4.7, −0.2; p = 0.034; d = 0.64).

Table 1. Baseline sociodemographic, health, and behavioural characteristics of participants by group.

Characteristics/Behaviours M (SD)	Total (*n* = 116)	Control Group (*n* = 36)	Pooled Intervention Group (*n* = 80)	Traditional Group (*n* = 41)	Enhanced Group (*n* = 39)
Age (years)	44.5 (10.4)	40.5 (10.7)	46.3 (9.8)	45.4 (10.2)	47.2 (9.4)
Sex, n (%)					
Female	82 (70.7)	25 (69.4)	57 (71.2)	30 (73.2)	27 (69.2)
Weight (kg)	90.7 (14.3)	92.5 (16.1)	89.8 (13.4)	88.9 (13.8)	90.8 (13.1)
BMI (kg/m^2)	31.7 (3.9)	31.9 (3.9)	31.7 (3.9)	31.4 (3.8)	31.9 (4.0)
Waist circumference (cm)	99.6 (11.0)	99.7 (11.7)	99.6 (10.8)	99.6 (9.0)	99.5 (12.5)
HbA1c (%)	5.4 (0.5)	5.3 (0.3)	5.5 (0.6)	5.5 (0.7)	5.5 (0.4)
Dietary intake					
Energy intake					
Total energy intake (kJ/d)	9683 (3146)	9153 (2810)	9922 (3274)	10,397 (2989)	9422 (3519)
Macronutrient intake					
Carbohydrate (%EI)	43.5 (6.8)	43.3 (8.3)	43.6 (6.0)	44.0 (5.3)	43.2 (6.7)
Fats (%EI)	33.3 (5.0)	33.8 (6.2)	33.0 (4.4)	33.7 (3.8)	32.2 (5.0)
Saturated fat (%EI)	14.0 (2.7)	14.4 (3.1)	13.8 (2.5)	14.3 (2.4)	13.2 (2.5)
Monounsaturated fat (%EI)	12.4 (2.2)	12.7 (2.8)	12.2 (1.92)	12.4 (1.6)	12.0 (2.2)
Polyunsaturated fat (%EI)	4.0 (1.0)	3.9 (1.0)	4.1 (1.0)	4.0 (0.9)	4.1 (1.2)
Protein (%EI)	18.0 (3.0)	18.0 (2.6)	18.0 (3.2)	17.9 (2.9)	18.2 (3.5)
Alcohol (%EI)	5.6 (7.1)	5.4 (7.3)	5.6 (7.0)	4.8 (6.5)	6.5 (7.4)
Micronutrient intake					
Sugars (g/d)	125.0 (57.7)	119.9 (56.4)	127.3 (58.5)	135.2 (54.0)	119.1 (62.6)
Fibre (g/d)	27.0 (8.9)	24.5 (8.4)	28.1 (9.0)	28.8 (8.1)	27.4 (9.9)
Sodium (mg/day)	2390.2 (858.1)	2310.0 (833.2)	2426.1 (871.9)	2598.6 (864.7)	2244.7 (852.9)
Nutrient-dense food intake					
Nutrient-dense foods (%EI)	58.3 (12.1)	56.1 (9.9)	59.3 (12.8)	59.5 (13.5)	59.1 (12.2)
Vegetables (%EI)	8.5 (3.9)	8.1 (3.3)	8.7 (4.2)	8.0 (4.1)	9.5 (4.1)
Vegetables (serves/day)	4.5 (1.7)	4.0 (1.3)	4.7 (1.7)	4.6 (1.6)	4.9 (1.9)
Fruits (%EI)	6.6 (4.2)	6.3 (4.2)	6.7 (4.1)	7.0 (4.2)	6.5 (4.2)
Fruits (serves/day)	1.8 (1.2)	1.8 (1.3)	1.8 (1.2)	2.0 (1.2)	1.7 (1.2)
Milk, yoghurt, cheese (%EI)	9.0 (5.3)	8.1 (4.8)	9.4 (5.5)	10.3 (5.5)	8.5 (5.4)
Breads, cereals, rice, pasta, noodles (%EI)	16.7 (5.8)	15.9 (6.6)	17.1 (5.4)	17.5 (5.7)	16.6 (5.2)
Lean meats, fish, poultry, eggs, nuts (%EI)	13.1 (6.2)	13.7 (6.1)	12.9 (6.2)	12.4 (5.9)	13.4 (6.8)
Protein alternatives, nuts, eggs, beans	4.4 (3.2)	4.4 (3.1)	4.4 (3.3)	4.2 (3.4)	4.5 (3.1)
Energy-dense food intake					
Energy-dense, nutrient-poor foods (%/day)	41.7 (12.1)	43.9 (9.9)	40.7 (12.8)	40.5 (13.5)	40.9 (12.2)
Fatty meats (%EI)	2.0 (1.6)	2.3 (2.0)	1.9 (1.4)	2.1 (1.6)	1.8 (1.3)

Table 1. *Cont.*

Characteristics/Behaviours M (SD)	Total (n = 116)	Control Group (n = 36)	Pooled Intervention Group (n = 80)	Traditional Group (n = 41)	Enhanced Group (n = 39)
Fried/takeaway foods (%EI)	9.1 (5.6)	10.1 (5.5)	8.7 (5.7)	8.1 (4.9)	9.3 (6.5)
Confectionery (%EI)	6.8 (6.0)	6.5 (5.3)	6.9 (6.3)	8.1 (7.3)	5.6 (5.0)
Baked sweet products (%EI)	5.6 (4.2)	4.6 (3.4)	6.0 (4.4)	6.8 (4.2)	5.1 (4.5)
Packaged snacks (%EI)	3.0 (3.0)	3.4 (3.4)	2.8 (2.9)	2.8 (2.7)	2.9 (3.1)
Sweetened drinks (%EI)	3.0 (5.4)	4.7 (8.4)	2.2 (3.1)	1.8 (2.3)	2.5 (3.9)
Diet quality					
Total diet quality (ARFS 0–73)	35.4 (9.0)	33.3 (10.0)	36.3 (8.4)	36.0 (8.0)	36.5 (9.0)
Vegetables (ARFS 0–21)	14.2 (4.0)	13.4 (4.5)	14.7 (3.8)	14.4 (3.6)	15.0 (3.9)
Fruits (ARFS 0–12)	5.4 (2.8)	5.3 (3.2)	5.5 (2.7)	5.6 (2.3)	5.3 (3.1)
Protein foods—meat/flesh (ARFS 0-7)	2.7 (1.4)	2.5 (1.4)	2.7 (1.3)	2.6 (1.0)	2.9 (1.6)
Protein foods—meat/flesh alternatives (ARFS 0-6)	2.2 (1.2)	2.1 (1.2)	2.2 (1.3)	2.1 (1.2)	2.3 (1.3)
Grains, breads, and cereals (ARFS 0–13)	5.4 (1.9)	4.8 (1.9)	5.7 (1.8)	5.8 (1.7)	5.6 (1.8)
Dairy foods (ARFS 0–11)	3.9 (1.8)	3.6 (1.8)	4.0 (1.8)	4.0 (1.6)	4.0 (2.0)
Water (AFRS 0–1)	0.5 (0.5)	0.5 (0.5)	0.5 (0.5)	0.5 (0.5)	0.5 (0.5)
Extras (ARFS 0–2)	1.1 (0.7)	1.2 (0.7)	1.0 (0.8)	1.1 (0.8)	1.0 (0.7)
Physical activity					
MVPA (min/w)	312.8 (297.8)	238.1 (239.2)	346.4 (316.3)	351.0 (357.7)	341.5 (270.6)
Sleep health					
Sleep quality score (PSQI Global score)	7.0 (3.0)	6.7 (3.1)	7.1 (3.0)	7.0 (3.1)	7.3 (2.8)
Sleep duration (min/d)	406.0 (58.3)	406.7 (51.9)	405.8 (61.3)	408.3 (60.3)	403.1 (63.0)

Abbreviations: %EI, percentage of energy intake; ARFS, Australian recommended food score; d, day; g, grams; kJ, kilojoules; mg, milligrams; MVPA, moderate-to-vigorous intensity physical activity; PSQI, Pittsburgh Sleep Quality Index; w, week.

Table 2. Between-group differences in dietary intake, physical activity, and sleep health at six months (n = 81).

Intake/Behaviour	Control Group (n = 21) M (SE)	Pooled Intervention Group (n = 60) M (SE)	Between-Group Difference (95% CI)	p-Value	Effect Size (Cohen's d)	Traditional Intervention Group (n =32) M (SE)	Enhanced Intervention Group (n =28) M (SE)	Between-Group Difference (95% CI)	p-Value	Effect Size (Cohen's d)
Energy intake										
Total energy intake (kJ/d)	9198 (399)	8187 (235)	−1011 (−1922, −101)	0.029	0.55	8276 (316)	8318 (339)	42 (−895, 979)	0.93	0.02
Macronutrient intake										
Carbohydrate (%EI)	41.6 (1.3)	42.0 (0.8)	0.4 (−2.5, 3.4)	0.771	0.07	42.6 (1.0)	41.9 (1.1)	−0.7 (−3.6, 2.1)	0.617	0.12
Total fats (%EI)	33.6 (0.9)	32.7 (0.6)	−0.9 (−3.1, 1.2)	0.381	0.21	32.6 (0.7)	33.2 (0.8)	0.6 (−1.5, 2.6)	0.596	0.14
Saturated fat (%EI)	13.5 (0.5)	13.2 (0.3)	−0.3 (−1.4, 0.7)	0.555	0.15	13.2 (0.3)	13.4 (0.4)	0.1 (−0.9, 1.1)	0.782	0.05
Monounsaturated fat (%EI)	12.6 (0.4)	12.3 (0.2)	−0.3 (−1.3, 0.6)	0.523	0.15	12.2 (0.3)	12.6 (0.4)	0.4 (−0.6, 1.4)	0.433	0.21
Polyunsaturated fat (%EI)	4.0 (0.2)	4.2 (0.1)	0.2 (−0.2, 0.7)	0.378	0.25	4.2 (0.1)	4.3 (0.2)	0.03 (−0.4, 0.5)	0.894	0.04

Table 2. *Cont.*

Intake/Behaviour	Control Group (n = 21) M (SE)	Pooled Intervention Group (n = 60) M (SE)	Between-Group Difference (95% CI)	p-Value	Effect Size (Cohen's d)	Traditional Intervention Group (n =32) M (SE)	Enhanced Intervention Group (n =28) M (SE)	Between-Group Difference (95% CI)	p-Value	Effect Size (Cohen's d)
Protein (%EI)	19.2 (0.7)	19.8 (0.4)	0.6 (−1.0, 2.2)	0.476	0.19	20.2 (0.6)	19.2 (0.6)	1.0 (−2.7, 0.7)	0.252	0.29
Alcohol (%EI)	5.6 (0.8)	5.6 (0.5)	0.05 (−1.8, 1.9)	0.961	0.01	4.6 (0.7)	5.6 (0.7)	1.0 (−1.0, 3.1)	0.305	0.26
Micronutrient intake										
Sugars (g/d)	113.6 (8.2)	103.2 (4.8)	−10.4 (−29.1, 8.2)	0.273	0.28	105.5 (5.5)	102.6 (5.9)	−2.8(−19.0, 13.4)	0.734	0.09
Fibre (g/d)	27.4 (1.2)	26.1 (0.7)	−1.3 (−4.0, 1.4)	0.332	0.24	27.6 (1.0)	25.9 (1.1)	−1.7 (−4.7, 1.2)	0.253	0.29
Sodium (mg/d)	2246.6 (122.1)	1933.4 (72.1)	−313.2 (−591.3, −35.0)	0.027	0.56	1983.8 (91.8)	1912.4 (98.6)	−71.4 (−344.3, 201.4)	0.608	0.13
Nutrient-dense food intake										
Nutrient-dense foods (%EI)	63.8 (2.1)	66.8 (1.2)	2.9 (−1.8, 7.7)	0.222	0.31	67.9 (1.8)	66.6 (2.0)	−1.2 (−6.6, 4.1)	0.654	0.12
Vegetables (%EI)	8.7 (0.8)	9.4 (0.5)	0.6 (−1.1, 2.4)	0.482	0.17	9.3 (0.7)	9.5 (0.8)	0.2 (−1.9, 2.2)	0.862	0.05
Vegetables (serves/d)	4.8 (0.3)	4.6 (0.2)	−0.2 (−1.0, 0.6)	0.703	0.13	4.9 (0.3)	4.5 (0.3)	−0.4 (−1.2, 0.5)	0.414	0.19
Fruits (%EI)	7.1 (0.9)	9.2 (0.5)	2.1 (0.1, 4.1)	0.04	0.53	9.9 (0.8)	8.3 (0.8)	−1.6 (−3.8, 0.6)	0.158	0.38
Fruits (serves/d)	1.8 (0.2)	2.2 (0.1)	0.3 (−0.2, 0.8)	0.233	0.25	2.4 (0.2)	1.8 (0.2)	−0.6 (−1.1, 0.02)	0.058	0.52
Milk, yoghurt, cheese (%EI)	10.5 (1.2)	10.0 (0.7)	−0.5 (−3.2, 2.1)	0.695	0.09	10.4 (0.7)	10.0 (0.8)	−0.4 (−2.6, 1.8)	0.727	0.09
Breads, cereals, rice, pasta, noodles (%EI)	17.0 (1.4)	17.3 (0.8)	0.3 (−2.9, 3.4)	0.862	0.05	16.9 (1.1)	18.7 (1.2)	1.8 (−1.5, 5.0)	0.287	0.27
Lean meats, fish, poultry, eggs, nuts (%EI)	14.9 (1.4)	15.6 (0.8)	0.7 (−2.5, 3.8)	0.674	0.11	15.7 (1.1)	14.6 (1.2)	−1.1 (−4.4, 2.1)	0.488	0.18
Protein alternatives, nuts, eggs, beans	5.2 (0.6)	5.6 (0.4)	0.4 (−1.0, 1.8)	0.581	0.14	5.5 (0.5)	5.8 (0.5)	0.3 (−1.0, 1.7)	0.618	0.11
Energy-dense food intake										
Energy-dense, nutrient-poor foods (%EI)	36.2 (2.1)	33.2 (1.2)	−2.9 (−7.7,1.8)	0.222	0.31	32.1 (1.8)	33.3 (2.0)	1.2 (−4.1, 6.6)	0.654	0.12
Fatty meats (%EI)	1.9 (0.3)	1.5 (0.2)	−0.3 (−0.9, 0.3)	0.31	0.25	1.4 (0.2)	1.4 (0.2)	−0.03 (−0.5, 0.4)	0.887	0.04
Fried/takeaway foods (%EI)	9.0 (0.8)	7.8 (0.5)	−1.2 (−3.0, 0.7)	0.214	0.34	8.4 (0.7)	7.2 (0.7)	−1.2 (−3.3, 0.8)	0.23	0.31
Confectionery (%EI)	4.4 (0.9)	5.0 (0.5)	0.6 (−1.4, 2.7)	0.537	0.15	4.1 (0.8)	6.3 (0.9)	2.2 (−0.2, 4.6)	0.076	0.47
Baked sweet products (%EI)	4.6 (0.7)	3.8 (0.4)	−0.8 (−2.3, 0.8)	0.338	0.25	4.5 (0.5)	3.6 (0.6)	−0.9 (−2.5, 0.7)	0.263	0.29
Packaged snacks (%EI)	2.4 (0.3)	1.7 (0.2)	−0.7 (−1.5, 0.02)	0.055	0.44	1.8 (0.2)	1.6 (0.3)	−0.2 (−0.9, 0.5)	0.563	0.13
Sweetened drinks (%EI)	2.0 (0.5)	1.1 (0.3)	−0.9 (−2.0, 0.2)	0.125	0.38	1.2 (0.3)	0.7 (0.3)	−0.5 (−1.2, 0.3)	0.214	0.32
Diet quality										
Total diet quality (ARFS 0-73)	36.2 (1.2)	36.7 (0.7)	0.5 (−2.1, 3.2)	0.687	0.09	37.6 (1.0)	37.2 (1.0)	−0.3 (−3.1, 2.4)	0.807	0.06
Vegetables (ARFS 0-21)	13.9 (0.5)	14.5 (0.3)	0.7 (−0.6, 1.9)	0.281	0.3	14.9 (0.5)	14.4 (0.5)	−0.5 (−1.8, 0.9)	0.483	0.18
Fruits (ARFS 0-12)	6.2 (0.5)	5.9 (0.3)	−0.3 (−1.5, 0.8)	0.551	0.13	6.3 (0.4)	5.6 (0.4)	−0.8 (−1.9, 0.4)	0.194	0.35
Protein foods—meat/flesh (ARFS 0-7)	2.5 (0.2)	3.0 (0.1)	0.5 (−0.1, 1.1)	0.08	0.42	3.0 (0.2)	2.9 (0.2)	−0.04 (−0.6, 0.5)	0.89	0.03
Protein foods—meat/flesh alternatives (ARFS 0-6)	2.1 (0.2)	2.4 (0.1)	0.3 (−0.2, 0.8)	0.221	0.38	2.3 (0.2)	2.5 (0.2)	0.2 (−0.3, 0.7)	0.346	0.26
Grains, breads, and cereals (ARFS 0–13)	5.5 (0.4)	5.5 (0.2)	−0.05 (−1.0, 0.9)	0.909	0.03	5.4 (0.3)	6.0 (0.3)	0.6 (−0.3, 1.5)	0.168	0.39
Dairy foods (ARFS 0–11)	3.9 (0.3)	3.7 (0.2)	−0.2 (−0.9, 0.6)	0.674	0.13	4.1 (0.2)	3.7 (0.3)	−0.3 (−1.0, 0.4)	0.37	0.19
Water (AFRS 0–1)	0.6 (0.1)	0.8 (0.04)	0.2 (−0.03, 0.3)	0.108	0.51	0.8 (0.1)	0.8 (0.1)	0.1 (−0.1, 0.3)	0.415	0.26
Extras (ARFS 0–2)	1.1 (0.1)	1.0 (0.1)	−0.1 (−0.4, 0.2)	0.388	0.13	0.7 (0.1)	1.1 (0.1)	0.4 (0.1, 0.7)	0.011	0.52
Physical activity										
MVPA (min/w)	393.1 (70.2)	455.8 (41.0)	62.7 (−98.2, 223.6)	0.445	0.19	535.5 (59.6)	402.6 (63.8)	−133.0 (−305.3, 39.4)	0.131	0.39
Sleep health										
Sleep quality score (PSQI Global score)	6.7 (0.7)	5.8 (0.4)	−0.8 (−2.4, 0.7)	0.292	0.25	6.1 (0.6)	5.7 (0.6)	−0.4 (−2.0, 1.3)	0.659	0.13
Sleep duration (min/d)	403.4 (11.9)	405.8 (7.0)	2.5 (−24.6, 29.5)	0.859	0.05	401.1 (10.0)	413.8 (10.7)	12.7 (−16.4, 41.8)	0.393	0.22

Abbreviations: %EI, percentage of energy intake; ARFS, Australian recommended food score; d, day; g, grams; kJ, kilojoules; m, mean; mg, milligrams; MVPA, moderate-to-vigorous intensity physical activity; PSQI, Pittsburgh Sleep Quality Index; SE, standard error; w, week. ***Note:*** Boldface indicates statistical significance ($p < 0.05$).

Table 3. Between-group differences in dietary intake, physical activity and sleep health at 12 months ($n = 54$).

Intake/Behaviour	Control Group ($n = 17$) M (SE)	Pooled Intervention Group ($n = 37$) M (SE)	Between-Group Difference (95% CI)	p-Value	Effect Size (Cohen's d)	Traditional Intervention Group ($n = 23$) M (SE)	Enhanced Intervention Group ($n = 14$) M (SE)	Between-Group Difference (95% CI)	p-Value	Effect Size (Cohen's d)
Energy intake										
Total energy intake (kJ/d)	9125 (450)	8211 (372)	−913 (−2033, 207)	0.11	0.47	8673 (490)	7949 (654)	−724 (−2389, 940)	0.394	0.29
Macronutrient intake										
Carbohydrate (%EI)	42.9 (1.9)	42.0 (1.0)	−0.8 (−5.2, 3.6)	0.715	0.11	43.6 (0.9)	40.5 (2.1)	−3.1 (−7.7, 1.5)	0.189	0.46
Total fats (%EI)	32.2 (1.1)	32.1 (0.7)	−0.1 (−2.7, 2.5)	0.931	0.02	31.6 (0.8)	33.3 (1.6)	1.7 (−1.8, 5.2)	0.341	0.32
Saturated fat (%EI)	13.3 (0.5)	12.9 (0.4)	−0.4 (−1.6, 0.9)	0.583	0.2	13.3 (0.5)	12.6 (0.6)	−0.7 (−2.3, 0.9)	0.41	0.3
Monounsaturated fat (%EI)	12.2 (0.5)	12.0 (0.3)	−0.1 (−1.0, 1.3)	0.811	0.05	11.5 (0.3)	12.8 (0.7)	1.3 (−0.3, 2.8)	0.107	0.55
Polyunsaturated fat (%EI)	3.6 (0.2)	4.1 (0.1)	0.4 (−0.1, 0.9)	0.119	0.39	3.9 (0.1)	4.4 (0.3)	0.5 (−1.2, 0.1)	0.103	0.56
Protein (%EI)	19.5 (1.0)	19.9 (0.5)	0.3 (−2.0, 2.6)	0.777	0.07	18.9 (0.5)	21.3 (1.1)	2.4 (0.1, 4.6)	**0.04**	0.74
Alcohol (%EI)	5.0 (1.0)	5.8 (0.5)	0.7 (1.4, 2.9)	0.508	0.19	5.7 (0.6)	4.6 (0.7)	−1.2 (−2.8, 0.5)	0.16	0.51
Micronutrient intake										
Sugars (g/d)	119.2 (13.1)	107.3 (5.8)	−11.9 (−40.0, 16.2)	0.407	0.24	114.6 (6.9)	99.5 (10.3)	−15.1 (−39.9, 9.6)	0.232	0.41
Fibre (g/d)	27.7 (1.4)	26.7 (1.1)	−1.0 (−4.5, 2.6)	0.586	0.16	27.4 (1.4)	27.8 (2.3)	0.3 (−5.3, 6.0)	0.908	0.04
Sodium (mg/d)	2265.3 (146.6)	1939.1 (93.0)	−326.2 (−662.6, 10.2)	0.057	0.56	2077.5 (130.2)	1814.7 (138.3)	−262.8 (−639.8, 114.1)	0.172	0.46
Nutrient-dense food intake										
Nutrient-dense foods (%EI)	62.9 (2.3)	66.0 (1.6)	3.1 (−2.5, 8.8)	0.281	0.31	63.9 (1.8)	71.3 (2.4)	7.4 (1.3, 13.5)	**0.017**	0.81
Vegetables (%EI)	9.6 (0.6)	10.1 (0.7)	0.4 (−1.3, 2.1)	0.619	0.13	9.4 (0.7)	11.1 (1.4)	1.7 (−1.6, 5.1)	0.312	0.34
Vegetables (serves/d)	5.0 (0.4)	4.9 (0.3)	−0.1 (−1.0, 0.7)	0.729	0.07	4.9 (0.4)	5.2 (0.4)	0.3 (−0.7, 1.4)	0.52	0.2
Fruits (%EI)	6.9 (0.8)	8.5 (0.6)	1.6 (−0.3, 3.5)	0.093	0.47	9.4 (0.8)	6.8 (1.0)	−2.6 (−5.3, 0.2)	0.066	0.63
Fruits (serves/d)	1.9 (0.2)	2.1 (0.1)	0.1 (−0.4, 0.7)	0.62	0.1	2.4 (0.2)	1.5 (0.2)	−0.8 (−1.5, −0.2)	**0.011**	0.9
Milk, yoghurt, cheese (%EI)	9.6 (1.2)	9.3 (1.1)	−0.3 (−3.4, 2.7)	0.824	0.06	10.2 (1.5)	8.9 (1.6)	−1.3 (−6.1, 3.5)	0.59	0.18
Breads, cereals, rice, pasta, noodles (%EI)	16.4 (1.4)	15.9 (1.0)	0.5 (−4.0, 2.9)	0.769	0.08	16.2 (1.2)	17.2 (2.1)	1.0 (−4.0, 6.0)	0.695	0.14
Lean meats, fish, poultry, eggs, nuts (%EI)	15.9 (2.0)	15.9 (1.0)	0.04 (−4.6, 4.5)	0.985	0.01	13.7 (0.9)	18.1 (1.9)	4.5 (0.4, 8.5)	**0.029**	0.76
Protein alternatives, nuts, eggs, beans	3.9 (0.5)	5.1 (0.6)	1.3 (−0.3, 2.9)	0.121	0.48	4.5 (0.5)	6.3 (1.4)	1.8 (−1.1, 4.7)	0.216	0.41
Energy-dense food intake										
Energy-dense, nutrient-poor foods (%EI)	37.1 (2.3)	34.0 (1.6)	−3.1 (−8.8, 2.5)	0.281	0.31	36.1 (1.8)	28.7 (2.4)	−7.4 (−13.5, −1.3)	**0.017**	0.81
Fatty meats (%EI)	1.5 (0.2)	1.6 (0.3)	0.05 (−0.8, 0.7)	0.883	0.04	1.8 (0.4)	1.1 (0.2)	−0.6 (−1.6, 0.3)	0.205	0.41
Fried/takeaway foods (%EI)	9.6 (0.9)	7.1 (0.6)	−2.4 (−4.7, −0.2)	0.034	0.64	8.4 (0.7)	4.8 (1.2)	−3.6 (−6.5, −0.7)	**0.015**	0.81
Confectionery (%EI)	4.8 (0.6)	4.5 (0.4)	−0.3 (−1.8, 1.2)	0.688	0.13	4.7 (0.6)	4.2 (0.7)	−0.4 (−2.2, 1.4)	0.64	0.15
Baked sweet products (%EI)	4.1 (0.5)	4.4 (0.4)	−0.3 (−1.1, 1.6)	0.696	0.13	5.5 (0.6)	3.5 (0.6)	−2.0 (−3.6, −0.4)	**0.014**	0.85
Packaged snacks (%EI)	2.8 (0.5)	1.7 (0.3)	−1.1 (−2.2, 0.1)	0.067	0.54	2.1 (0.4)	1.0 (0.3)	−1.1 (−2.2, −0.03)	**0.043**	0.75
Sweetened drinks (%EI)	2.9 (1.1)	1.8 (0,4)	−1.1 (−3.4, 1.1)	0.322	0.29	1.2 (0.4)	2.3 (0.8)	1.1 (−0.5, 2.7)	0.194	0.47
Diet quality										
Total diet quality (ARFS 0–73)	36.9 (1.5)	37.5 (1.2)	0.6 (−3.2, 4.4)	0.752	0.09	38.6 (0.8)	36.8 (2.7)	−1.8 (−7.4, 3.8)	0.532	0.22
Vegetables (ARFS 0–21)	15.3 (0.9)	15.7 (0.5)	0.4 (−1.6, 2.3)	0.725	0.12	15.6 (0.4)	16.0 (1.3)	0.4 (−2.3, 3.0)	0.779	0.1
Fruits (ARFS 0–12)	5.4 (0.5)	5.8 (0.4)	0.4 (−0.9, 1.7)	0.589	0.17	6.0 (0.4)	5.5 (1.0)	−0.5 (−2.6, 1.6)	0.645	0.15
Protein foods—meat/flesh (ARFS 0–7)	2.8 (0.2)	3.0 (0.1)	0.2 (−0.4, 0.7)	0.529	0.2	3.0 (0.2)	2.8 (0.2)	−0.2 (−0.8, 0.5)	0.632	0.23
Protein foods—meat/flesh alternatives (ARFS 0–6)	1.9 (0.2)	2.4 (0.2)	0.5 (−0.1, 1.1)	0.082	0.49	2.4 (0.2)	2.3 (0.4)	−0.1 (−1.0, 0.8)	0.811	0.07
Grains, breads, and cereals (ARFS 0–13)	5.7 (0.5)	5.3 (0.3)	−0.4 (−1.4, 0.7)	0.516	0.23	5.7 (0.3)	5.0 (0.4)	−0.7 (−1.8, 0.4)	0.229	0.4
Dairy foods (ARFS 0–11)	3.9 (0.4)	3.8 (0.2)	−0.1 (−0.9, 0.8)	0.852	0.07	4.5 (0.3)	3.2 (0.3)	−1.2 (−2.2, −0.3)	**0.011**	0.81
Water (AFRS 0–1)	0.7 (0.1)	0.7 (0.1)	0.03 (−0.2, 0.2)	0.805	0.09	0.6 (0.1)	0.8 (0.1)	0.2 (0.01, 0.4)	**0.037**	0.68
Extras (ARFS 0–2)	1.0 (0.1)	0.8 (0.1)	−0.2 (−0.5, 0.1)	0.241	0.29	0.8 (0.1)	0.9 (0.2)	0.1 (−0.4, 0.6)	0.614	0.17
Physical activity, M (SD)	395.1 (117.4)	433.8 (53.2)	38.7 (−218.5, 296.0)	0.768	0.09	525.2 (76.9)	358.3 (70.2)	−166.9 (−370.2, 36.4)	0.108	0.55

Table 3. *Cont.*

Intake/Behaviour	Control Group (n = 17) M (SE)	Pooled Intervention Group (n = 37) M (SE)	Between-Group Difference (95% CI)	p-Value	Effect Size (Cohen's d)	Traditional Intervention Group (n = 23) M (SE)	Enhanced Intervention Group (n = 14) M (SE)	Between-Group Difference (95% CI)	p-Value	Effect Size (Cohen's d)
MVPA (min/w)										
Sleep health, M (SD)										
Sleep quality score (PSQI Global score)	5.8 (0.6)	5.1 (0.5)	−0.7 (−2.3, 0.9)	0.423	0.26	5.1 (0.7)	5.1 (0.6)	−0.02 (−1.8, 1.7)	0.977	0.01
Sleep duration (min/d)	420.8 (12.7)	422.2 (9.2)	1.4 (−30.2, 33.0)	0.931	0.03	425.5 (13.8)	422.9 (9.0)	−2.6 (−35.9, 30.6)	0.877	0.05

Abbreviations: %EI, percentage of energy intake; ARFS, Australian recommended food score; d, day; g, grams; kJ, kilojoules; m, mean; mg, milligrams; MVPA, moderate-to-vigorous intensity physical activity; PSQI, Pittsburgh Sleep Quality Index; SE, standard error; w, week. ***Note:*** Boldface indicates statistical significance ($p < 0.05$).

At six months, there were no differences in intake of nutrient-dense foods, energy-dense nutrient-poor foods, and diet quality (Table 2) between the traditional and enhanced groups (d = 0.05–0.52). However, at 12 months, significant differences were observed between the traditional and enhanced groups for eight dietary outcomes, with the enhanced group consuming a higher percentage of energy intake from nutrient-dense foods (+7.4 %EI; 95% CI 1.3, 13.5; p = 0.017; d = 0.81), protein (+2.4 %EI; 95% CI 0.1, 4.6; p = 0.040; d = 0.74), and lean meats (+4.5 %EI; 95% CI 0.4, 8.5; p = 0.029; d = 0.76), and lower percentage of energy intake from energy-dense nutrient-poor foods (−7.4 %EI; 95% CI −13.5, −1.3; p = 0.017; d = 0.81), fried/takeaway foods (−3.6 %EI; 95% CI −6.5, −0.7; p = 0.015; d = 0.81), baked sweet products (−2.0 %EI; 95% CI −3.6, −0.4; p = 0.014; d = 0.85), and packaged snacks (−1.1 %EI; 95% CI −2.2, −0.03; p = 0.043; d = 0.75). The enhanced group also reported significantly lower serves of fruit than the traditional group at 12 months (−0.8 serves; 95% CI −1.5, −0.2; p = 0.011; d = 0.90) (Table 3).

4. Discussion

The primary objective was to evaluate whether a multi-component m-Health weight loss intervention in adults with overweight and obesity that targeted dietary, physical activity, and sleep behaviours was effective at improving dietary intake over six months, and after 12 month follow-up, compared with a wait-list control. Results indicated that the pooled intervention, relative to the control, was effective in the short term (six months) in achieving significantly lower total energy and sodium intakes and increased fruit intake. Whilst the pooled intervention group maintained these improvements at 12 months, the differences between the pooled intervention and the control groups were not significant. The secondary objective was to evaluate whether the intervention targeting dietary, physical activity, and sleep behaviours (enhanced) was more effective at improving dietary intake than the dietary and physical activity intervention only (traditional). The results of this study show that at six months there were no significant differences in dietary intake observed between the intervention groups. However, at 12 months, the enhanced group reported a significantly higher percentage of energy intake from nutrient-dense foods, in particular lean meats, and significantly lower percentage of energy intake from energy-dense nutrient-poor foods. The enhanced group reported lower daily energy intake at 12 months than at six months, while traditional group reported increased energy intake between six and 12 months.

The significant improvements in energy, sodium, and fruit intake achieved by the pooled intervention group compared with the control at six months align with results from other dietary interventions [11,57–59]. This suggests that adults with overweight or obesity can change these aspects of dietary intake in the short term when provided with evidenced-based advice. Reduced energy intake and increased fruit intake are associated with reduced risk of weight gain [60], and lower energy intake may assist in achieving clinically significant weight loss [61]. In turn, this may reduce multiple cardiometabolic risk factors including cholesterol, hypertension, and insulin sensitivity [62]. In addition, the reduced sodium intake to less than 2000 mg per day achieved by the pooled intervention group lowers cardiovascular disease risk [63], which is a leading cause of death [64].

This study observed that at 12 months, the pooled intervention group reported a non-statistically significant difference in energy intake relative to the control. This non-significant difference may be attributed to an increase in energy intake by the traditional group between six and 12 months. This indicates that on average, traditional group participants had difficulty in maintaining reduced energy intake beyond the active intervention period. Consequently, longer-term support such as face-to-face/telephone check-ins and social support may be needed, given that accountability has been reported as a strong facilitator of motivation and compliance with weight management behaviour change [65,66]. However, the enhanced group reported a positive shift in energy intake from nutrient-dense and energy-dense foods at 12 months. The two intervention groups did not differ in sleep quality at 12 months, but the enhanced group was the only group to receive the sleep intervention, and it also had the largest improvement in sleep quality from baseline to 12 months. Given that sleep duration is related to sleep quality [67] and short sleep duration influences dietary intake [27], the improved sleep quality of the enhanced group may have contributed to the reduced intake of energy-dense nutrient-poor foods. Although to date, research has focused on how sleep duration and not sleep quality influences dietary intake, and further research is needed to clarify this.

Overall, the results demonstrate that the sleep intervention did not produce significant improvements in indicators of sleep health relative to the traditional intervention. This may be because at baseline, both intervention groups reported sleep duration of ≈6.8 h per night, which was close to meeting the recommendation of 7–9 h per night for adults aged 18–64 years [68]. Relative to the much shorter sleep duration reported in meta-analyses that had an effect on dietary intake (≤5.5 h per night) [18,26,27], it may be that the participants' sleep was not short enough and not improved enough during the intervention period to affect dietary behaviours (e.g., frequency of meals/snacks) and dietary intake. Furthermore, participants in the enhanced and traditional groups reported mean (SD) PSQI global scores of 7.3 (2.8) and 7.0 (3.1) at baseline, respectively. A PSQI global score of ≤5 indicates 'good' sleep quality, and a global score of >5 indicates 'poor' sleep quality, with a score of 21 representing the most severe sleep difficulty [53]. Therefore, it may be that the participants' sleep quality, although classified as poor, was not impaired enough, and not improved enough in the enhanced intervention group to affect dietary intake. In this study, participants were recruited based on body weight status, rather than on physical activity, diet, and sleep behaviours. Future studies may benefit from specifically recruiting adults with overweight and obesity who are poor sleepers.

To the authors' knowledge, this study is the only multiple-behaviour-change weight loss RCT to include a sleep health component and report dietary outcomes in the longer-term (i.e., 12 months). To our knowledge, existing RCTs have limited their investigation to the effect of extending sleep duration on a small number of dietary outcomes, and the studies were all ≤6 weeks in length [29–31,69]. This highlights the unique contribution of the current study, as it is the only published weight loss RCT to include a sleep health component that also measured a large number of dietary outcomes over a longer period. Additional longer-term (>6 months) studies are required to evaluate effects of improved sleep health on dietary intake, as sleep health is multi-dimensional, and changes in behaviour may take some time to become apparent [67]. In terms of overall dietary patterns, there was no significant difference observed between the groups in diet quality at six or 12 months, which is consistent with another sleep intervention [29]. The dietary intervention in this study primarily focused on increased intake of lower energy, nutrient-dense foods (e.g., fruits and vegetables) and reduced intake of energy-dense, nutrient-poor foods and drinks to achieve a reduction in daily energy intake, as reported by the enhanced group at 12 months. The dietary advice was less focused on increasing the food variety, and future dietary interventions are encouraged to focus on this to promote improvements in overall diet quality.

This study had a number of strengths including the RCT design, assessor blinding, validated outcome measures, comprehensive set of dietary outcomes assessed, and generalisability. This study also had some limitations. The study was powered to detect a difference between groups in the primary outcome of weight only, which increased the probability of a type 2 error. As multiple comparisons were performed, the chance of a type 1 error was also increased. The 95% confidence intervals are relatively wide for the between-group differences in total energy intake, sodium intake, and the percentage of energy intake from fruit at six months, and for a number of outcomes at 12 months, which indicates high variability in the between-group differences. As such, these results should be interpreted with some caution. A further limitation is the use of self-reported questionnaires and FFQs, which can result in misreporting, recall bias, measurement error, and vulnerability to social desirability bias [70]. However, the same tool was used to assess diet in all participants at all time points, although the AES could not detect changes in portion sizes, only consumption frequency.

5. Conclusions

This multi-component m-Health diet, physical activity, and sleep behaviour weight loss intervention significantly reduced energy intake and sodium intake, and it increased fruit intake in adults with overweight and obesity in the short term (six months). The enhanced group, who received an additional sleep health intervention, reported improved dietary intake relative to the traditional group in the longer term (12 months). Further adequately powered studies using longer-term sleep interventions are needed to examine whether improving sleep health is an effective strategy for improving dietary intake and weight management.

Supplementary Materials: The following are available online at https://www.mdpi.com/article/10.3390/nu13072468/s1, Figure S1: CONSORT diagram describing study design and flow of participants.

Author Contributions: M.J.D. conceptualised the study. S.F., T.L.B. and C.E.C. contributed to the development of the dietary intervention and the assessment methodology. A.T.R., B.M. and M.J.D. contributed to the development of the physical activity and sleep interventions. S.F., T.L.B., A.T.R., B.M. and M.J.D. contributed to the intervention development and design. M.J.D. and S.F. developed the data analysis plan and conducted the analyses. S.F. drafted the manuscript. All authors provided a critical review of the manuscript, and all authors edited and approved the final version of the manuscript prior to submission. All authors have read and agreed to the published version of the manuscript.

Funding: M.J.D. is supported by a Career Development Fellowship (APP1141606) from the National Health and Medical Research Council (NHMRC). M.J.D. (ID 100029) was supported by Future Leaders Fellowships from the National Heart Foundation of Australia. T.L.B. is supported by a NHMRC investigator grant (G1801414). C.E.C. was supported by a NHMRC Senior Research Fellowship (APP1108095). This project is partly supported by a Diabetes Australia General Grant (Y17G-DUNM) and the National Heart Foundation of Australia (105584; 100629).

Institutional Review Board Statement: The reporting of this study adheres to the Consolidated Standards of Reporting Trials (CONSORT) guidelines. Ethical approval was granted by the Human Research Ethics Committee of The University of Newcastle (H-2017-0039), and the trial was prospectively registered with the Australian New Zealand Clinical Trials Registry (ACTRN12617000735358; U1111-1219-2050).

Informed Consent Statement: Written informed consent was obtained from all subjects involved in this study.

Data Availability Statement: Study data are available from the corresponding author (M.J.D.) upon request.

Acknowledgments: The authors would like to thank Megan Whatnall, Amelia Richards, Levi Wade, Sally Barker, and Melina Cooper for their assistance.

Conflicts of Interest: The authors declare no conflict of interest. The funders had no role in the study design, data collection, analysis, decision to publish, or the preparation of this manuscript.

References

1. Chooi, Y.C.; Ding, C.; Magkos, F. The epidemiology of obesity. *Metabolism* **2019**, *92*, 6–10. [CrossRef] [PubMed]
2. OECD Health Policy Studies. *The Heavy Burden of Obesity, The Economics of Prevention*; OECD Publishing: Paris, French, 2019.
3. Collins, C.; Young, A.; Hodge, A. Diet Quality Is Associated with Higher Nutrient Intake and Self-Rated Health in Mid-Aged Women. *J. Am. Coll. Nutr.* **2008**, *27*, 146–157. [CrossRef] [PubMed]
4. Australian Bureau of Statistics. National Health Survey: First Results, 2017–18: Fruit and Vegetable Consumption. Available online: https://www.abs.gov.au/ausstats/abs@.nsf/Lookup/by%20Subject/4364.0.55.001~{}2017-18~{}Main%20Features~{}Fruit%20and%20vegetable%20consumption~{}105 (accessed on 6 May 2019).
5. O'Brien, K.; Hutchesson, M.; Jensen, M.; Morgan, P.; Callister, R.; Collins, C. Participants in an online weight loss program can improve diet quality during weight loss: A randomized controlled trial. *Nutr. J.* **2014**, *13*, 82. [CrossRef] [PubMed]

6. Australian Government National Health and Medical Research Council. *Clinical Practice Guidelines for the Management of Overweight and Obesity in Adults, Adolescents and Children in Australia*; Australian Government National Health and Medical Research Council: Canberra, Australia, 2013.
7. Harris, J.; Felix, L.; Miners, A.; Murray, E.; Michie, S.; Ferguson, E.; Free, C.; Lock, K.; Landon, J.; Edwards, P. Adaptive e-learning to improve dietary behaviour: A systematic review and cost-effectiveness analysis. *Health Technol. Assess.* **2011**, *15*. [CrossRef] [PubMed]
8. Olson, C. Behavioral nutrition interventions using e- and m-health communication technologies: A narrative review. *Annu. Rev. Nutr.* **2016**, *36*, 647–664. [CrossRef] [PubMed]
9. Browne, S.; Minozzi, S.; Bellisario, C.; Sweeney, M.; Susta, D. Effectiveness of interventions aimed at improving dietary behaviours among people at higher risk of or with chronic non-communicable diseases: An overview of systematic reviews. *Eur. J. Clin. Nutr.* **2019**, *73*, 9–23. [CrossRef]
10. Ambrosini, G.; Solis-Trapala, I.; Ahern, A.; Fuller, N.; Holzapfel, C.; Hauner, H.; Caterson, I.; Jebb, S. Greater improvements in diet quality among overweight participants following a group-based commercial weight loss programme than those receiving support to lose weight in primary care. *Nutr. J.* **2018**, *17*, 1–11. [CrossRef]
11. Hutchesson, M.; Collins, C.; Morgan, P.; Watson, J.F.; Guest, M.; Callister, R. Changes to dietary intake during a 12-week commercial web-based weight loss program: A randomized controlled trial. *Eur. J. Clin. Nutr.* **2014**, *68*, 64–70. [CrossRef]
12. Collins, C.E. Dietary strategies for successful weight loss and maintenance: More evidence required. *J. Am. Diet. Assoc.* **2011**, *111*, 1822–1825. [CrossRef]
13. Dulloo, A.G.; Miles-Chan, J.L.; Montani, J.-P. Nutrition, movement and sleep behaviours: Their interactions in pathways to obesity and cardiometabolic diseases. *Obes. Rev.* **2017**, *18*, 3–6. [CrossRef]
14. Chaput, J.P. Sleep patterns, diet quality and energy balance. *Physiol. Behav.* **2014**, *134*, 86–89. [CrossRef] [PubMed]
15. Nedeltcheva, A.; Kilkus, J.; Imperial, J.; Kasza, K.; Schoeller, D.; Penev, P. Sleep curtailment is accompanied by increased intake of calories from snacks. *Am. J. Clin. Nutr.* **2009**, *89*, 126–133. [CrossRef] [PubMed]
16. Spaeth, A.M.; Dinges, D.F.; Goel, N. Effects of experimental sleep restriction on weight gain, caloric intake, and meal timing in healthy adults. *Sleep* **2013**, *36*, 981–990. [CrossRef] [PubMed]
17. Markwald, R.R.; Melanson, E.L.; Smith, M.R.; Higgins, J.; Perreault, L.; Eckel, R.H.; Wright, K.P., Jr. Impact of insufficient sleep on total daily energy expenditure, food intake, and weight gain. *Proc. Natl. Acad. Sci. USA* **2013**, *110*, 5695–5700. [CrossRef]
18. Zhu, B.; Shi, C.; Park, C.G.; Zhao, X.; Reutrakul, S. Effects of sleep restriction on metabolism-related parameters in healthy adults: A comprehensive review and meta-analysis of randomized controlled trials. *Sleep Med. Rev.* **2019**, *45*, 18–30. [CrossRef]
19. St-Onge, M.-P.; McReynolds, A.; Trivedi, Z.B.; Roberts, A.L.; Sy, M.; Hirsch, J. Sleep restriction leads to increased activation of brain regions sensitive to food stimuli. *Am. J. Clin. Nutr.* **2012**, *95*, 818–824. [CrossRef]
20. St-Onge, M.P.; Wolfe, S.; Sy, M.; Shechter, A.; Hirsch, J. Sleep restriction increases the neuronal response to unhealthy food in normal-weight individuals. *Int. J. Obes.* **2014**, *38*, 411–416. [CrossRef]
21. Fang, Z.; Spaeth, A.M.; Ma, N.; Zhu, S.; Hu, S.; Goel, N.; Detre, J.A.; Dinges, D.F.; Rao, H. Altered salience network connectivity predicts macronutrient intake after sleep deprivation. *Sci. Rep.* **2015**, *5*, 8215. [CrossRef]
22. Greer, S.; Goldstein, A.; Walker, M. The impact of sleep deprivation on food desire in the human brain. *Nat. Commun.* **2013**, *4*, 2259. [CrossRef]
23. Broussard, J.; Kilkus, J.; Delebecque, F.; Abraham, V.; Day, A.; Whitmore, H.; Tasali, E. Elevated ghrelin predicts food intake during experimental sleep restriction. *Obesity* **2016**, *24*, 132–138. [CrossRef]
24. Taheri, S.; Lin, L.; Austin, D.; Young, T.; Mignot, E. Short sleep duration is associated with reduced leptin, elevated ghrelin, and increased body mass index. *PLoS Med.* **2004**, *1*, e62. [CrossRef]
25. Nedeltcheva, A.V.; Scheer, F.A.J.L. Metabolic effects of sleep disruption, links to obesity and diabetes. *Curr. Opin. Endocrinol. Diabetes Obes.* **2014**, *21*, 293–298. [CrossRef]
26. Al Khatib, H.K.; Harding, S.V.; Darzi, J.; Pot, G.K. The effects of partial sleep deprivation on energy balance: A systematic review and meta-analysis. *Eur. J. Clin. Nutr.* **2017**, *71*, 614–624. [CrossRef]
27. Fenton, S.; Burrows, T.L.; Skinner, J.A.; Duncan, M.J. The influence of sleep health on dietary intake: A systematic review and meta-analysis of intervention studies. *J. Hum. Nutr. Diet.* **2021**, *34*, 273–285. [CrossRef]
28. Pizinger, T.; Aggarwal, B.; St-Onge, M.-P. Sleep extension in short sleepers: An evaluation of feasibility and effectiveness for weight management and cardiometabolic disease prevention. *Front. Endocrinol.* **2018**, *9*. [CrossRef]
29. Al Khatib, H.K.; Hall, W.L.; Creedon, A.; Ooi, E.; Masri, T.; McGowan, L.; Harding, S.V.; Darzi, J.; Pot, G.K. Sleep extension is a feasible lifestyle intervention in free-living adults who are habitually short sleepers: A potential strategy for decreasing intake of free sugars? A randomized controlled pilot study. *Am. J. Clin. Nutr.* **2018**, *107*, 43–53. [CrossRef] [PubMed]
30. Haack, M.; Serrador, J.; Cohen, D.; Simpson, N.; Meier-Ewert, H.; Mullington, J. Increasing sleep duration to lower beat-to-beat blood pressure: A pilot study. *J. Sleep Res.* **2013**, *22*, 295–304. [CrossRef] [PubMed]
31. Tasali, E.; Kahn, E.; Hoddy, K.; Kilkus, J.; Wroblewski, K.; Schoeller, D.A. Sleep Extension Reduces Energy Intake in Free-Living Overweight Adults: A Randomized Controlled Study. In Proceedings of the D109. SRN: Outcomes and Impact of Sleep and Respiratory Disorders, Dallas, TX, USA, 22 May 2019. [CrossRef]
32. Burrows, T.; Fenton, S.; Duncan, M. Diet and sleep health: A scoping review of intervention studies in adults. *J. Hum. Nutr. Diet.* **2020**, *33*, 308–329. [CrossRef]

33. Duncan, M.J.; Fenton, S.; Brown, W.J.; Collins, C.E.; Glozier, N.; Kolt, G.S.; Holliday, E.G.; Morgan, P.J.; Murawski, B.; Plotnikoff, R.C.; et al. Efficacy of a multi-component m-Health weight-loss intervention in overweight and obese adults: A randomised controlled trial. *Int. J. Environ. Res. Public Health* **2020**, *17*, 6200. [CrossRef] [PubMed]
34. Duncan, M.J.; Brown, W.J.; Burrows, T.L.; Collins, C.E.; Fenton, S.; Glozier, N.; Kolt, G.S.; Morgan, P.J.; Hensley, M.; Holliday, E.G.; et al. Examining the efficacy of a multicomponent m-Health physical activity, diet and sleep intervention for weight loss in overweight and obese adults: Randomised controlled trial protocol. *BMJ Open* **2018**, *8*, e026179. [CrossRef] [PubMed]
35. Schulz, K.F.; Altman, D.G.; Moher, D. CONSORT 2010 Statement: Updated guidelines for reporting parallel group randomised trials. *BMJ* **2010**, *340*, c332. [CrossRef]
36. Abraham, C.; Michie, S. A taxonomy of behavior change techniques used in interventions. *Health Psychol.* **2008**, *27*, 379–387. [CrossRef]
37. Bandura, A. The primacy of self-regulation in health promotion. *Appl. Psychol. Health Well Being* **2005**, *54*, 245–254. [CrossRef]
38. Teixeira, P.J.; Marques, M.M. Health Behavior Change for Obesity Management. *Obes. Facts* **2017**, *10*, 666–673. [CrossRef]
39. Burrows, T.L.; Hutchesson, M.J.; Rollo, M.E.; Boggess, M.M.; Guest, M.; Collins, C.E. Fruit and vegetable intake assessed by food frequency questionnaire and plasma carotenoids: A validation study in adults. *Nutrients* **2015**, *7*, 3240–3251. [CrossRef]
40. Collins, C.E.; Watson, J.F.; Guest, M.; Boggess, M.M.; Duncanson, K.; Pezdirc, K.; Rollo, M.; Hutchesson, M.J.; Burrows, T.L. Reproducibility and comparative validity of a food frequency questionnaire for adults. *Clin. Nutr.* **2014**, *33*, 906–914. [CrossRef]
41. Australian Government National Health and Medical Research Council. Eat for Health Educator Guide. Available online: https://www.eatforhealth.gov.au/sites/default/files/files/the_guidelines/n55b_eat_for_health_educators_guide.pdf (accessed on 20 May 2019).
42. Collins, C.E.; Neve, M.; Palmer, M.; Wicks, L.; Williams, L.; Baines, S.; Burrows, T.; Hure, A. *DAA Best Practice Guidelines for the Treatment of Overweight and Obesity in Adults*; Report to Inform the 2011 Revision of the 2005 Guidelines; Dietitians Association of Australia: Canberra, Australia, 2012.
43. Duncan, M.; Vandelanotte, C.; Kolt, G.S.; Rosenkranz, R.R.; Caperchione, C.M.; George, E.S.; Ding, H.; Hooker, C.; Karunanithi, M.; Maeder, A.J.; et al. Effectiveness of a web- and mobile phone-based intervention to promote physical activity and healthy eating in middle-aged males: Randomized controlled trial of the ManUp study. *J. Med. Internet Res.* **2014**, *16*, e136. [CrossRef] [PubMed]
44. Williams, R.L.; Rollo, M.E.; Schumacher, T.; Collins, C. Diet quality scores of Australian adults who have completed the Healthy Eating Quiz. *Nutrients* **2017**, *9*, 880. [CrossRef] [PubMed]
45. Australian Government Department of Health. Australia's Physical Activity and Sedentary Behaviour Guidelines and the Australian 24-hour Movement Guidelines. Available online: https://www1.health.gov.au/internet/main/publishing.nsf/Content/health-pubhlth-strateg-phys-act-guidelines#npa1864 (accessed on 20 May 2019).
46. Murawski, B.; Plotnikoff, R.C.; Rayward, A.T.; Oldmeadow, C.; Vandelanotte, C.; Brown, W.J.; Duncan, M.J. Efficacy of an m-Health Physical Activity and Sleep Health Intervention for Adults: A Randomized Waitlist-Controlled Trial. *Am. J. Prev. Med.* **2019**, *57*, 503–514. [CrossRef] [PubMed]
47. Rayward, A.T.; Murawski, B.; Duncan, M.J.; Holliday, E.G.; Vandelanotte, C.; Brown, W.J.; Plotnikoff, R.C. Efficacy of an m-health physical activity and sleep intervention to improve sleep quality in middle-aged adults: The Refresh Study randomized controlled trial. *Ann. Behav. Med.* **2020**, *54*, 1–14. [CrossRef] [PubMed]
48. Collins, C.E.; Burrows, T.; Rollo, M.E.; Boggess, M.M.; Watson, J.F.; Guest, M.; Duncanson, K.; Pezdirc, K.; Hutchesson, M.J. The comparative validity and reproducibility of a diet quality index for adults: The Australian Recommended Food Score. *Nutrients* **2015**, *7*, 785–798. [CrossRef]
49. Bush, K.; Kivlahan, D.R.; McDonell, M.B.; Fihn, S.D.; Bradley, K.A. The AUDIT alcohol consumption questions (AUDIT-C): An effective brief screening test for problem drinking. Ambulatory Care Quality Improvement Project (ACQUIP). Alcohol Use Disorders Identification Test. *Arch. Intern. Med.* **1998**, *158*, 1789–1795. [CrossRef] [PubMed]
50. Timperio, A.; Salmon, J.; Crawford, D.V. Validity and reliability of a physical activity recall instrument among overweight and non-overweight men and women. *J. Sci. Med. Sport* **2003**, *6*, 477–491. [CrossRef]
51. Brown, W.J.; Trost, S.; Bauman, A.; Mummery, K.; Owen, N. Test-retest reliability of four physical activity measures used in population surveys. *J. Sci. Med. Sport* **2004**, *7*, 205–215. [CrossRef]
52. Reeves, M.M.; Marshall, A.L.; Owen, N.; Winkler, E.A.; Eakin, E.G. Measuring physical activity change in broad-reach intervention trials. *J. Phys. Act. Health* **2010**, *7*, 194–202. [CrossRef] [PubMed]
53. Buysse, D.J.; Reynolds, C.F.; Monk, T.H.; Berman, S.R.; Kupfer, D.J. The Pittsburgh sleep quality index: A new instrument for psychiatric practice and research. *Psychiatry Res.* **1989**, *28*, 193–213. [CrossRef]
54. Backhaus, J.; Junghanns, K.; Broocks, A.; Riemann, D.; Hohagen, F. Test–retest reliability and validity of the Pittsburgh Sleep Quality Index in primary insomnia. *J. Psychosom. Res.* **2002**, *53*, 737–740. [CrossRef]
55. StataCorp. *Stata Statistical Software: Release 15*; StataCorp LLC: College Station, TX, USA, 2017.
56. Cohen, J. *Statistical Power Analyses for the Behavioral Sciences*; Academic Press: New York, NY, USA, 1988.
57. Aguiar, E.; Morgan, P.; Collins, C.; Plotnikoff, R.; Young, M.; Callister, R. Efficacy of the Type 2 Diabetes Prevention Using LifeStyle Education Program RCT. *Am. J. Prev. Med.* **2015**, *50*. [CrossRef]
58. Guyenet, S. Impact of Whole, Fresh Fruit Consumption on Energy Intake and Adiposity: A Systematic Review. *Front. Nutr.* **2019**, *6*. [CrossRef]

59. Blomfield, R.; Collins, C.; Hutchesson, M.; Young, M.D.; Jensen, M.; Callister, R.; Morgan, P. Impact of self-help weight loss resources with or without online support on the dietary intake of overweight and obese men: The SHED-IT randomised controlled trial. *Obes. Res. Clin. Pract.* **2014**, *8*, e476–e487. [CrossRef]
60. Schlesinger, S.; Neuenschwander, M.; Schwedhelm, C.; Hoffmann, G.; Bechthold, A.; Boeing, H.; Schwingshackl, L. Food Groups and Risk of Overweight, Obesity, and Weight Gain: A Systematic Review and Dose-Response Meta-Analysis of Prospective Studies. *Adv. Nutr.* **2019**, *10*, 205–218. [CrossRef]
61. American Dietetic Association. Position of the American Dietetic Association: Weight Management. *J. Am. Diet. Assoc.* **2009**, *109*, 330–346. [CrossRef]
62. Kraus, W.; Bhapkar, M.; Huffman, K.; Pieper, C.; Krupa Das, S.; Redman, L.; Villareal, D.; Rochon, J.; Roberts, S.; Ravussin, E.; et al. 2 years of calorie restriction and cardiometabolic risk (CALERIE): Exploratory outcomes of a multicentre, phase 2, randomised controlled trial. *Lancet Diabetes Endocrinol.* **2019**, *7*, 673–683. [CrossRef]
63. Heart Foundation. Salt & Heart Healthy Eating: Position Statement. Available online: https://www.heartfoundation.org.au/getmedia/7fdc9882-713b-405f-8e54-8de413e0b44d/190729_Nutrition_Position_Statement_-_Salt2.pdf (accessed on 18 July 2021).
64. World Health Organization. Raised Blood Pressure: Situation and Trends. Global Health Observatory. Available online: https://www.who.int/gho/ncd/risk_factors/blood_pressure_prevalence/en/ (accessed on 18 July 2021).
65. Carels, R.A.; Burmeister, J.M.; Koball, A.M.; Oehlhof, M.W.; Hinman, N.; LeRoy, M.; Bannon, E.; Ashrafioun, L.; Storfer-Isser, A.; Darby, L.A.; et al. A randomized trial comparing two approaches to weight loss: Differences in weight loss maintenance. *J. Health Psychol.* **2014**, *19*, 296–311. [CrossRef] [PubMed]
66. Metzgar, C.J.; Preston, A.G.; Miller, D.L.; Nickols-Richardson, S.M. Facilitators and barriers to weight loss and weight loss maintenance: A qualitative exploration. *J. Hum. Nutr. Diet.* **2015**, *28*, 593–603. [CrossRef]
67. Buysse, D. Sleep Health: Can We Define It? Does It Matter? *Sleep* **2014**, *37*, 9–17. [CrossRef] [PubMed]
68. Hirshkowitz, M.; Whiton, K.; Albert, S.M.; Alessi, C.; Bruni, O.; DonCarlos, L.; Hazen, N.; Herman, J.; Katz, E.S.; Kheirandish-Gozal, L.; et al. National Sleep Foundation's sleep time duration recommendations: Methodology and results summary. *Sleep Health* **2015**, *1*, 40–43. [CrossRef] [PubMed]
69. So-Ngern, A.; Chirakalwasan, N.; Saetung, S.; Thakkinstian, A.; Reutrakul, S. Effects of sleep extension on glucose metabolism in chronically sleep-deprived individuals. *J. Sleep Res.* **2019**, *27* (Suppl. 1), 314. [CrossRef]
70. National Institutes of Health, National Cancer Institute. Dietary Assessment Primer. Food Frequency Questionnaire at a Glance. Available online: http://dietassessmentprimer.cancer.gov/profiles/questionnaire/ (accessed on 18 July 2021).

nutrients

MDPI

Article

Diet-Related Phototoxic Reactions in Psoriatic Patients Undergoing Phototherapy: Results from a Multicenter Prospective Study

Alessia Pacifico [1], Rosalynn R. Z. Conic [2], Antonio Cristaudo [1], Sergio Garbarino [3], Marco Ardigò [1], Aldo Morrone [1], Paolo Iacovelli [1], Sara di Gregorio [4], Paolo Daniele Maria Pigatto [4,5], Ayman Grada [6], Steven Richard Feldman [7], Egeria Scoditti [8], Khalaf Kridin [9,†] and Giovanni Damiani [4,5,10,*,†]

1 Clinical Dermatology Department, San Gallicano Dermatological Institute, IRCCS, 00144 Rome, Italy; alessia.pacifico@gmail.com (A.P.); antonio.cristaudo@libero.it (A.C.); ardigo.marco@gmail.com (M.A.); aldomorrone54@gmail.com (A.M.); paolo.iacovelli@ifo.gov.it (P.I.)
2 Department of Preventive Medicine, Maryland University, Baltimore, MD 21201, USA; ruzica.conic@gmail.com
3 Department of Neuroscience, Rehabilitation, Ophthalmology, Genetics and Maternal/Child Sciences, Polyclinic Hospital San Martino Istituto di Ricovero e Cura a Carattere Scientifico (IRCCS), University of Genoa, 16132 Genoa, Italy; sgarbarino.neuro@gmail.com
4 Clinical Dermatology, IRCCS Istituto Ortopedico Galeazzi, 20161 Milan, Italy; saradigregorio95@gmail.com (S.d.G.); paolo.pigatto@unimi.it (P.D.M.P.)
5 Department of Biomedical, Surgical and Dental Sciences, University of Milan, 20122 Milan, Italy
6 Department of Dermatology, Boston University School of Medicine, Boston, MA 02118, USA; grada@bu.edu
7 Wake Forest School of Medicine, Department of Dermatology, Winston-Salem, NC 27157, USA; sfeldman@wakehealth.edu
8 Institute of Clinical Physiology, National Research Council (CNR), 73100 Lecce, Italy; egeria.scoditti@ifc.cnr.it
9 Lübeck Institute of Experimental Dermatology, University of Lübeck, 23538 Lübeck, Germany; dr_kridin@hotmail.com
10 PhD Degree Program in Pharmacological Sciences, Department of Pharmaceutical and Pharmacological Sciences, University of Padua, 35131 Padua, Italy
* Correspondence: dr.giovanni.damiani@gmail.com; Tel.: +39-02-662141
† These authors contributed equally to this work.

Citation: Pacifico, A.; Conic, R.R.Z.; Cristaudo, A.; Garbarino, S.; Ardigò, M.; Morrone, A.; Iacovelli, P.; di Gregorio, S.; Pigatto, P.D.M.; Grada, A.; et al. Diet-Related Phototoxic Reactions in Psoriatic Patients Undergoing Phototherapy: Results from a Multicenter Prospective Study. *Nutrients* **2021**, *13*, 2934. https://doi.org/10.3390/nu13092934

Academic Editor: Winston Craig

Received: 18 July 2021
Accepted: 19 August 2021
Published: 25 August 2021

Abstract: Vegans and vegetarians often consume foods containing photosensitizers capable of triggering phytophotodermatitis. The potential effect of vegan and vegetarian diets on the response of psoriatic patients undergoing phototherapy is not well characterized. We assessed clinical outcomes of vegan, vegetarian and omnivore adult psoriatic patients undergoing band ultraviolet B phototherapy (NB-UVB). In this multicenter prospective observational study, we enrolled 119 adult, psoriatic patients, of whom 40 were omnivores, 41 were vegetarians and 38 were vegans, with phototherapy indication. After determining the minimum erythemal dose (MED), we performed NB-UVB sessions for 8 weeks. The first irradiation dosage was 70.00% of the MED, then increased by 20.00% (no erythema) or by 10.00% (presence of erythema) until a maximum single dose of 3 J/cm^2 was reached and constantly maintained. All the enrolled patients completed the 8 weeks of therapy. Severe erythema was present in 16 (42.11%) vegans, 7 (17.07%) vegetarians and 4 (10.00%) omnivores ($p < 0.01$). MED was lowest among vegans (21.18 ± 4.85 J/m^2), followed by vegetarians (28.90 ± 6.66 J/m^2) and omnivores (33.63 ± 4.53 J/m^2, $p < 0.01$). Patients with severe erythema were more likely to have a high furocumarin intake (OR 5.67, 95% CI 3.74–8.61, $p < 0.01$). Vegans consumed the highest amount of furocumarin-rich foods. A model examining erythema, adjusted for gender, age, skin type, MED, phototherapy type, number of phototherapies and furocumarin intake, confirmed that vegans had a lower number of treatments. Vegans had more frequent severe erythema from NB-UVB, even after adjustment of the phototherapy protocol for their lower MED. Assessing diet information and adapting the protocol for vegan patients may be prudent.

Keywords: psoriasis; diet; vegans; vegetarians; omnivores; phototherapy; NB-UVB; efficacy; precision medicine; exposures; exposome; inflammation

1. Introduction

For more than a decade, psoriasis has been considered as a systemic, inflammatory disease with cutaneous [1,2] and musculoskeletal manifestations [3] enriched by a wide range of comorbidities [4], ranging from respiratory [5–8] to cardiometabolic ones [9,10].

While obesity and metabolic syndrome are risk factors for psoriasis, the contribution of diet to psoriasis and its treatment outcomes are still not completely defined [11]. Vegans and vegetarians often consume vegetables containing furocumarins (i.e., celery, parsnip, carrot, parsley, citrus and figs), natural photosensitizers and triggers for phytophotodermatitis in both humans and animals [12–18]. Furocumarins are characterized by a coumarin structure with a furan ring and are traditionally classified into linear ("psoralen type") and angular ("angelicin type") types [19]. Furocumarins are well absorbed from food sources and rapidly distributed into several tissues, including skin [19], so phototoxic reactions due to the ingestion of food containing such photosensitizers can occur [20].

Narrow-band ultraviolet B phototherapy (NB-UVB) represents a useful pre-biologics therapeutic option for psoriatic patients [21]. NB-UVB phototherapy may induce psoriasis clearance through pleiotropic effects on human cutaneous immunity, including the activation of apoptosis, DNA damage response and repair pathways, cell cycle control/differentiation and inflammation regulation [22]. Circadian rhythmicity disruption is also involved in psoriasis pathophysiology [23] and may influence NB-UVB response [22,24,25]. Furthermore, the concept that diet influences both inflammation and peripheral clocks is well established [16,17,26,27], but the impact of vegan and vegetarian diets on phototherapy remains neglected. Due to the increasing number of vegetarians and vegans in the general population, we explored the effect of these diets in psoriatic patients undergoing NB-UVB.

2. Materials and Methods

2.1. Study Design

This was a multicenter prospective observational study involving two primary referral phototherapy centers (IRCCS San Gallicano Hospital-Rome, and IRCCS Istituto Ortopedico Galeazzi-Milan, Italy). Patients fulfilling inclusion criteria were enrolled, evaluated (T0) and followed up for 8 weeks (T1). Vegans, vegetarians and omnivores were matched for age, gender, skin phenotype and Psoriasis Area Severity Index (PASI). Saint Rafael Hospital (OSR) local ethical committee approved in 28 May 2021 the study protocol 176/int/2020 and the current study represents a post-hoc analysis. Each patient signed an written informed consent.

2.2. Inclusion/Exclusion Criteria

We enrolled adult patients (≥18 years) with plaque psoriasis and Fitzpatrick skin type II-IV with an indication for NB-UVB phototherapy. Psoriatic erythroderma was excluded because it is an absolute contraindication to phototherapy. Patients with psoriatic arthritis that refuse systemic treatments but underwent NB-UVB were included.

Conversely, patients were excluded in case of (a) previous history of skin cancer or chemotherapy (<5 years before); (b) ongoing therapies potentially aggravating psoriasis or recognized as photosensitizing; (c) undergoing topical anti-psoriatic treatments in the previous 2 weeks or systemic ones in the previous 4 weeks; or (d) acute or chronic infectious comorbidities.

2.3. MED Evaluation and Erythema Quantification

MED testing was performed on normal dorsal skin of each enrolled patient before starting phototherapy using a Multiport UV Solar Simulator 601 (Solar Light CO.INC: Philadelphia, PA, USA) [28]. MED testing was repeated after 8 weeks of phototherapy treatment to evaluate photoadaptation. Before and after the phototherapy course, each patient was evaluated for the erythema index with a skin reflectance measuring instrument (Mexameter MX16, Courage & Khazaka Electronics, Cologne, Germany).

MED was calculated more than 6 hours after the last meal to prevent furocumarins absorption peak [19].

2.4. Phototherapy Protocol

NB-UVB was delivered by a PUVA Combi Light PCL 8000 phototherapy booth (Heverlee, Belgium) equipped with 48 Phillips® TL100 W/01 tubes in both centers, and both used the same dosimeter. The fluence in the center was 15 mW/cm^2, as measured with a Waldmann Variocontrol dosimeter (WaldmannMedizintechnik GmbH, Villingen-Schwenningen, Germany). The initial irradiation dose was chosen based on a subject's minimal erythema dose (MED).

The first irradiation dosage was 70.00% of the MED, which was then increased by 20.00% (in the absence of erythema) or by 10% (in case of a minimal perceptible onset erythema) until a maximum single dose of 3.00 J/cm^2 was reached, after which the dose was constantly maintained. Irradiations were administered three times weekly for up to 8 weeks. In the case of patients susceptible to develop a severe erythema, dosing was reduced to two treatments per week.

2.5. Dietary Evaluation

During the dermatological visit, dietary information was recorded and patients were classified as vegans, vegetarians or omnivores using the following definitions:

- Vegans where patients "ate only all kinds of fruits, vegetables, nuts, grains, seeds, beans and pulses" [29];
- Vegetarians or "fully vegetarians" where patients "never ate meat, poultry and fish, or ate these foods less than once a month" [30];
- Omnivores if they patients not represented by the previous classifications.

We asked patients to list and weigh each food they consumed in a diet diary for a week. From this information, we identified the approximative quantities of polyphenols [31], carotenoids [32], astaxanthins [33] and furocumarins [34] consumed using converting tables already present in the literature. Then, we divided the obtained quantitatives for each evaluated substance into tertiles and named the first tertile as "low" ("none" in case of no intake), the second one as "intermediate" and the third one as "high" consumption.

2.6. Statistical Analysis

Demographic and food intake characteristics between omnivores, vegetarians and vegans were compared using the t-test for continuous variables and the chi-square test for categorical variables. Association between erythema and diet was further examined using a linear regression adjusting for gender, age, skin type, MED, phototherapy type, number of therapies and furocumarin use.

3. Results

3.1. Clinical Data and Demographics

We enrolled 119 psoriatic patients (40 omnivores, 41 vegetarians and 38 vegans) and all of them completed the 8 weeks of therapy. There were no differences in gender ($p = 0.68$), age ($p = 0.79$), skin type ($p = 0.43$), number of phototherapy treatments (0.40) or baseline (T0) PASI ($p = 0.55$).

Remarkably, after 8 weeks of NB-UVB, there was a statistically significant difference in PASI between omnivores, vegans and vegetarians (1.95 vs. 8.87 vs. 4.59, $p < 0.01$) (Table 1).

Table 1. Socio-demographic and clinical characteristics of omnivores, vegetarians and vegans.

Population Characteristics	Omnivores (n = 40)	Vegans (n = 38)	Vegetarians (n = 41)	p
Male, n (%)	19 (47.50)	18 (47.37)	16 (39.02)	0.65
Skin Phototypes				
II	4 (10.00)	9 (23.68)	8 (19.51)	
III	17 (42.50)	16 (42.11)	19 (46.34)	0.43
IV	19 (47.50)	12 (31.58)	14 (34.15)	
V	0 (0.00)	1 (2.63)	0 (0.0))	
Age (average (SD), years)	39.27 (9.24)	40.66 (8.28)	39.80 (8.96)	0.79
PASI (average (SD))				
T0	12.78 (2.83)	12.39 (2.95)	13.15 (3.27)	0.55
T1	1.95 (4.09)	8.87 (4.31)	4.59 (5.24)	<0.01
MED (average (SD), mJ/cm^2)	33.62 (4.53)	21.18 (4.85)	28.90 (6.66)	<0.01
Treatments (average (SD)	15.07 (4.28)	12.13 (6.41)	14.61 (5.38)	0.40
Erythema (%)				
Absent	19 (47.50)	0 (0.00)	14 (34.15)	
Mild	17 (42.50)	10 (26.32)	14 (34.15)	<0.001
Moderate	0 (0.00)	12 (31.58)	6 (14.63)	
Severe	4 (10.00)	16 (42.11)	7 (17.07)	
PsA, n (%)	11 (27.50)	11 (28.95)	11 (26.83)	0.98
DAPSA (average (SD))	14.75 (2.99)	16.00 (2.05)	12.82 (2.32)	0.21

DAPSA: Disease Activity Index for PSoriatic Arthritis, MED: minimal erythematous dose, PASI: Psoriasis Area Severity Index, PsA: psoriatic arthritis, SD: standard deviation.

3.2. Erythema and MED

Severe erythema was present in 16 (42.11%) vegans, 7 (17.07%) vegetarians and 4 (10.00%) omnivores ($p < 0.01$). In total, 19 (47.50%) omnivores and 14 (34.15%) vegetarians experienced no erythema, whilst all vegans experienced some degree of erythema. Patients with severe erythema underwent fewer phototherapy sessions. MED was lowest among vegans (21.18 $\pm$ 4.85 J/m^2), followed by vegetarians (28.90 $\pm$ 6.66 J/m^2) and omnivores (33.63 $\pm$ 4.53 J/m^2, $p < 0.01$).

3.3. Diet Photoactives and Their Impact

Interestingly, 21 (55.26%) vegans consumed high doses of furocumarins, followed by 4 (10.00%) vegetarians and no omnivores ($p < 0.01$). Astaxanthines consumption was high in about 23 (29.11%) vegans and vegetarians, but only in 1 omnivore ($p < 0.01$). High polyphenol consumption was reported by 14 (36.84%) vegans, 11 (26.83%) of vegetarians and only 1 omnivore ($p < 0.01$). There was no difference in carotenoid intake between vegetarians, vegans and omnivores ($p = 0.31$) (Table 2). Only two omnivores consumed a high number of vegetables.

A model examining erythema, adjusted for gender, age, skin type, MED, phototherapy type, number of phototherapies and furocumarins intake, confirmed that vegans had a lower number of treatments. Patients with a high furocumarins intake displayed a 5.67-fold risk (95%CI 3.74–8.61, $p < 0.001$) of developing erythema, followed by medium (OR 1.96 (1.52–2.53)) and low (OR 4.43 (2.93–6.71)) intakes, compared to no furocumarin intake. All the main foods rich in furocumarins consumed by our population achieved a statistically significant difference between omnivores, vegans and vegetarians (see Table 3)

Table 2. Different intakes of the main photoactives between omnivores, vegetarians and vegans.

Level of Photoactives Ingested	Omnivores (*N* = 40)	Vegans (*N* = 38)	Vegetarians (*N* = 41)	*p*
Furocumarins, *N* (%)				
None	29 (72.50)	0 (0.00)	14 (34.15)	
Low	11 (27.50)	10 (26.32)	16 (39.02)	<0.01
Intermediate	0 (0.00)	7 (18.42)	7(17.07)	
High	0 (0.00)	21 (55.26)	4 (9.76)	
Carotenoids, *N* (%)				
None	4 (10.00)	9 (23.68)	8 (19.51)	
Low	17 (42.50)	16 (42.11)	19 (46.34)	0.31
Intermediate	19 (47.50)	12 (31.58)	14 (34.15)	
High	0 (0.00)	1 (2.63)	0 (0.0)	
Astaxanthines, *N* (%)				
None	15 (37.50)	0 (0.00)	0 (0.00)	
Low	16 (40.00)	13 (34.21)	17 (41.46)	<0.01
Intermediate	8 (20.00)	14 (36.84)	12 (29.27)	
High	1 (2.50)	11 (28.95)	12 (29.27)	
Polyphenols, *N* (%)				
None	17 (42.50)	0 (0.00)	0 (0.00)	
Low	18 (45.00)	19 (50.00)	16 (39.02)	<0.01
Intermediate	4 (10.00)	5 (13.16)	14 (34.15)	
High	1 (2.50)	14 (36.84)	11 (26.83)	

Table 3. Main foods rich in furocumarins consumed by the studied population.

Specific Foods Intake	Omnivores (*n* = 40), g/week	Vegans (*n* = 38), g/week	Vegetarians (*n* = 41), g/week	*p*
Parsley (average (SD)	1.0 (2.8)	601.3 (467.4)	35.3 (80.0)	<0.001
Grapefruit (average (SD)	7.5 (26.7)	1611.8 (1122.1)	113.9 (223.6)	<0.001
Lime (average (SD)	2.5 (11.0)	277.6 (245.3)	38.3 (60.9)	<0.001
Lemon (average (SD)	5.6 (17.4)	960.5 (495.0)	34.1 (74.5)	<0.001
Celeriac (average (SD)	0 (0)	717.1 (948.3)	67.1 (187.6)	<0.001
Parsnip (average (SD)	0 (0)	1063.2 (920.1)	23.2 (65.3)	<0.001
Celery (average (SD)	3.8 (13.3)	1355.3 (553.3)	169.5 (304.9)	<0.001
Orange (average (SD)	13.8 (40.8)	1726.3 (615.4)	402.2 (458.8)	<0.001
Cilantro (aver-age (SD)	0.3 (1.6)	96.2 (145.4)	6.7 (13.9)	<0.001
Carrots (average (SD)	8.2 (33.6)	1815.8 (711.1)	525.6 (785.4)	<0.001

4. Discussion

Several anecdotal cases had focused on vegetarian diet in humans with discordant results in establishing a potential cause-effect link between furocumarin-rich foods intake and a potential photosensitivity modification, conversely we found that a higher dietary intake of furocumarins in vegan and vegetarian diets is associated with greater skin sensitivity to NB-UVB phototherapy in psoriatic patients. Due to the increased prevalence of vegans and vegetarians among psoriatic patients [35], dermatologists started to evaluate potential differences in clinical outcomes and therapeutic management [36,37]. Although the Mediterranean diet, which is rich in vegetables, is regarded as beneficial for psoriatic patients [1], the impact of the single-food ingredients on phototherapy is entirely unknown. Furthermore, several foods included in the Mediterranean diet, such as celery, parsnip, carrot, parsley, citrus and figs, contain furocumarins, a natural, well-known photosensitizer able to trigger phytophotodermatitis in both humans and animals [20,38]. Since 100 g of parsnip or celery may contain up to 4–5 mg furocumarins, and a normal US diet only

1.3 mg furocumarins per day, a diet rich in some vegetables may contain up to 13 mg per day in the case of vegans and vegetarians, potentially sustaining photosensitivity [39,40].

Thus, our results support the notion that diet has a clinically meaningful impact on phototherapy management in psoriasis patients. In particular, clinicians treating psoriatic patients with phototherapy should include diet information in medical history. Vegans and vegetarians with psoriasis displayed higher photosensitivity than omnivores, and deserve *ad hoc* phototherapy management [28].

Phototoxicity has been widely proven in UVA wavelengths, since, in this UV range, psoralen DNA monoadducts are efficiently induced, but the greater the quantity ingested with a diet might convert the monoadducts to crosslinks [41,42]. Maximal phototoxic skin reactions in humans are induced by wavelengths between 334 and 425 nm, which perfectly falls into UVA spectrum (320–340 nm); however, in the presence of psoralens, such as furocumarins, phototoxicity may also appear in the spectrum of the less erythematogenic NB-UVB (311–313 nm).

To further empower this statement, small amounts of dietary intake of furocumarins were reported to trigger photoxicity in both PUVA and NB-UVB [20,36].

We found an increased erythemal response to NB-UVB in subjects with high consumptions of furocumarins-rich foods; this is in contrast with Beattie et al., who found no effect when analyzing UVA-related photosensitivity [42].

Our results regarding erythemal response in psoriatic NB-UVB users undergoing different diet regimens and furocumarins intakes could be explained both quantitatively, since vegans and vegetarians consume more food containing furocumarins than omnivores, and qualitatively, since furocumarins bioavailability is affected by storage (fresh vs. frozen) and cooking (fresh vs. cooked) methods [41]. Interestingly, vegans and vegetarians consume foods containing high quantities of both antioxidants (astaxanthines and polyphenols) and furocumarins, but antioxidants may not counteract furocumarins-related photosensitivity/phototoxicity in our cohort of psoriatic patients. It is noteworthy that furocumarins have been shown to induce the secretion of melatonin in humans [43]; melatonin is a known antioxidant and anti-inflammatory hormone that exerts beneficial properties in the skin, and it is lower in psoriatic patients compared to healthy controls [44]. Although melatonin plasma levels were not measured in our cohort, this specific effect by furocumarins on melatonin, coupled with the effect on phototherapy response demonstrated here, may be of interest in the management of disorders, including psoriasis, which is itself associated with abnormalities in circadian rhythms [43].

This pilot study presents limitations, including a small sample size and a lack of skin metabolomic differences between the three considered groups. Furthermore, we used a solar simulator since the monochromator, the most precise instrument to measure MED on the market, was not present in the involved centers [45,46]. As current Italian guidelines did not suggest a specific instrument to measure MED [28], we used a solar simulator that, together with the monochromator, is internationally regarded as the gold standard [28].

5. Conclusions

In our study we found a different diet-related effect to NB-UVB, and this aspect acquired paramount importance, since vegans and vegetarians experience more erythema during NB-UVB, thus limiting the total number of phototherapy sessions in the considered period. Dermatologists should consider dietary assessment in patients' clinical evaluation, and eventually also adapt phototherapy protocols to diet characteristics, since they seems to moderately influence patient management and clinical outcomes. Our clinical recommendations are summarized in Table 4.

Further studies warranted to assess the impact of circadian rhythmicity on NB-UVB phototherapy sessions in clinical practice.

Table 4. Clinical recommendations based on diet for NB-UVB phototherapy in patients with psoriasis.

NB-UVB CLINICAL RECOMMENDATIONS BASED ON DIET	
Omnivores	The starting dose is established after MED evaluation and corresponds to 70% of the MED; then, the dose is increased by 20% (if no erythema) or by 10% (in case of erythema) up to a maximum dosage of 3 J/cm^2
Vegetarians and Vegans	The starting dose is established after MED evaluation and, in case of a low MED (20–25 mJ/cm^2), corresponds to 40% of the MED; then, the dose is increased by 10% (if no erythema) up to a maximum single dose of 2.5 J/cm^2. In case of erythema the dose is maintained constant

MED: minimal erythematous dose.

Author Contributions: Conceptualization, A.P. and G.D.; methodology, R.R.Z.C.; software, R.R.Z.C.; validation, A.P., S.d.G., A.G., E.S. and K.K.; formal analysis, R.R.Z.C. and K.K.; investigation, A.P., A.C., M.A., A.M., P.D.M.P. and G.D.; resources, S.R.F., S.G., E.S., K.K. and G.D.; data curation, R.R.Z.C., S.d.G. and G.D.; writing—original draft preparation, A.P., R.R.Z.C. and G.D.; writing—review and editing, A.P., R.R.Z.C., A.C., S.G., M.A., A.M., P.I., S.d.G., P.D.M.P., A.G., S.R.F., S.G., K.K. and G.D.; visualization, R.R.Z.C. and K.K.; supervision, G.D.; project administration, A.P., A.C. and A.M.; funding acquisition, G.D. All authors have read and agreed to the published version of the manuscript.

Funding: This research received no external funding.

Institutional Review Board Statement: Saint Rafael Hospital (OSR) local ethical committee approved in 28 May 2021 the study protocol 176/int/2020 and the current study represents a post-hoc analysis.

Informed Consent Statement: Informed consent was obtained from all subjects involved in the study.

Data Availability Statement: Data available on request due to ethical restrictions. The data presented in this study are available on request from the corresponding author.

Acknowledgments: We thank our data-manger, Ilaria Controne, for her expertise in nutrition.

Conflicts of Interest: The authors declare no conflict of interest.

References

1. Ben Abdallah, H.; Johansen, C.; Iversen, L. Key Signaling Pathways in Psoriasis: Recent Insights from Antipsoriatic Therapeutics. *Psoriasis (Auckland)* **2021**, *11*, 83–97.
2. Marzano, A.V.; Damiani, G.; Genovese, G.; Gattorno, M. A dermatologic perspective on autoinflammatory diseases. *Clin. Exp. Rheumatol.* **2018**, *36* (Suppl. 110), 32–38.
3. Stober, C. Pathogenesis of psoriatic arthritis. *Best Pract. Res. Clin. Rheumatol.* **2021**, *35*, 101694. [CrossRef]
4. Yan, D.; Blauvelt, A.; Dey, A.K.; Golpanian, R.S.; Hwang, S.T.; Mehta, N.N.; Myers, B.; Shi, Z.-R.; Yosipovitch, G.; Bell, S.; et al. New Frontiers in Psoriatic Disease Research, Part II: Comorbidities and Targeted Therapies. *J. Investig. Dermatol.* **2021**. [CrossRef] [PubMed]
5. Santus, P.; Rizzi, M.; Radovanovic, D.; Airoldi, A.; Cristiano, A.; Conic, R.; Petrou, S.; Pigatto, P.D.M.; Bragazzi, N.; Colombo, D.; et al. Psoriasis and Respiratory Comorbidities: The Added Value of Fraction of Exhaled Nitric Oxide as a New Method to Detect, Evaluate, and Monitor Psoriatic Systemic Involvement and Therapeutic Efficacy. *Biomed Res. Int.* **2018**, *2018*, 3140682. [CrossRef] [PubMed]
6. Damiani, G.; Pacifico, A.; Rizzi, M.; Santus, P.; Bridgewood, C.; Bragazzi, N.; Adawi, M.; Watad, A. Patients with psoriatic arthritis have higher levels of FeNO than those with only psoriasis, which may reflect a higher prevalence of a subclinical respiratory involvement. *Clin. Rheumatol.* **2020**, *39*, 2981–2988. [CrossRef] [PubMed]
7. Damiani, G.; Radaeli, A.; Olivini, A.; Calvara-Pinton, P.; Malerba, M. Increased airway inflammation in patients with psoriasis. *Br. J. Dermatol.* **2016**, *175*, 797–799. [CrossRef] [PubMed]
8. Malerba, M.; Damiani, G.; Radaeli, A.; Ragnoli, B.; Olivini, A.; Calzavara-Pinton, P. Narrowband ultraviolet B phototherapy in psoriasis reduces proinflammatory cytokine levels and improves vitiligo and neutrophilic asthma. *Br. J. Dermatol.* **2015**, *173*, 1544–1545. [CrossRef]
9. Conic, R.R.; Damiani, G.; Schrom, K.P.; Ramser, A.E.; Zheng, C.; Xu, R.; McCormick, T.S.; Cooper, K.D. Psoriasis and Psoriatic Arthritis Cardiovascular Disease Endotypes Identified by Red Blood Cell Distribution Width and Mean Platelet Volume. *J. Clin. Med.* **2020**, *9*, 186. [CrossRef]
10. Seth, D.; Ehlert, A.N.; Golden, J.B.; Damiani, G.; McCormick, T.S.; Cameron, M.J.; Cooper, K.D. Interaction of Resistin and Systolic Blood Pressure in Psoriasis Severity. *J. Investig. Dermatol.* **2020**, *140*, 1279–1282.e1. [CrossRef]

11. Barrea, L.; Balato, N.; Di Somma, C.; Macchia, P.E.; Napolitano, M.; Savanelli, M.C.; Esposito, K.; Colao, A.; Savastano, S. Nutrition and psoriasis: Is there any association between the severity of the disease and adherence to the Mediterranean diet? *J. Transl. Med.* **2015**, *13*, 18. [CrossRef] [PubMed]
12. Ford, A.R.; Siegel, M.; Bagel, J.; Cordoro, K.M.; Garg, A.; Gottlieb, A.; Green, L.J.; Gudjonsson, J.E.; Koo, J.; Lebwohl, M.; et al. Dietary Recommendations for Adults with Psoriasis or Psoriatic Arthritis from the Medical Board of the National Psoriasis Foundation: A Systematic Review. *JAMA Dermatol.* **2018**, *154*, 934–950. [CrossRef]
13. Phan, C.; Touvier, M.; Kesse-Guyot, E.; Adjibade, M.; Hercberg, S.; Wolkenstein, P.; Chosidow, O.; Ezzedine, K.; Sbidian, E. Association between Mediterranean Anti-inflammatory Dietary Profile and Severity of Psoriasis: Results from the NutriNet-Santé Cohort. *JAMA Dermatol.* **2018**, *154*, 1017–1024. [CrossRef]
14. Adawi, M.; Damiani, G.; Bragazzi, N.L.; Bridgewood, C.; Pacifico, A.; Conic, R.R.; Morrone, A.; Malagoli, P.; Pigatto, P.D.M.; Amital, H.; et al. The Impact of Intermittent Fasting (Ramadan Fasting) on Psoriatic Arthritis Disease Activity, Enthesitis, and Dactylitis: A Multicentre Study. *Nutrients* **2019**, *11*, 601. [CrossRef] [PubMed]
15. Damiani, G.; Watad, A.; Bridgewood, C.; Pigatto, P.D.M.; Pacifico, A.; Malagoli, P.; Bragazzi, N.L.; Adawi, M. The Impact of Ramadan Fasting on the Reduction of PASI Score, in Moderate-To-Severe Psoriatic Patients: A Real-Life Multicenter Study. *Nutrients* **2019**, *11*, 277. [CrossRef]
16. Kocic, H.; Damiani, G.; Stamenkovic, B.; Tirant, M.; Jovic, A.; Tiodorovic, D.; Peris, K. Dietary compounds as potential modulators of microRNA expression in psoriasis. *Ther. Adv. Chronic Dis.* **2019**, *10*, 2040622319864805. [CrossRef] [PubMed]
17. Damiani, G.; Bragazzi, N.L.; McCormick, T.S.; Pigatto, P.D.M.; Leone, S.; Pacifico, A.; Tiodorovic, D.; Di Franco, S.; Alfieri, A.; Fiore, M. Gut microbiota and nutrient interactions with skin in psoriasis: A comprehensive review of animal and human studies. *World J. Clin. Cases* **2020**, *8*, 1002–1012. [CrossRef]
18. Melough, M.M.; Chun, O.K. Dietary furocoumarins and skin cancer: A review of current biological evidence. *Food Chem. Toxicol.* **2018**, *122*, 163–171. [CrossRef]
19. Melough, M.M.; Cho, E.; Chun, O.K. Furocoumarins: A review of biochemical activities, dietary sources and intake, and potential health risks. *Food Chem. Toxicol.* **2018**, *113*, 99–107. [CrossRef]
20. Boffa, M.J.; Gilmour, E.; Ead, R.D. Celery soup causing severe phototoxicity during PUVAtherapy. *Br. J. Dermatol.* **1996**, *135*, 334. [CrossRef] [PubMed]
21. Pacifico, A.; Damiani, G.; Iacovelli, P.; Conic, R.R.; Scarabello, A.; Filoni, A.; Malagoli, P.; Bragazzi, N.; Pigatto, P.D.; Morrone, A. Photoadaptation to ultraviolet B TL01 in psoriatic patients. *J. Eur. Acad. Dermatol. Venereol.* **2020**, *34*, 1750–1754. [CrossRef]
22. Addison, R.; Weatherhead, S.C.; Pawitri, A.; Smith, G.R.; Rider, A.; Grantham, H.J.; Cockell, S.J.; Reynolds, N.J. Therapeutic wavelengths of ultraviolet B radiation activate apoptotic, circadian rhythm, redox signalling and key canonical pathways in psoriatic epidermis. *Redox Biol.* **2021**, *41*, 101924. [CrossRef] [PubMed]
23. Ando, N.; Nakamura, Y.; Aoki, R.; Ishimaru, K.; Ogawa, H.; Okumura, K.; Shibata, S.; Shimada, S.; Nakao, A. Circadian Gene Clock Regulates Psoriasis-Like Skin Inflammation in Mice. *J. Investig. Dermatol.* **2015**, *135*, 3001–3008. [CrossRef]
24. Rezzani, R.; Rodella, L.F.; Favero, G.; Damiani, G.; Paganelli, C.; Reiter, R.J. Attenuation of ultraviolet A-induced alterations in NIH3T3 dermal fibroblasts by melatonin. *Br. J. Dermatol.* **2014**, *170*, 382–391. [CrossRef] [PubMed]
25. Damiani, G.; Bragazzi, N.L.; Garbarino, S.; Chattu, V.K.; Shapiro, C.M.; Pacifico, A.; Malagoli, P.; Pigatto, P.D.M.; Conic, R.R.Z.; Tiodorovic, D.; et al. Psoriatic and psoriatic arthritis patients with and without jet-lag: Does it matter for disease severity scores? Insights and implications from a pilot, prospective study. *Chronobiol. Int.* **2019**, *36*, 1733–1740. [CrossRef]
26. Duszka, K.; Wahli, W. Peroxisome Proliferator-Activated Receptors as Molecular Links between Caloric Restriction and Circadian Rhythm. *Nutrients* **2020**, *12*, 3476. [CrossRef] [PubMed]
27. Martínez-Lozano, N.; Tvarijonaviciute, A.; Ríos, R.; Barón, I.; Scheer, F.; Garaulet, M. Late Eating Is Associated with Obesity, Inflammatory Markers and Circadian-Related Disturbances in School-Aged Children. *Nutrients* **2020**, *12*, 2881. [CrossRef] [PubMed]
28. Russo, F.; Vispi, M.; Sirna, R.; Mancini, V.; Bagnoni, G.; Bartoli, L.; Bellini, M.; Brandini, L.; Caproni, M.; Castelli, A.; et al. Tuscan consensus on the use of UVBnb phototherapy in the treatment of psoriasis. *G. Ital. Dermatol. Venereol.* **2019**, *154*, 99–105. [CrossRef]
29. The Vegan Society. So What Do Vegans Eat? Available online: www.vegansociety.com/go-vegan/definition-veganism (accessed on 28 June 2021).
30. Jaacks, L.M.; Kapoor, D.; Singh, K.; Narayan, K.V.; Ali, M.K.; Kadir, M.M.; Mohan, V.; Tandon, N.; Prabhakaran, D. Vegetarianism and cardio- metabolic disease risk factors: Differences between South Asian and US adults. *Nutrition* **2016**, *32*, 975–984. [CrossRef]
31. Phenol-Explorer. Database on Polyphenol Content in Foods. Available online: http://phenol-explorer.eu/foods (accessed on 18 August 2021).
32. Food Science Graz. Available online: https://foodscience.tugraz.at (accessed on 18 August 2021).
33. Fernández-García, E.; Carvajal-Lérida, I.; Jarén-Galán, M.; Garrido-Fernández, J.; Pérez-Gálvez, A.; Hornero-Méndez, D. Carotenoids bioavailability from foods: From plant pigments to efficient biological activities. *Food Res. Int.* **2012**, *46*, 438–450. [CrossRef]
34. Fernández-García, E.; Carvajal-Lérida, I.; Jarén-Galán, M.; Garrido-Fernández, J.; Pérez-Gálvez, A.; Hornero-Méndez, D. Identification and Quantitation of Furocoumarins in Popularly Consumed Foods in the U.S. Using QuEChERS Extraction Coupled with UPLC-MS/MS Analysis. *J. Agric. Food Chem.* **2017**, *65*, 5049–5055.

35. Afifi, L.; Danesh, M.J.; Lee, K.M.; Beroukhim, K.; Farahnik, B.; Ahn, R.S.; Yan, D.; Singh, R.K.; Nakamura, M.; Koo, J.; et al. Dietary Behaviors in Psoriasis: Patient-Reported Outcomes from a U.S. National Survey. *Dermatol. Ther.* **2017**, *7*, 227–242. [CrossRef] [PubMed]
36. Fusano, M.; Fusano, I.; Galimberti, M.G.; Bencini, M.; Bencini, P.L. Comparison of Postsurgical Scars between Vegan and Omnivore Patients. *Dermatol. Surg.* **2020**, *46*, 1572–1576. [CrossRef] [PubMed]
37. Fusano, M.; Zane, C.; Calzavara-Pinton, P.; Bencini, P.L. Photodynamic therapy for actinic keratosis in vegan and omnivore patients: The role of diet on skin healing. *J. Dermatolog. Treat.* **2019**, *32*, 78–83. [CrossRef]
38. Puig, L.; de Moragas, J.M. Enhancement of PUVA phototoxic effects following celery ingestion: Cool broth also can burn. *Arch. Dermatol.* **1994**, *130*, 809–810. [CrossRef]
39. Wolters, M. Die Bedeutung der Ernährung und begleitender Faktoren für die Psoriasis [The significance of diet and associated factors in psoriasis]. *Hautarzt* **2006**, *57*, 999–1004. [CrossRef]
40. Wagstaff, D.J. Dietary exposure to furocumarins. *Regul. Toxicol. Pharmacol.* **1991**, *14*, 261–272. [CrossRef]
41. Ortiz-Rodríguez, L.A.; Reichardt, C.; Hoehn, S.J.; Jockusch, S.; Crespo-Hernández, C.E. Detection of the thietane precursor in the UVA formation of the DNA 6-4 photoadduct. *Nat. Commun.* **2020**, *11*, 3599. [CrossRef]
42. Beattie, P.E.; Wilkie, M.J.V.; Smith, G.; Ferguson, J.; Ibbottson, S.H. Can dietary furocumarin ingestion enhance the erythemal response during high dose UVA1 therapy? *J. Am. Acad. Dermatol.* **2007**, *56*, 84–87. [CrossRef]
43. Souetre, E.; Salvati, E.; Belugou, J.L.; Krebs, B.; Darcourt, G. 5-Methoxypsoralen as a specific stimulating agent of melatonin secretion in humans. *J. Clin. Endocrinol. Metab.* **1990**, *71*, 670–674. [CrossRef]
44. Mozzanica, N.; Tadini, G.; Radaelli, A.; Negri, M.; Pigatto, P.; Morelli, M.; Frigerio, U.; Finzi, A.; Esposti, G.; Rossi, D. Plasma melatonin levels in psoriasis. *Acta Derm.-Venereol.* **1988**, *68*, 312–316. [PubMed]
45. Mackenzi, L.A.; Frainbel, W. Construction and Development of a Grating Monochromator and Its Application to Study of Reaction of Skin to Light. *Br. J. Dermatol.* **1973**, *89*, 251–264. [CrossRef] [PubMed]
46. Moseley, H.; Ferguson, J. Which light source should be used for the investigation of clinical phototoxicity: Monochromator or solar simulator? *Photodermatol. Photoimmunol. Photomed.* **2010**, *26*, 3–6. [CrossRef] [PubMed]

nutrients

MDPI

Article

A Higher Intake of Energy at Dinner Is Associated with Incident Metabolic Syndrome: A Prospective Cohort Study in Older Adults

Ygor Hermenegildo-López [1,2], Carolina Donat-Vargas [1,3,4], Helena Sandoval-Insausti [5], Belén Moreno-Franco [6,7], Monserrat Rodríguez-Ayala [1], Jimena Rey-García [1,8], José Ramón Banegas [1], Fernando Rodríguez-Artalejo [1,3] and Pilar Guallar-Castillón [1,3,*]

1 Department of Preventive Medicine and Public Health, School of Medicine, Universidad Autónoma de Madrid-IdiPaz, CIBERESP (CIBER of Epidemiology and Public Health), 28029 Madrid, Spain; ygorhl@gmail.com (Y.H.-L.); carolina.donat@imdea.org (C.D.-V.); ms.rodriguezayala@gmail.com (M.R.-A.); jimena.reygarcia@gmail.com (J.R.-G.); joseramon.banegas@uam.es (J.R.B.); fernando.artalejo@uam.es (F.R.-A.)
2 Centro de Transfusión de la Comunidad de Madrid, Blood Donation Department, 28032 Madrid, Spain
3 IMDEA-Food Institute, CEI UAM+CSIC, 28049 Madrid, Spain
4 Unit of Nutritional Epidemiology, Institute of Environmental Medicine, Karolinska Institutet, 17177 Stockholm, Sweden
5 Department of Nutrition, Harvard T.H. Chan School of Public Health, Boston, MA 02115, USA; helenagabar@gmail.com
6 Department of Microbiology, Pediatrics, Radiology and Public Health, Universidad de Zaragoza, 50009 Zaragoza, Spain; mbmoreno@posta.unizar.es
7 Fundación Instituto de Investigación Sanitaria de Aragón (IIS Aragón), 50009 Zaragoza, Spain
8 Department of Internal Medicine, Ramón y Cajal University Hospital, 28034 Madrid, Spain
* Correspondence: mpilar.guallar@uam.es; Tel.: +34-914975480

Citation: Hermenegildo-López, Y.; Donat-Vargas, C.; Sandoval-Insausti, H.; Moreno-Franco, B.; Rodríguez-Ayala, M.; Rey-García, J.; Banegas, J.R.; Rodríguez-Artalejo, F.; Guallar-Castillón, P. A Higher Intake of Energy at Dinner Is Associated with Incident Metabolic Syndrome: A Prospective Cohort Study in Older Adults. *Nutrients* **2021**, *13*, 3035. https://doi.org/10.3390/nu13093035

Academic Editors: Egeria Scoditti and Sergio Garbarino

Received: 7 August 2021
Accepted: 25 August 2021
Published: 30 August 2021

Abstract: A higher energy intake (EI) at night has been associated with a higher risk of obesity, while a higher EI at lunch may protect against weight gain. This study examined the association between EI throughout the day and incident metabolic syndrome (MetS) among older adults. A cohort of 607 individuals aged $\geq$ 60 free from MetS at baseline was followed from 2008–2010 until 2015. At baseline, habitual EI was assessed on six eating occasions: breakfast, mid-morning snack, lunch, afternoon snack, dinner, and snacking. MetS was defined according to the harmonized definition. Statistical analyses were performed with logistic regression and adjusted for the main confounders, including total EI, diet quality, and physical activity/sedentary behavior. During follow-up, 101 new MetS cases occurred. Compared to the lowest sex-specific quartile of EI at dinner, the OR (95% confidence interval) for incident MetS were: 1.71 (0.85–3.46) in the second, 1.70 (0.81–3.54) in the third, and 2.57 (1.14–5.79) in the fourth quartile (p-trend: 0.034). Elevated waist circumference and triglycerides were the MetS components that most contributed to this association. A higher EI at dinner was associated with a higher risk of MetS in older adults. Reducing EI at dinner might be a simple strategy to prevent MetS.

Keywords: metabolic syndrome; chronobiology; timing of food, older adults; intake of energy; food intake; dinner intake; eating occasions

1. Introduction

The understanding of circadian rhythm and its impact on health has increased substantially in the last decade. The circadian system is regulated by a central clock located in the suprachiasmatic nucleus and various peripheral tissues located in other brain regions, as well as in the liver, pancreas, gut, white adipose tissue, and skeletal muscle, which act as peripheral clocks. The suprachiasmatic nucleus acts via the autonomic nervous system, and is responsible for the circadian secretion of hormones, as well as the regulation of the temperature throughout the day [1]. Neural and endocrine, as well as behavioral, functions have 24 h rhythms and are major determinants of human metabolism [2].

Some external exposures act as synchronizers of the central and the peripheral clocks. These synchronizers are also known as time-givers ("zeitgebers" in German) [3]. The most relevant of them are the cycles of light and dark exposure [4]. However, there is increasing evidence that nutritional timing, specifically the time of day when energy is consumed, is also part of this regulation [2,5].

The interest in chronobiology has increased from the 1990s due to the identification of relevant genetic determinants of circadian rhythmicity (e.g., the CLOCK gene and its association with obesity) [6,7]. Along with this, the detrimental influence of the misalignment of clocks (such as eating late, shift work, or jet lag) on metabolic risk has contributed further to the emerging body of evidence [8,9].

A higher energy intake (EI) at night is associated with a higher risk of obesity [10], while a higher EI at lunch protects against weight gain [11]. In addition, it has been suggested that avoiding eating late could be a strategy to help prevent obesity and the metabolic syndrome (MetS) [5]. Accordingly, this study aimed to assess, for the first time in the literature, the prospective association between the distribution of EI throughout the day and MetS among community-living older adults in Spain who are at high risk of developing MetS.

2. Materials and Methods

2.1. Study Design and Participants

Data at baseline were taken from the ENRICA Study, which is a representative sample of the noninstitutionalized Spanish population aged 18 and over, recruited in 2008–2010 [12]. Those aged 60 years or older formed the Seniors-ENRICA-1 cohort which was established at baseline with 3518 participants who were followed up until 2015.

Baseline information was collected in three stages: a telephone interview to obtain data on health status, lifestyle, morbidity, and use of health services; a first home visit, performed by a nurse, to obtain blood samples; and a second home visit by trained staff to obtain information on diet and to perform a physical examination. In 2015, the information was updated for those participants who were still alive and agreed to a new blood collection, constituting a subsample of 1821 older adults. Of these, we excluded 676 individuals with MetS at baseline, 72 with missing values that were unable to estimate MetS at baseline, 453 lacking data to calculate the incidence of MetS in 2015, and 13 with missing values for covariates. Thus, the final analytical sample comprised 607 individuals.

All study participants gave written informed consent. The Clinical Research Ethics Committee of the La Paz University Hospital approved the study.

2.2. Dietary Assessment

Food consumption in the previous year was collected using a validated computerized dietary history [13]. This tool allowed for the collection of 860 foods, as well as 184 recipes for dishes commonly eaten in Spain, and includes 127 sets of digitized photographs to better estimate the size of food portions. It automatically converts the food into nutrients and energy using standard Spanish food composition tables [14].

Participants were asked to report all the food consumed and when it was eaten at least once every 15 days. Participants were questioned about food consumed at each eating occasion, as follows: "What do you usually eat for breakfast, lunch, dinner, etc.?". A total of six eating occasions were considered: breakfast, mid-morning snack, lunch, afternoon snack, dinner, and snacking (food consumed between the previous occasions, including before bedtime and when going out for a drink). To facilitate reporting of the food eaten at lunch and dinner, we asked about the first and second courses, desserts, and beverages consumed, as well as about bread and wine. The mean time to complete the diet history was 54 min.

2.3. Metabolic Syndrome

MetS is defined, according to the harmonized definition [15], as having at least three of the following five criteria: abdominal obesity (waist circumference of ≥102 cm in men and ≥88 cm in women); fasting blood glucose of ≥100 mg/dL or receiving antidiabetic drugs; systolic/diastolic blood pressure of ≥130/ 85 mmHg or receiving antihypertensive drugs; serum triglycerides of ≥150 mg/dL; and serum HDL-cholesterol of <40 mg/dL in men and <50 mg/dL in women.

Waist circumference was measured with a flexible inelastic belt-type tape at the midpoint between the last rib and the iliac crest, at the end of a normal exhalation [16]. Blood glucose was determined in 12 h fasting blood samples by the glucose oxidase method [17]. Blood pressure was measured using standard procedures with a validated automatic blood pressure device [18]. Serum triglycerides were measured by the glycerol phosphate oxidase method, and serum HDL-cholesterol by direct elimination/catalase method.

2.4. Potential Confounders

At baseline, study participants reported sociodemographic and lifestyle variables, including age, sex, level of education (primary or less, secondary, or university), smoking status (never, former, current), and being an ex-drinker. Physical activity at leisure time and in the household was obtained with the questionnaire used in the EPIC-cohort of Spain and was expressed in metabolic equivalents (MET-h/week) [19]. Participants also reported the number of hours spent watching TV and sleeping (summing up sleeping time at night and during the day). Weight and height were measured using standard procedures, and the body mass index (BMI) was calculated dividing weight by squared height (kg/m^2). Dieting was evaluated with the question: "Are you on a weight loss diet?". Diet quality was assessed using the MEDAS index of adherence to the Mediterranean diet. To obtain a high MEDAS index indicates a high adherence to the Mediterranean diet [20]. Finally, participants reported if they had been diagnosed with coronary heart disease, chronic respiratory disease, cancer at any site, osteoarthritis, or arthritis.

2.5. Statistical Analysis

Logistic regression models were built to assess the risk of incident MetS. The main independent variables were the percentage of EI at each eating occasion, and the study associations were expressed with odds ratios (OR) and their 95% confidence interval (CI). The percentage of EI was modelled as sex-specific quartiles, and the lowest quartile was the reference group. Two logistic models were built. Model 1 was adjusted for sex, age, education level, and total EI (kcal/day). Model 2 was additionally adjusted for smoking status, ex-drinker status, leisure-time physical activity, physical activity in the household, hours/day spent watching TV, total sleeping time (hours/day), BMI, dieting, ethanol intake (g/day), and MEDAS score, as well as the baseline chronic diagnosed diseases: coronary disease, chronic respiratory disease, cancer at any site, osteoarthritis, or arthritis.

The same type of analysis was performed for each component of the MetS. In addition, an isocaloric substitution model was built to evaluate the effect of replacing the percentage of EI at breakfast by same amount of EI at dinner on the risk of incident MetS [21]. Similar analyses were performed for replacing EI% at lunch or at any other occasion with the same EI% at dinner.

All analyses were performed with STATA software v.13.1 (College Station, TX, USA: StataCorp LP). *p*-Values of <0.05 were considered as statistically significant.

3. Results

Among the 607 participants, 312 (51.4%) were women, and mean age was 67.3 years (SD 5.3). The main eating occasions were breakfast, lunch, and dinner, with an EI% of 17.1%, 41.9%, and 27.8%, respectively (Table 1). Compared with participants in the lowest quartile of EI% at breakfast, those in the highest quartile were less frequently current smokers, had less total EI, spent less time watching TV, had a lower BMI, and showed lower ethanol intake. Participants in the highest quartile of EI% at lunch were less educated, with lower total EI, and higher adherence to a Mediterranean diet, and those in the highest quartile of EI% at dinner were more frequently current smokers (Table 2).

Table 1. Energy intake at each eating occasion, and percentage of individuals who skipped each eating occasion, among the study participants at baseline. Seniors-ENRICA-1 cohort study, 2008–2010. *N* = 607.

		Energy Intake (kcal/day)		
Eating Occasions	**% of Energy Intake**	**Mean**	**SD**	**Skipping Eating Occasion (%)**
Breakfast	17.1	339.2	210.5	0.7
Mid-morning snack	4.2	89.0	151.8	41.5
Lunch	41.9	840.4	284.7	0.0
Afternoon snack	4.0	82.3	118.2	38.4
Dinner	27.8	563.2	235.7	0.3
Snacking	5.0	108.3	165.9	37.7
Total energy	100.0	2022.5	559.4	-

Table 2. Baseline characteristics of the participants in the Seniors-ENRICA-1 cohort study according to sex-specific quartiles of the percentage of energy intake at breakfast, lunch, and dinner. *N* = 607.

	Breakfast †		Lunch †		Dinner †	
	Q1	**Q4**	**Q1**	**Q4**	**Q1**	**Q4**
Sex, % of women	51.3	51.7	51.3	51.7	51.3	51.7
Age (years), mean	67.0 (5.6)	67.6 (6.2)	66.9 (5.9)	67.7 (5.6)	67.4 (5.8)	67.2 (5.6)
	Breakfast †		**Lunch †**		**Dinner †**	
Level of education, %						
Primary or less	42.1	41.1	35.5	45.7	39.5	42.4
Secondary	31.6	31.8	30.9	31.8	30.9	31.1
University	26.3	27.2	33.6	22.5 *	29.6	26.5
Smoking status, %						
Never smoker	53.3	66.2	53.3	58.9	62.5	53.0
Former smoker	32.2	27.8	33.6	33.1	30.3	30.5
Current smoker	14.5	6.0 **	13.2	8.0	7.2	16.6 **
Ex-drinker, %	11.2	6.0	12.5	6.0	7.9	6.0
Energy (kcal/day), mean	2095 (562)	1951 ** (591)	2184 (637)	1924 *** (501)	2016 (609)	2001 (516)
Physical activity during leisure time (METs h/week), mean	23.2 (14.6)	25.7 (18.1)	24.1 (17.2)	22.3 (16.6)	23.9 (18.5)	24.4 (15.3)
Physical activity in the household (METs h/week), mean	37.8 (31.2)	35.6 (28.2)	37.4 (30.9)	38.6 (28.8)	38.0 (32.2)	33.8 (28.4)
Watching TV (h/week), mean	18.2 (11.6)	15.4 * (10.4)	15.4 (10.3)	15.9 (11.2)	16.4 (12.6)	17.7 (10.1)
Sleeping time (hour/day), mean	7.2 (1.2)	7.2 (1.3)	7.1 (1.5)	7.3 (1.5)	7.2 (1.4)	7.2 (1.3)
Body mass index (kg/m2), mean	27.5 (3.4)	26.8 * (3.6)	26.7 (3.3)	27.6 (4.0)	26.7 (3.6)	27.2 (3.2)
Dieting, %	10.5	6.0	7.2	9.3	6.6	8.6
Ethanol intake (g/day), mean	15.6 (22.1)	8.6 ** (15.1)	9.8 (17.9)	12.7 (16.7)	11.2 (16.8)	14.6 (22.2)
MEDAS score, mean	7.8 (1.8)	7.4 (1.7)	6.8 (2.0)	7.8 *** (1.7)	7.3 (1.9)	7.6 (1.6)
Coronary disease, %	0	2.0	0.7	0	1.3	0
Chronic respiratory disease, %	9.2	6.6	5.9	8.0	6.6	6.0
Cancer, %	2.6	2.0	2.0	1.3	2.0	2.7
Osteoarthritis, %	33.6	38.4	36.8	41.7	37.5	38.4
Arthritis, %	12.5	7.3	11.2	6.0	7.2	13.9

Q1, quartile 1 (lowest); Q4, quartile 4 (highest); MET, metabolic equivalents; MEDAS, Mediterranean Diet Adherence Screener. * p for trend <0.05, ** p for trend < 0.01, *** p for trend < 0.001. † Cut-off points for the quartiles in men: breakfast Q1: (0–11.0), Q2: (11.0–15.2), Q3: (15.2–21.0), Q4: (21.2–62.1); lunch Q1: (14.8–36.5), Q2: (36.6–43.4), Q3: (43.4–49.6), Q4: (49.7–71.0; dinner Q1: (0–22.4), Q2: (22.5–28.6), Q3: (28.6–34.1), Q4: (34.1–59.8). Cut-off points for the quartiles in women: breakfast Q1: (0.1–11.5), Q2: (11.7–17.0), Q3: (17.1–21.5), Q4: (21.5–67.4); lunch Q1: (11.0–34.0), Q2: (34.1–40.8), Q3: (40.8–46.0), Q4: (46.1–78.4); dinner Q1: (0–21.0), Q2: (21.1–27.4), Q3: (27.5–32.0), Q4: (32.0–53.8).

At the end of the follow-up period, 101 (16.6%) of the participants developed MetS. Of them, 50 were men and 51 were women. The risk of MetS increased across quartiles of EI% at dinner: the OR (95% CI) of MetS was 1 for the lowest quartile, 1.71 (0.85–3.46) for the second, 1.70 (0.81–3.54) for the third, and 2.57 (1.14–5.79) for the highest quartile (P-trend: 0.034). No associations were found for EI% any other eating occasion (Table 3).

Table 3. Odds ratios (95% confidence interval) for incident metabolic syndrome (2008/10 to 2015) according to sex-specific quartiles of the percentage of energy at each eating occasion among participants in the Seniors-ENRICA-1 cohort study. $N = 607$.

	N/Cases	Model 1 OR (95% CI)	Model 2 OR (95% CI)
Breakfast †	607/101		
Quartile 1 (lowest)	152/25	Ref	Ref
Quartile 2	152/33	1.51 (0.83–2.73)	1.46 (0.78–2.75)
Quartile 3	152/26	1.12 (0.58–2.17)	1.28 (0.64–2.55)
Quartile 4 (highest)	151/17	0.79 (0.35–1.80)	0.84 (0.35–1.97)
p for trend		0.619	0.815
Mid-morning snack †	607/101		
Quartile 1 (lowest)	252/43	Ref	Ref
Quartile 2	119/23	1.32 (0.73–2.41)	1.26 (0.67–2.37)
Quartile 3	118/18	0.98 (0.51–1.86)	0.97 (0.50–1.91)
Quartile 4 (highest)	118/17	1.14 (0.55–2.36)	1.09 (0.51–2.34)
p for trend		0.838	0.910
Lunch †	607/101		
Quartile 1 (lowest)	152/22	Ref	Ref
Quartile 2	152/24	0.99 (0.51–1.91)	1.05 (0.53–2.08)
Quartile 3	152/24	0.99 (0.50–1.98)	1.10 (0.53–2.29)
Quartile 4 (highest)	151/31	1.62 (0.73–3.58)	1.71 (0.73–3.97)
p for trend		0.300	0.258
Afternoon snack †	607/101		
Quartile 1 (lowest)	233/39	Ref	Ref
Quartile 2	125/25	1.34 (0.74–2.41)	1.31 (0.71–2.43)
Quartile 3	125/19	0.96 (0.51–1.81)	1.12 (0.58–2.17)
Quartile 4 (highest)	124/18	1.07 (0.53–2.15)	1.05 (0.50–2.19)
p for trend		0.978	0.871
Dinner †	607/101		
Quartile 1 (lowest)	152/16	Ref	Ref
Quartile 2	152/27	1.76 (0.89–3.46)	1.71 (0.85–3.46)
Quartile 3	152/26	1.69 (0.84–3.41)	1.70 (0.81–3.54)
Quartile 4 (highest)	151/32	2.31 (1.06–5.03) *	2.57 (1.14–5.79) *
p for trend		0.054	0.034
Snacking †	607/101		
Quartile 1 (lowest)	229/38	Ref	Ref
Quartile 2	127/22	1.09 (0.60–2.00)	1.17 (0.62–2.19)
Quartile 3	126/24	1.30 (0.71–2.37)	1.15 (0.61–2.16)
Quartile 4 (highest)	125/17	1.00 (0.48–2.10)	1.07 (0.49–2.34)
p for trend		0.723	0.748

* $p < 0.05$. Model 1 was mutually adjusted for the percentage of energy intake consumed at each eating occasion as appropriate, as well as sex, age, level of education (primary or less, secondary, or university), and total energy intake (kcal/day). Model 2 was adjusted as per model 1, plus smoking status (never, former, and current smokers), ex-drinker status, leisure-time physical activity (METs h/week), physical activity in the household (METs h/week), watching TV (hours/week), sleeping time (hours/day), body mass index (kg/m^2), dieting, ethanol intake (g(day), Mediterranean Diet Adherence Screener score, coronary diseases, chronic respiratory disease, cancer, osteoarthritis, and arthritis. † Cut-off points for the quartiles in men: breakfast Q1: (0–11.0), Q2: (11.0–15.2), Q3: (15.2–21.0), Q4: (21.2–62.1); mid-morning snack Q1: (0), Q2: (0.1–3.0), Q3: (3.1–8.9), Q4: (9.0–51.8); lunch Q1: (14.8–36.5), Q2: (36.6–43.4), Q3: (43.4–49.6), Q4: (49.7–71.0); afternoon snack Q1: (0), Q2: (0–2.9), Q3: (3.0–6.6), Q4: (6.7–34.9); dinner Q1: (0–22.4), Q2: (22.5–28.6), Q3: (28.6–34.1), Q4: (34.1–59.8); snacking Q1: (0), Q2: (0.2–3.9), Q3: (3.9–8.2), Q4: (8.3–40.2). Cut-off points for the quartiles in women: breakfast Q1: (0.1–11.5), Q2: (11.7–17.0), Q3: (17.1–21.5), Q4: (21.5–67.4); mid-morning snack Q1: (0), Q2: (0.2–3.3), Q3: (3.3–6.8), Q4: (6.9–32.1); lunch Q1: (11.0–34.0), Q2: (34.1–40.8), Q3: (40.8–46.0), Q4:(46.1–78.4); afternoon snack Q1: (0), Q2: (0.3–3.5), Q3: (3.5–7.5), Q4: (7.5–31.3); dinner Q1: (0–21.0), Q2: (21.1–27.4), Q3: (27.5–32.0), Q4: (32.0–53.8); snacking Q1: (0), Q2: (0.2–3.7), Q3: (3.7–9.7), Q4: (9.7–53.8).

Isocaloric replacement of EI at breakfast with the same amount of EI at dinner was associated with a higher risk of MetS, so that the OR (95% CI) of MetS was 2.73 (1.31–5.68; *p*-trend: 0.011) for the highest vs. lowest quartile of EI at dinner. Likewise, replacement of EI at any other time of the day with the same percentage of EI at dinner was associated with a higher risk of MetS (OR: 2.42; 95% CI: 1.22–4.81; *p*-trend: 0.019) (Table 4).

With regard to the five components of the MetS, we found that participants in the highest vs. lowest quartile of EI at dinner had a higher risk of abdominal obesity (OR 2.15; (95% CI: 1.08–4.25; *p*-trend: 0.013), as well as a significant tendency for elevated triglyceridemia (*p*-trend: 0.025) (Table 5).

Table 4. Odds ratios (95% confidence interval) for incident metabolic syndrome among participants in the Seniors-ENRICA-1 cohort study (2008/10 to 2015) when isocaloric substitution of breakfast, lunch, or all other occasions of energy intake for energy intake at dinner. *N* = 607.

	N/Events	Model 1 OR (95% CI)	Model 2 OR (95% CI)
Isocaloric substitution of energy consumed at breakfast for dinner	607/101		
Quartile 1 (lowest)	152/16	Ref	Ref
Quartile 2	152/27	1.79 (0.92–3.52)	1.76 (0.87–3.55)
Quartile 3	152/26	1.76 (0.89–3.49)	1.75 (0.85–3.57)
Quartile 4 (highest)	151/32	2.53 (1.26–5.07) **	2.73 (1.31–5.68) **
p for trend		0.014	0.011
Isocaloric substitution of energy consumed at lunch for dinner			
Quartile 1 (lowest)	152/16	Ref	Ref
Quartile 2	152/27	1.72 (0.87–3.37)	1.70 (0.84–3.43)
Quartile 3	152/26	1.58 (0.80–3.15)	1.57 (0.76–3.22)
Quartile 4 (highest)	151/32	1.97 (0.97–4.01)	2.17 (1.03–4.54) *
p for trend		0.095	0.065
Isocaloric substitution of energy consumed at all other occasions for dinner			
Quartile 1 (lowest)	152/16	Ref	Ref
Quartile 2	152/27	1.79 (0.92–3.49)	1.76 (0.88–3.53)
Quartile 3	152/26	1.70 (0.87–3.32)	1.66 (0.82–3.36)
Quartile 4 (highest)	151/32	2.26 (1.18–4.35) *	2.42 (1.22–4.81) *
p for trend		0.024	0.019

* $p < 0.05$, ** $p < 0.01$. Models 1 and 2 were adjusted as in Table 3.

Table 5. Odds ratios (95% confidence interval) for the incidence of each component of metabolic syndrome according to sex-specific quartiles of percentage of energy intake at dinner among participants in the Seniors-ENRICA-1 cohort study (2008/10 to 2015).

	Model 1		Model 2	
	N/Events	OR (95% CI)	*N*/Events	OR (95% CI)
Abdominal Obesity	647/176		647/176	
Quartile 1 (lowest)	162/36	Ref	162/36	Ref
Quartile 2	162/40	1.18 (0.69–2.02)	162/40	1.10 (0.61–2.00)
Quartile 3	162/48	1.62 (0.93–2.81)	162/48	1.82 (0.98–3.36)
Quartile 4 (highest)	161/52	1.99 (1.06–3.74) *	161/52	2.15 (1.08–4.25) *
p for trend		0.020		0.013
Hyperglycemia/diabetes	834/150		834/150	
Quartile 1 (lowest)	209/37	Ref	209/37	Ref
Quartile 2	209/29	0.75 (0.44–1.30)	209/29	0.75 (0.43–1.30)

Table 5. *Cont.*

	Model 1		Model 2	
	N/Events	OR (95% CI)	N/Events	OR (95% CI)
Quartile 3	209/38	1.04 (0.61–1.77)	209/38	1.00 (0.58–1.73)
Quartile 4 (highest)	207/46	1.40 (0.75–2.59)	207/46	1.34 (0.72–2.50)
p for trend		0.239		0.308
Arterial hypertension	241/102		241/102	
Quartile 1 (lowest)	61/28	Ref	61/28	Ref
Quartile 2	60/23	0.81 (0.37–1.74)	60/23	0.79 (0.36–1.76)
Quartile 3	61/23	0.73 (0.33–1.58)	61/23	0.75 (0.33–1.70)
Quartile 4 (highest)	59/28	0.87 (0.36–2.11)	59/28	0.86 (0.33–2.22)
p for trend		0.656		0.671
Hypertriglyceridemia	1141/98		1141/98	
Quartile 1 (lowest)	286/24	Ref	286/24	Ref
Quartile 2	285/18	0.80 (0.42–1.52)	285/18	0.82 (0.42–1.60)
Quartile 3	286/28	1.53 (0.82–2.87)	286/28	1.66 (0.87–3.19)
Quartile 4 (highest)	284/28	1.92 (0.93–3.97)	284/28	2.07 (0.98–4.35)
p for trend		0.038		0.025
Low HDL-cholesterol	1109/156		1109/156	
Quartile 1 (lowest)	278/36	Ref	278/36	Ref
Quartile 2	277/43	1.20 (0.74–1.96)	277/43	1.18 (0.71 -1.95)
Quartile 3	277/40	1.06 (0.64–1.77)	277/40	1.12 (0.66 –1.92)
Quartile 4 (highest)	277/37	0.86 (0.47–1.57)	277/37	0.89 (0.48–1.67)
p for trend		0.626		0.773

* $p < 0.05$. Models 1 and 2 were adjusted as in Table 3.

4. Discussion

In this prospective cohort of older adults, a higher percentage of EI at dinner was associated with an increased risk of MetS. Replacing the EI consumed at breakfast or at any other eating occasion with the same amount of energy at dinner also increased the risk of the MetS. Elevated waist circumference and triglycerides were the MetS components that most contributed to this association. In Spain, a country where dinner is usually eaten later than in other countries (generally after 9:00 p.m. [22]), to eat more energy at this eating occasion was detrimental and associated with the development of the MetS.

Previous studies are in line with these findings. However, the exposures were measured differently, and it is difficult to establish comparisons. The association between meal timing and MetS has been previously studied in Japan and Korea. In a cross-sectional study of Japanese people aged 20–75 years, late-night dining habits were associated with a higher risk of the MetS [23]. In another cross-sectional study of Korean adults aged 19 or older (KNHANES Study), night eating (≥25% of total EI at night) was associated with a higher frequency of MetS among males [24]. Finally, in a prospective study of Japanese adults aged 40–54 years, a positive association between dining immediately before going to bed and the risk of incident MetS did not reach statistical significance; however, women who had both the habits of dining just before going to bed and of snacking after dinner had an increased risk of MetS [25].

Our results are in line with current knowledge on nutritional chronobiology [3,5], as well as with previous findings in observational studies [23]. However, the different ways to assess eating timing in relation to MetS, which mainly depends on data availability, make the studies difficult to compare. For instance, some studies considered only two time bands, daytime and evening eating [9,24], while we considered six eating occasions, including snacking, during each 24 h period. Likewise, in some previous studies, the exposure was an intake of at least 25% of energy after 9:00 p.m. [24], or as having dinner just before going to bed [25].

Regarding MetS components, there is some evidence that chronobiology and timing in EI are associated with obesity [10,26]. Specifically, a higher EI during the night was associated with a higher risk of obesity [26]. Of note is that abdominal obesity is a central component of MetS, and can lead to lipid abnormalities, alterations of glucose metabolism, and elevated blood pressure [27]. However, although energy intake timing seems to be relevant in the development of obesity, some inconsistencies have also been observed concerning energy intake at dinner. A previous study conducted in participants 18 years old and over found that while a higher energy intake at lunch was associated with a lower risk of weight gain, no association was found between the energy consumed at dinner and weight gain (>3 kg after 3 years of follow-up) [11].

Regarding hypertriglyceridemia, an experiment with mice found that a high-fat diet at the end of their active period was associated with hypertriglyceridemia, weight gain, increased body adiposity, decreased glucose tolerance, and hyperleptinemia [28]. In addition, another study with mice that were CLOCK gene mutants (with hyperphagia during their sleep/inactivity phase) showed increased weight gain and hypertriglyceridemia [29]. In humans, timing misalignments (i.e., shift work) have been frequently studied, and an association with hypertriglyceridemia was also found; a study among night-shift nurses showed a higher EI at night as well as higher levels of triglycerides, LDL-cholesterol, and total cholesterol. Night shift was also associated with higher levels of HDL-cholesterol [9].

In our analyses, EI at dinner was not associated with other components of the MetS, such as hyperglycemia, high blood pressure, or low HDL-cholesterol. However, these associations were found in some cross-sectional studies. For example, night-shift nurses showed elevated fasting glucose [9]. In addition, in the KNHANES Study, an association between eating at night and reduced HDL-cholesterol was found among men [24].

Although having a regular breakfast has been suggested to decrease adiposity [30], in our analyses no association was found between EI at breakfast and risk of MetS. However, skipping breakfast, which could influence energy balance throughout the day, was very rare (0.7%) in our sample. Furthermore, our results did not show an association between snacking and MetS, although the association was positive, again not reaching statistical significance. Of note is that snacking varies greatly among different populations [31], and in our sample it accounted for only 5% of total EI.

The increasing body of evidence on the influence of EI throughout the day and its impact on metabolic health makes the association between EI timing and the development of MetS plausible [2]. For example, it has been suggested that insulin (the hormone that is altered in MetS) is involved in the post-prandial synchronization of the circadian clocks throughout the body [32]. Another factor that could be acting is the proximity of food intake to the nocturnal rise in melatonin. Calorie intake closer to this nocturnal rise was associated with increased adiposity and impaired glucose homeostatic [33]. It is also known that circadian clocks influence both appetite and energy expenditure, which are ultimately linked to metabolic disorders [34]. Finally, the present study also contributes to the epidemiological and population-based support for the influence of energy timing on cardiometabolic risk management [31].

4.1. Importance and Practical Consequences

Our findings suggest that reducing %EI at dinner may serve to lower the risk of MetS. This is important for a number of reasons. First, because of the high prevalence of MetS syndrome (>40%) in our older population [35]. Second, because MetS has been associated with an increased risk of diabetes [36,37] and is a major contributor to the epidemic of cardiovascular disease in the Western world [38]. Third, because MetS does not easily revert. Fourth, because metabolic conditions associated with misalignments do not seem to have a clear genetic component and are based on unhealthy lifestyles [5]. Fifth, because misalignments are becoming more frequent due to long working days and the use of electronic devices at bedtime before sleeping. Finally, because changes in EI timing might be easier to follow than just caloric restrictions. Similar approaches have been proposed to prevent obesity [39,40].

4.2. Strengths and Limitations

The main strength of this study is its longitudinal design. However, although the cohort was initially representative of the older adult population in Spain, attrition bias does not allow us to ensure representativeness. Moreover, analyses were adjusted for a significant number of potential confounders, such as total energy, physical activity during leisure time and in the household, watching TV, sleeping, dieting, and alcohol consumption. In addition, certified personnel performed all measurements following standardized protocols.

The main limitation is that diets were self-reported, so recall bias or social desirability bias cannot be ruled out. In addition, despite each participant reporting the food consumed on each eating occasion, the time of the meals was not collected. A diet history validation was performed for food groups with many nutrients and total energy [13], but a specific validation for eating occasions was not performed. On the other hand, the specific time of the six eating occasions was not recorded, although these are quite characteristic of the dietary habits in the Spanish population. Additionally, the influence of sleep preferences was not considered in the analyses. Finally, we cannot rule out some selection bias due to the exclusion of participants with lacking data.

5. Conclusions

In this study of older adults, a higher percentage of EI at dinner was associated with a higher incidence of MetS, mostly due to abdominal obesity and hypertriglyceridemia. These findings suggest that a reduction of energy intake at dinner may be useful to prevent MetS. However, the efficacy of this strategy must be proven in future clinical trials with populations.

Author Contributions: Conceptualization, P.G.-C. and F.R.-A.; methodology, P.G.-C.; formal analysis, Y.H.-L.; data cu-ration, Y.H.-L.; writing—original draft preparation, Y.H.-L. and P.G.-C.; writing—review and editing, C.D.-V., B.M.-F., H.S.-I., M.R.-A., J.R.-G., J.R.B. and F.R.-A. All authors have read and agreed to the published version of the manuscript.

Funding: This research was funded by Instituto Carlos III, State Secretary of R+D+I, and FEDER/FSE, grant number 17/1709, 19/319, 19/665 and 20/00144 and by CIBERESP, Instituto de Salud Carlos III.

Institutional Review Board Statement: The study was conducted according to the guidelines of the Declaration of Helsinki, and approved by The Clinical Research Ethics Committee of the La Paz University Hospital.

Informed Consent Statement: Informed consent was obtained from all subjects involved in the study.

Data Availability Statement: The datasets generated during and/or analyzed during the current study are available from the corresponding author on reasonable request.

Acknowledgments: We wish to thank Rocío Ordóñez Ortega who assisted with the collection and analysis of data.

Conflicts of Interest: The authors declare no conflict of interest.

References

1. Albrecht, U. Timing to Perfection: The Biology of Central and Peripheral Circadian Clocks. *Neuron* **2012**, *74*, 246–260. [CrossRef]
2. Johnston, J.D.; Ordovás, J.M.; Scheer, F.A.; Turek, F.W. Circadian Rhythms, Metabolism, and Chrononutrition in Rodents and Humans. *Adv. Nutr.* **2016**, *7*, 399–406. [CrossRef]
3. Flanagan, A.; Bechtold, D.A.; Pot, G.K.; Johnston, J.D. Chrono-nutrition: From molecular and neuronal mechanisms to human epidemiology and timed feeding patterns. *J. Neurochem.* **2021**, *157*, 53–72. [CrossRef] [PubMed]
4. Skene, D.J.; Arendt, J. Human circadian rhythms: Physiological and therapeutic relevance of light and melatonin. *Ann. Clin. Biochem. Int. J. Lab. Med.* **2006**, *43 Pt 5*, 344–353. [CrossRef]
5. Lopez-Minguez, J.; Gómez-Abellán, P.; Garaulet, M. Timing of Breakfast, Lunch, and Dinner. Effects on Obesity and Metabolic Risk. *Nutrients* **2019**, *11*, 2624. [CrossRef] [PubMed]
6. Garaulet, M.; Gomez-Abellan, P. Chronobiology and obesity. *Nutr. Hosp.* **2013**, *28* (Suppl. 5), 114–120.

7. Li, S.; Lin, J.D. Molecular control of circadian metabolic rhythms. *J. Appl. Physiol.* **2009**, *107*, 1959–1964. [CrossRef]
8. Mattson, M.P.; Allison, D.; Fontana, L.; Harvie, M.; Longo, V.D.; Malaisse, W.J.; Mosley, M.; Notterpek, L.; Ravussin, E.; Scheer, F.; et al. Meal frequency and timing in health and disease. *Proc. Natl. Acad. Sci. USA* **2014**, *111*, 16647–16653. [CrossRef] [PubMed]
9. Molzof, H.E.; Wirth, M.D.; Burch, J.B.; Shivappa, N.; Hebert, J.R.; Johnson, R.L.; Gamble, K.L. The impact of meal timing on cardiometabolic syndrome indicators in shift workers. *Chronobiol. Int.* **2017**, *34*, 337–348. [CrossRef]
10. Wang, J.B.; Patterson, R.E.; Ang, A.; Emond, J.A.; Shetty, N.; Arab, L. Timing of energy intake during the day is associated with the risk of obesity in adults. *J. Hum. Nutr. Diet.* **2014**, *27* (Suppl. 2), 255–262. [CrossRef]
11. Hermenegildo, Y.; Lopez-Garcia, E.; Garcia-Esquinas, E.; Perez-Tasigchana, R.F.; Rodriguez-Artalejo, F.; Guallar-Castillon, P. Distribution of energy intake throughout the day and weight gain: A population-based cohort study in Spain. *Br. J. Nutr.* **2016**, *115*, 2003–2010. [CrossRef]
12. Rodríguez-Artalejo, F.; Graciani, A.; Guallar-Castillón, P.; León-Muñoz, L.M.; Zuluaga, M.C.; López-García, E.; Gutiérrez-Fisac, J.L.; Taboada, J.M.; Aguilera, M.T.; Regidor, E.; et al. Rationale and Methods of the Study on Nutrition and Cardiovascular Risk in Spain (ENRICA). *Rev. Esp. Cardiol.* **2011**, *64*, 876–882. [CrossRef] [PubMed]
13. Guallar-Castillón, P.; Sagardui-Villamor, J.; Balboa-Castillo, T.; Sala-Vila, A.; Astolfi, M.J.A.; Pelous, M.D.S.; León-Muñoz, L.M.; Graciani, A.; Laclaustra, M.; Benito, C.; et al. Validity and Reproducibility of a Spanish Dietary History. *PLoS ONE* **2014**, *9*, e85074. [CrossRef] [PubMed]
14. Farrán, A.Z.R.; Cervera, P. *Tablas de Composición de Alimentos del CESNID*; McGraw-Hill Interamericana: Madrid, Spain, 2003.
15. Alberti, K.G.; Eckel, R.H.; Grundy, S.M.; Zimmet, P.Z.; Cleeman, J.I.; Donato, K.A.; Fruchart, J.-C.; James, W.P.T.; Loria, C.M.; Smith, S.C., Jr.; et al. Harmonizing the metabolic syndrome: A joint interim statement of the International Diabetes Federation Task Force on Epidemiology and Prevention; National Heart, Lung, and Blood Institute; American Heart Association; World Heart Federation; International Atherosclerosis Society; and International Association for the Study of Obesity. *Circulation* **2009**, *120*, 1640–1645. [PubMed]
16. Gutiérrez-Fisac, J.L.; Guallar-Castillon, P.; León-Muñoz, L.M.; Graciani, A.; Banegas, J.R.; Rodríguez-Artalejo, F. Prevalence of general and abdominal obesity in the adult population of Spain, 2008–2010: The ENRICA study. *Obes. Rev.* **2011**, *13*, 388–392. [CrossRef]
17. Navarro-Vidal, B.; Banegas, J.R.; Leon-Munoz, L.M.; Rodriguez-Artalejo, F.; Graciani, A. Achievement of cardiometabolic goals among diabetic patients in Spain. A nationwide population-based study. *PLoS ONE* **2013**, *8*, e61549. [CrossRef] [PubMed]
18. Banegas, J.R.; Graciani, A.; de la Cruz-Troca, J.J.; Leon-Munoz, L.M.; Guallar-Castillon, P.; Coca, A.; Ruilope, L.M.; Rodríguez-Artalejo, F. Achievement of cardiometabolic goals in aware hypertensive patients in Spain: A nationwide population-based study. *Hypertension* **2012**, *60*, 898–905. [CrossRef]
19. Pols, M.A.; Peeters, P.H.; Ocke, M.C.; Slimani, N.; Bueno-de-Mesquita, H.B.; Collette, H.J. Estimation of reproducibility and relative validity of the questions included in the EPIC Physical Activity Questionnaire. *Int. J. Epidemiol.* **1997**, *26* (Supp . 1), S181–S189. [CrossRef]
20. Schröder, H.; Fitó, M.; Estruch, R.; Martínez-González, M.A.; Corella, D.; Salas-Salvadó, J.; Lamuela-Raventós, R.; Ros, E.; Salaverría, I.; Fiol, M.; et al. A Short Screener Is Valid for Assessing Mediterranean Diet Adherence among Older Spanish Men and Women. *J. Nutr.* **2011**, *141*, 1140–1145. [CrossRef]
21. Hu, F.B.; Stampfer, M.J.; Rimm, E.; Ascherio, A.; Rosner, B.A.; Spiegelman, D.; Willett, W.C. Dietary Fat and Coronary Heart Disease: A Comparison of Approaches for Adjusting for Total Energy Intake and Modeling Repeated Dietary Measurements. *Am. J. Epidemiol.* **1999**, *149*, 531–540. [CrossRef]
22. Medina, F. *Food Culture in Spain*; Greenwood Press: Westport, CT, USA, 2005; p. 90.
23. Kutsuma, A.; Nakajima, K.; Suwa, K. Potential Association between Breakfast Skipping and Concomitant Late-Night-Dinner Eating with Metabolic Syndrome and Proteinuria in the Japanese Population. *Scientifica* **2014**, *2014*, 253581. [CrossRef]
24. Ha, K.; Song, Y. Associations of meal timing and frequency with obesity and metabolic syndrome among Korean adults. *Nutrients* **2019**, *11*, 2437. [CrossRef]
25. Yoshida, J.; Eguchi, E.; Nagaoka, K.; Ito, T.; Ogino, K. Association of night eating habits with metabolic syndrome and its components: A longitudinal study. *BMC Public Health* **2018**, *18*, 1366. [CrossRef]
26. Xiao, Q.; Garaulet, M.; Scheer, F.A.J.L. Meal timing and obesity: Interactions with macronutrient intake and chronotype. *Int. J. Obes.* **2019**, *43*, 1701–1711. [CrossRef]
27. Grundy, S.M. What is the contribution of obesity to the metabolic syndrome? *Endocrinol. Metab. Clin. N. Am.* **2004**, *33*, 267–282. [CrossRef] [PubMed]
28. Bray, M.S.; Tsai, J.-Y.; Villegas-Montoya, C.; Boland, B.; Blasier, Z.; Egbejimi, O.; Kueht, M.; Young, M.E. Time-of-day-dependent dietary fat consumption influences multiple cardiometabolic syndrome parameters in mice. *Int. J. Obes.* **2010**, *34*, 1589–1598. [CrossRef] [PubMed]
29. Turek, F.W.; Joshu, C.; Kohsaka, A.; Lin, E.; Ivanova, G.; McDearmon, E.; Laposky, A.; Losee-Olson, S.; Easton, A.; Jensen, D.R.; et al. Obesity and Metabolic Syndrome in Circadian Clock Mutant Mice. *Science* **2005**, *308*, 1043–1045. [CrossRef]
30. Gibney, M.J.; Barr, S.I.; Bellisle, F.; Drewnowski, A.; Fagt, S.; Livingstone, B.; Masset, G.; Moreiras, G.V.; Moreno, L.A.; Smith, J.; et al. Breakfast in Human Nutrition: The International Breakfast Research Initiative. *Nutrients* **2018**, *10*, 559. [CrossRef] [PubMed]
31. St-Onge, M.-P.; Ard, J.; Baskin, M.L.; Chiuve, S.E.; Johnson, H.M.; Kris-Etherton, P.; Varady, K.; American Heart Association Obesity Committee of the Council on Lifestyle and Cardiometabolic Health; Council on Cardiovascular Disease in the Young;

Council on Clinical Cardiology. Meal Timing and Frequency: Implications for Cardiovascular Disease Prevention: A Scientific Statement From the American Heart Association. *Circulation* **2017**, *135*, e96–e121. [CrossRef]

32. Crosby, P.; Hamnett, R.; Putker, M.; Hoyle, N.P.; Reed, M.; Karam, C.J.; Maywood, E.S.; Stangherlin, A.; Chesham, J.E.; Hayter, E.; et al. Insulin/IGF-1 Drives PERIOD Synthesis to Entrain Circadian Rhythms with Feeding Time. *Cell* **2019**, *177*, 896–909.e20. [CrossRef]
33. McHill, A.W.; Phillips, A.J.; Czeisler, C.A.; Keating, L.; Yee, K.; Barger, L.K.; Garaulet, M.; Scheer, F.; Klerman, E.B. Later circadian timing of food intake is associated with increased body fat. *Am. J. Clin. Nutr.* **2017**, *106*, 1213–1219. [CrossRef]
34. Page, A.J.; Christie, S.; Symonds, E.; Li, H. Circadian regulation of appetite and time restricted feeding. *Physiol. Behav.* **2020**, *220*, 112873. [CrossRef] [PubMed]
35. Guallar-Castillón, P.; Gil-Montero, M.; León-Muñoz, L.M.; Graciani, A.; Bayán-Bravo, A.; Taboada, J.M.; Banegas, J.R.; Rodríguez-Artalejo, F. Magnitude and Management of Hypercholesterolemia in the Adult Population of Spain, 2008–2010: The ENRICA Study. *Rev. Esp. Cardiol.* **2012**, *65*, 551–558. [CrossRef] [PubMed]
36. Oda, E. Metabolic syndrome: Its history, mechanisms, and limitations. *Acta Diabetol.* **2011**, *49*, 89–95. [CrossRef] [PubMed]
37. Samson, S.L.; Garber, A.J. Metabolic syndrome. *Endocrinol. Metab. Clin. N. Am.* **2014**, *43*, 1–23. [CrossRef] [PubMed]
38. Sherling, D.H.; Perumareddi, P.; Hennekens, C.H. Metabolic Syndrome. *J. Cardiovasc. Pharmacol. Ther.* **2017**, *22*, 365–367. [CrossRef]
39. Berrea, L.; Frias-Toral, E.; Aprano, S.; Castellucci, B.; Pugliese, G.; Rodriguez-Veintimilla, D.; Vitale, G.; Gentilini, D.; Colao, A.; Savastano, S.; et al. The clock diet: A practical nutrition guide to manage obesity through chrononutrition. *Minerva Med.* **2021**. [CrossRef]
40. Horn, C.; Laupsa-Borge, J.; Andersen, A.I.O.; Dyer, L.; Revheim, I.; Leikanger, T.; Næsheim, N.T.; Storås, I.; Johannessen, K.K.; Mellgren, G.; et al. Meal patterns associated with energy intake in people with obesity. *Br. J. Nutr.* **2021**, *12*, 1–48. [CrossRef]

nutrients

MDPI

Article

Eucaloric Balanced Diet Improved Objective Sleep in Adolescents with Obesity

Oussama Saidi [1,2,3], Emmanuelle Rochette [1,4,5], Giovanna Del Sordo [1], Éric Doré [2,3], Étienne Merlin [4,5], Stéphane Walrand [6] and Pascale Duché [1,*]

1 Laboratory Impact of Physical Activity on Health (IAPS), Toulon University, F-83000 Toulon, France; oussama.saidi@univ-tln.fr (O.S.); e_rochette@chu-clermontferrand.fr (E.R.); giovanna.del@hotmail.fr (G.D.S.)
2 Laboratory of Adaptations to Exercise under Physiological and Pathological Conditions (AME2P), Clermont Auvergne University, F-63170 Clermont-Ferrand, France; eric.dore@uca.fr
3 Regional Center for Human Nutrition (CRNH Auvergne), F-63000 Clermont-Ferrand, France
4 Department of Pediatrics, Clermont-Ferrand University Hospital, F-63000 Clermont-Ferrand, France; e_merlin@chu-clermontferrand.fr
5 INSERM, CIC 1405, CRECHE Unit, CHU Clermont-Ferrand, Clermont Auvergne University, F-63000 Clermont-Ferrand, France
6 INRAE, UNH, CHU Clermont-Ferrand, Clermont Auvergne University, F-63000 Clermont-Ferrand, France; stephane.walrand@inra.fr
* Correspondence: pascale.duche@univ-tln.fr; Tel.: +33-(0)-652-1838-91

Abstract: Background: A better understanding of the influence of energy balance on sleep in adolescents, particularly those with obesity, could help develop strategies to optimize sleep in these populations. The purpose of this study was to investigate sleep under ad libitum-vs-controlled diets adjusted to energy requirement (eucaloric) among adolescents with obesity and their normal weight controls. Methods: Twenty-eight male adolescents aged between 12 and 15 years, $n = 14$ adolescents with obesity (OB: BMI $\geq$ 90th centile) and $n = 14$ normal weight age matched controls (NW), completed an experimental protocol comprising ad libitum or eucaloric meals for three days, in random order. During the third night of each condition, they underwent in home polysomnography (PSG). Results: An interaction effect of energy intake (EI) was detected ($p < 0.001$). EI was higher during ad libitum compared to the eucaloric condition ($p < 0.001$) and in OB compared to NW ($p < 0.001$) in the absence of any substantial modification to macronutrient proportions. Analyses of energy intake distribution throughout the day showed a significant interaction with both a condition and group effect during lunch and dinner. Sleep improvements were noted in OB group during the eucaloric condition compared to ad libitum with reduced sleep onset latency and N1 stage. Sleep improvements were correlated to reduced EI, especially during the evening meal. Conclusion: Simply adjusting dietary intake to energy requirement and reducing the energy proportion of the evening meal could have therapeutic effects on sleep in adolescents with obesity. However, positive energy balance alone cannot justify worsened sleep among adolescents with obesity compared to normal weight counterparts.

Keywords: obesity; polysomnography; energy balance; energy intake; evening meal; youth

Citation: Saidi, O.; Rochette, E.; Del Sordo, G.; Doré, É.; Merlin, É.; Walrand, S.; Duché, P. Eucaloric Balanced Diet Improved Objective Sleep in Adolescents with Obesity. *Nutrients* **2021**, *13*, 3550. https://doi.org/10.3390/nu13103550

Academic Editors: Egeria Scoditti and Sergio Garbarino

Received: 23 August 2021
Accepted: 8 October 2021
Published: 10 October 2021

1. Introduction

Striking data from all over the world has reported commonly poor sleep patterns among adolescents [1,2]. A problem compounded by an array of endogenous and exogenous factors forming the so-called "Perfect Storm" of both altered sleep duration and quality [3]. Furthermore, obesity has been associated with sleep disturbances through several pathogenetic pathways [4]. Several clinical studies underlined how sleep affects both components of energy balance [5,6]. Although the effect on energy expenditure remains uncertain, a growing body of evidence demonstrates that sleep alteration increases energy intake with a more pronounced craving for energy-dense foods through numerous hormonal, neuroendocrine, cognitive, metabolic, and behavioral pathways [7–9] and these findings have also been noted in youth [10,11]. However, it appears that a potential bidirectional effect links sleep to energy balance and that in turn both nutrition and physical activity could exert ongoing feedback on biological clock and sleep physiology.

Although the effect of physical activity on sleep is well documented in both adult and young subjects [12,13], the effect of nutrition has been disregarded for a long time. However, a growing interest in this topic has emerged over the last decade and several reviews have been published [14–17]. Cross-sectional studies found numerous associations between dietary patterns and sleep [18–20]. However, fewer studies included objective sleep measurement. Spaeth et al., (2017) reported that in healthy adults, dietary intake was associated with sleep measured by polysomnography (PSG). Greater protein intake and lower carbohydrate intake were associated with more time spent in rapid eye movement (REM) sleep, while higher fiber was associated with increased slow wave sleep (SWS) [21]. In another study, higher nocturnal fat intake was associated with increased wake after sleep onset (WASO) and reduced sleep efficiency (SE) [22]. Besides these descriptive observational studies, some clinical trials attempt to report how dietary manipulation may affect sleep staging and quality [23–25]. However, the overwhelming majority of these studies were conducted in laboratory settings among healthy adults and potentially good sleepers who may be resistant to the effects of nutrition on sleep. Moreover, physical activity was not assessed during these interventions and may represent a potential confounder.

Adolescents, especially those with obesity, seem to be vulnerable to sleep disturbances [26]. Unfortunately, few interventional lifestyle studies, intended to improve sleep, were conducted among these populations. Yet, both nutrition and physical activity could offer basic options to improve sleep during adolescence. This is of significant importance especially for those with obesity since the latest recommendations have encouraged the consideration of sleep in the management of pediatric obesity [27]. A more comprehensive understanding of how energy balance affects sleep in these populations would be of considerable interest in the development of future dietary strategies promoting better sleep and effective weight loss intervention. Therefore, in this study we sought to examine objectively measured sleep under two different conditions (1) controlled balanced diet adjusted to energy requirement (eucaloric), and (2) diet offered ad libitum among adolescents with obesity and their age-matched normal weight peers. In other words, we intended to examine if an adolescent with obesity in neutral energy balance and balanced macronutrient proportions would have better or worse sleep quality than a normal weight adolescent in positive energy balance and vice versa. We hypothesize that a eucaloric balanced diet would improve sleep quality in both groups. However, given that adolescents with obesity may experience more sleep disturbances, a greater improvement was expected in this group according to the ceiling and floor effect.

2. Materials and Methods

2.1. Subjects

A total of 28 male adolescents aged between 12–15 years, and over Tanner stage 3, either with obesity (n = 14) (BMI $\geq$ 90th centile) recruited through the Pediatric Obesity Centre (Tza Nou, La Bourboule, France) or normal weight age matched controls (n = 14) recruited in collaboration with the Pediatric Department, CHU (Clermont-Ferrand, France) participated in this study. No participants suffered from any diagnosed major sleep disorders, e.g., narcolepsy, obstructive sleep apnea (OSA), or obesity comorbidities, e.g., diabetes, or used any medication or therapy that may interfere with sleep outcomes, e.g., antidepressants or benzodiazepines. Moreover, they were screened for depression using the Kutcher Adolescent Depression Scale (KADS) (LeBlanc et al., 2002). All included participants presented a score below 6 in this questionnaire indicating a low risk of depression. Adolescents with obesity were living in "Tza Nou" center (Obesity Treatment Center for Children and Adolescents). Normal weight peers were recruited from a boarding school to ensure comparable living conditions between groups. The investigators traveled to both sites with the laboratory equipment to ensure an ambulatory running of the study. The present study received approval from the relevant Institutional Ethics Review Board (Clinical trial No. NCT04041934). The research protocol was in line with the principles of the Helsinki Declaration. Information and consent forms were distributed to parents and adolescents prior to the study launch.

We underline that although we carefully included adolescents without diagnosed OSA, it turned out that some participants from the OB group ($n = 4$) presented undiagnosed OSA. During the habituation night, they presented an Apnea Hypopnea Index (AHI) of (4–5) episodes$\cdot h^{-1}$. Yet, during the experimental protocol, the value for AHI was mildly >5 episodes$\cdot h^{-1}$ potentially indicating the presence of OSA.

2.2. *Study Design and Procedure*

Participants were assessed for anthropometric and body composition. They filled in questionnaires in paper format regarding their circadian typology: the Horne–Östberg Morningness–Eveningness Questionnaire (MEQ), sleep quality: Pittsburgh Sleep Quality Index (PSQI), and sleepiness: Epworth Sleepiness Scale (ESS). One week prior to the launch of experimental sessions, time in bed (TIB) was fixed (from 22:00 to 7:00) in order to avoid the potential effect of an irregular bedtime schedule on circadian rhythm and sleep quality. During this week, participants resting metabolic rate (RMR) was assessed by indirect calorimetry. They wore accelerometers to assess physical activity and energy expenditure. The investigators also equipped participants with portable PSG in order to familiarize them sleeping with the device.

Subsequently, adolescents participated in the experimental protocol comprising two three-day sessions in random order, separated by a 10-day washout: one session was eucaloric, and the other ad libitum (Figure 1). A manipulation check was performed to ensure the washout period was effective at taking all participants back to baseline sleep and dietary intake between the two sets of conditions (eucaloric vs. ad libitum). No violations occurred. Physical activity was continuously monitored by accelerometry. In order to reduce biases that might overwhelm the relationship between sleep and dietary intake, participants were asked to maintain the same habitual daily activities as in the previous week and to refrain from any structured vigorous physical activities at least 48 h before the experimental protocol. Moreover, all sessions took part on the same days of the week and no scholastic commitments were held during the sleep assessment nights. The use of electronic media (smartphone, laptop...) in the evening was prohibited and evening bedtime was fixed. During night 3, participants were instructed to go to their room at 21:00. The investigators equipped them with PSG devices and the light was switched off at 22:00. In this manner they were given an ideal time of 9 h for sleep (from 22:00 to 07:00).

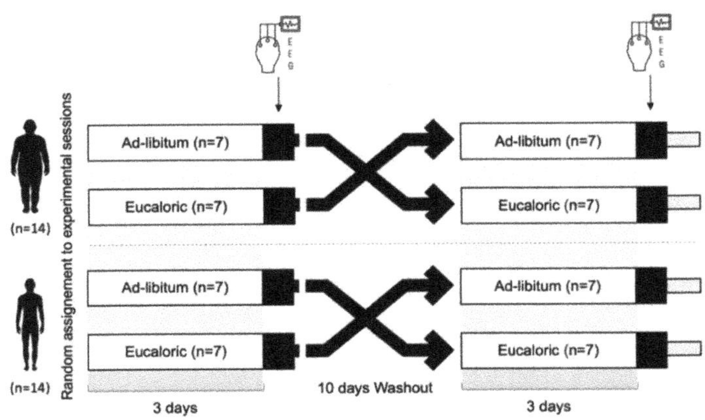

Figure 1. Study design.

2.3. Baseline Evaluations

2.3.1. Anthropometry and Body Composition

Barefoot height was measured using a portable stadiometer (TANITA, HR001, Japan). Body mass (BM) was measured using a digital scale (TANITA, BC-545N, Japan). Subsequently, Body Mass Index (BMI) was calculated as follows; BMI = body mass divided by height squared expressed in kg/m^2. Fat mass, and fat free mass were determined using dual-energy X-ray absorptiometry (QDR4500A scanner, Hologic, Waltham, MA, USA).

2.3.2. Circadian Typology and Subjective Sleep

The Horne–Östberg Morningness–Eveningness Questionnaire (MEQ) was used to assess circadian typology based on habitual waking and bedtimes as well as the times of day at which an individual prefers to 'perform' [28]. The PSQI was used to evaluate the quality of sleep during the last month based on 19 items covering seven clinically-relevant components of sleep difficulties. Each item is weighted on a scale, and a global score of sleep quality is calculated by adding up the seven component scores, producing an overall PSQI score ranging from 0–21, where a score > 5 indicates poor sleep quality [29]. The Epworth Sleepiness Scale (ESS) was used to measure daytime sleepiness [30]. The ESS consists of a simple self-report questionnaire with eight questions and responses on a 4-point Likert scale (0–3). An ESS score > 10 indicates excessive daytime sleepiness and the possibility of a high risk of underlying sleep disorders.

2.3.3. Resting Metabolic Rate

Resting metabolic rate (RMR) was measured in the morning, in a fasted state, using indirect calorimetry (Metamax 3B portable gas analyzer, Cortex, Leipzig, Germany). Before each test, the equipment was calibrated in accordance with the manufacturer's recommendations. Participants were placed in a supine position in a thermoneutral environment (22–25 °C room temperature) for 45 min before starting the measurements. After achieving a steady state, O_2 consumption and CO_2 production, standardized for temperature, barometric pressure, and humidity, were recorded at 1 min intervals for 20–45 min and averaged over the whole measurement period. Resting metabolic rate (in kcal/day) and respiratory quotient (ratio of CO2/O_2) were calculated thereafter.

2.3.4. Habitual Physical Activity and Energy Expenditure

Accelerometers are a non-intrusive and cost-effective tool that provide accurate and reliable measurement of physical activity and energy expenditure in youth under free living conditions. ActiGraph accelerometers (GT3X+, Actigraph LLC, Pensacola, FL, USA) were initialized to record data in 60-s epochs and worn for 3-consecutive days (one week prior to experimental sessions) on the dominant hip as recommended by the manufacturer. The device was only removed when water contact was possible (bathing, swimming). Data were analyzed using the manufacturer's software (Actilife 6.0, Pensacola, FL, USA). Physical activity was assessed based on the translation of count per min (CPM) into time (min/day) spent at various levels of movement intensity. Thresholds of CPM for sedentary, light, moderate, and vigorous levels of physical activities were defined according to Trost et al., (2011) [31]. Adolescents' energy expenditure was individually calculated based on RMR and daily metabolic equivalent of task values derived from the accelerometer's raw data.

2.4. Experimental Sessions

2.4.1. Dietary Intervention

All meals conformed to the adolescents' tastes as determined by a food questionnaire completed at baseline (Table S1). Disliked items were excluded in order to limit leftovers and prevent uncontrolled snacking outside planned meals during the eucaloric session. Top rated items were avoided to limit excessive intake and reflect habitual dietary patterns during the ad libitum session. Dietary intake during eucaloric sessions was planned by a registered dietician and was not designed to produce weight loss. The diet met the mean estimated energy requirement of adolescent males. Energy requirement was established individually based on subject's characteristics as well as physical activity and energy expenditure tracked on the previous week with the following macronutrient distribution: 18:52:30 (Protein:Carbohydrates:Fat). Individual dietary intake was calculated in advance and meal trays were prepared for each subject. Energy intake (EI) was distributed as follows: (25% from breakfast, 30% from lunch, 30% from dinner, and 15% from snacks). Participants were encouraged to finish their entire meals. If not, food leftovers were weighed and recorded, then they were subtracted from the initially calculated food intake. During the ad libitum session food was offered as buffet meals and the participants were asked to eat until they were satisfied. Food consumption was weighed and recorded by investigators. Energy intake and macronutrient distribution were calculated using a professional computerized nutrient analysis program (Bilnut 4.0 SCDA Nutrisoft software, France) and Ciqual tables (year-2018 version). Three meals and a snack were offered daily for both sessions at regular time intervals: 7:30 (breakfast), 12:00 (lunch), 17:30 (snack), and 19:00 (dinner). Caffeinated beverages such as colas, coffee, or tea were not allowed during the experimental sessions.

2.4.2. Polysomnography

The Sleep Profiler PSG2 (Advanced Brain Monitoring, Carlsbad, CA, USA) is an ambulatory alternative to laboratory polysomnography approved by the Food and Drugs Administration. The system provides access to 13-channels: electro-encephalography (EEG), electro-oculography (EOG), and electro-myography (EMG) of front-polar sites (AF7-AF8, AF7-Fpz, and AF8-Fpz) allowing the characterization of sleep staging and quality as well as wireless oximetry, nasal pressure/airflow, chest and abdomen respiratory effort, forehead and finger pulse rate, head movement and position, and quantitative snoring. The obtained records were uploaded to the Sleep Profiler portal where automated algorithms were applied to the signals. Auto-staging was performed based on the ratios of the power spectral densities and auto-detection of cortical and microarousals, sleep spindles, and ocular activity [32,33]. After the processing of sleep recordings, an experienced sleep expert reviewed each recording in order to confirm the accuracy of the auto-sleep staging and customize editing if needed. Sleep study featured the following outcomes: total sleep time (TST), sleep latency (SL), wake up after sleep (WASO), sleep efficiency (SE), arousal index, as well as sleep architecture (Wake, N1, N2, N3, REM sleep) according to recommendations of the American Academy of Sleep Medicine (AASM).

2.5. *Statistical Analysis*

Data analyses were performed using SPSS, (Version 26, SPSS, Inc, Chicago, IL, USA) Pre-experimental differences in characteristics between adolescents with obesity and their matched controls with regard to anthropometry, body composition, circadian typology and subjective sleep, as well as resting metabolic rate, physical activity, and energy expenditure were assessed using t-tests. Repeated measures analysis of variance (ANOVA) were performed with a 2 (weight status as between-subject factor) $\times$ 2 (conditions as within-subject factor) design, in order to assess differences between groups (obese vs. normal weight) and conditions (ad libitum vs. eucaloric) on each dietary intake variable (total energy intake, %EI from proteins, %EI from lipids and %EI from CHO), energy intake distribution throughout the day (breakfast, lunch, snack and dinner) as well as sleep outcomes (TST, SE, SL, Wake, WASO, REM, N1, N2, N3, and arousals). When significance was obtained, post hoc pairwise comparisons were computed using the Bonferroni method to examine differences between the two conditions (ad libitum vs. eucaloric) for the obese and normal weight groups separately. To test whether the presence of OSA influences the results obtained, additional analyses were conducted while excluding participants with potential OSA. All mentioned effect sizes were obtained using Cohen's d. Pearson's correlations were performed to examine the relationships between Δ energy intake and Δ sleep outcomes (ad libitum—eucaloric).

3. Results

3.1. Subject Characteristics

Twenty-eight adolescents participated in this study, OB (n = 14) were adolescents with obesity (BMI ≥ 90th centile) and NW (n = 14) were their normal weight age matched controls. Their mean age was 14.0 ± 0.9 years. Detailed characteristics of the two groups are presented in Table 1. The two groups differed in body weight, BMI, and percentage of body fat (all $p < 0.001$). OB presented poorer sleep quality as indicated by PSQI score ($p < 0.001$), as well as higher sleepiness estimated by ESS score ($p < 0.001$). No differences were found in circadian typology ($p = 0.210$). No differences were found in time spent on sedentary, light, and moderate to vigorous physical activity (all $p > 0.05$). However, RMR measured by indirect calorimetry was higher in OB (OB = 2035 ± 364 vs. NW = 1631 ± 106 kcal·day^{-1}, $p < 0.001$), which resulted in a higher total energy expenditure (TEE) in OB compared to NW (OB = 2864 ± 369 vs. NW = 2460 ± 148 kcal·day^{-1}, $p < 0.001$).

Table 1. Descriptive characteristics, subjective sleep quality, daytime sleepiness, and perceived stress of the sample population.

	OB Mean (SD)	NW Mean (SD)	p Value
Anthropometry & body composition			
Age (year)	14.0 (0.9)	14.0 (0.9)	-
Height (cm)	163.0 (11.7)	164.7 (7.8)	0.635
Weight (kg)	88.7 (23.8)	56.27 (6.14)	<0.001
BMI (kg·m^{-2})	33.1 (7.1)	20.71 (1.58)	<0.001
FM (%)	36.9 (5.7)	14.0 (3.2)	<0.001
FFM (kg)	53.4 (12.3)	48.2 (4.7)	0.164
Circadian typology and subjective sleep			
Circadian typology (MEQ_{score})	29.7 (3.4)	32.9 (5.9)	0.210
Sleep quality ($PSQI_{score}$)	8.4 (3.0)	4.2 (1.9)	<0.001
Sleepiness (ESS_{score})	9.3 (3.8)	4.3 (2.0)	<0.001
Physical activity, RMR and TEE			
Sedentary (min·day^{-1})	441 (46)	436 (62)	0.667
Light (min·day^{-1})	240 (36)	235 (43)	0.603
MVPA (min·day^{-1})	50 (27)	63 (39)	0.454
RMR (kcal·day^{-1})	2035 (364)	1631 (106)	<0.001
TEE (kcal·day^{-1})	2864 (369)	2460 (148)	<0.001

BMI: body mass index; ESS: Epworth Sleepiness Scale; FFM: fat free mass; FM: fat mass; MEQ: Horne–Östberg Morningness–Eveningness questionnaire; MVPA: moderate to vigorous physical activity; NW: normal weight controls; OB: adolescents with obesity; PSQI: Pittsburgh sleep quality index; RMR: resting metabolic rate; SD: standard deviation; TEE: Total energy expenditure.

3.2. Energy Balance and Macronutrient Intake Outcomes

Figure 2 shows energy intake (EI), energy expenditure (EE), and energy balance (EB) among OB and NW during ad libitum and eucaloric sessions. An interaction effect of EI was detected ($p < 0.001$). Overall, EI was higher during ad libitum compared to eucaloric ($p < 0.001$) and in OB compared to NW ($p < 0.001$). Post-hoc analysis showed a significant increase of EI among OB during ad libitum compared to eucaloric (+653 kcal, $p < 0.001$). However, the increase of EI in NW did not reach statistical significance (+125.8 kcal, $p = 0.24$). Mean values of EE did not differ from pre-experimental measurement. No differences were detected between group nor condition. Thereby, a significant interaction effect of EB was obtained ($p < 0.001$). As for EI, post-hoc analysis showed a significant increase in EB only in OB but not NW group.

Energy derived from each macronutrient is included in Table S2, interaction effects were noted for protein ($p < 0.001$) and CHO ($p = 0.007$). However, macronutrient distribution did not differ between conditions (all $p > 0.05$). The results were similar after excluding subject with OSA from the analyses (Table S3).

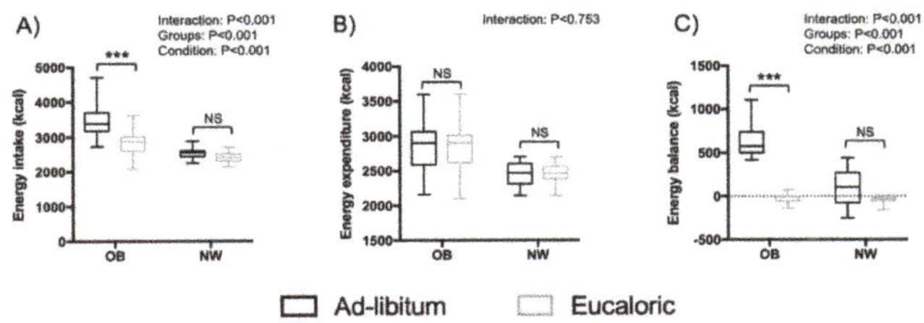

Figure 2. Energy intake (**A**), energy expenditure (**B**) and energy balance (**C**) among OB and NW during ad libitum and eucaloric sessions. OB: adolescents with obesity, NW: normal weight age matched controls, NS: non-significant, ***: significant difference in Post hoc pairwise comparisons (ad libitum vs. eucaloric) with $p < 0.001$.

3.3. Distribution of Energy Intake on the Four Meals

Giving that the proportions of all macronutrients were not different between conditions, in Figure 3 we focused on EI distribution throughout the offered four meals among OB and NW between the two sessions. Significant interaction effects were found for EI during lunch and dinner. Consumption of energy intake was higher in OB compared to NW and in ad libitum compared to the eucaloric condition.

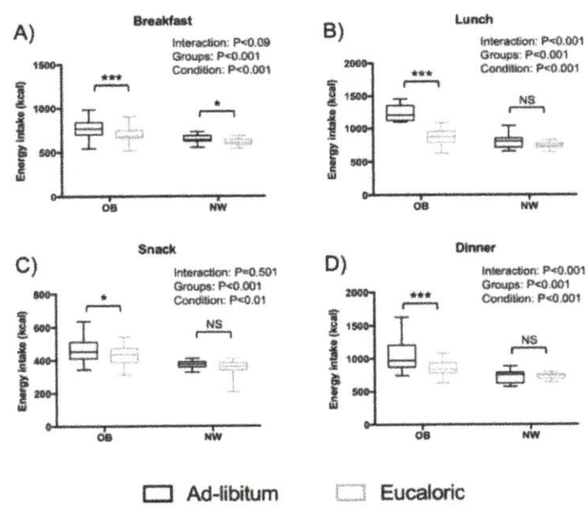

Figure 3. Distribution of energy intake throughout the four meals among adolescents with obesity and their matched normal weight controls during ad libitum and eucaloric sessions. *: significant difference in Post hoc pairwise comparisons (ad libitum vs. eucaloric) with $p < 0.05$. ***: significant difference in Post hoc pairwise comparisons (ad libitum vs. eucaloric) with $p < 0.001$.

3.4. Sleep Outcomes

Polysomnography outcomes are shown in Table 2. A significant interaction effect was only detected in N1 sleep ($p < 0.03$). However, trends toward significance were also obtained for SL ($p < 0.065$) with a moderate effect size. There were significant differences between conditions in SL ($p < 0.001$) and N1 ($p < 0.003$). However, no differences in SL and N1 were found between groups. Post hoc tests comparing these outcomes between sessions for each group revealed no differences among NW. However, SL and N1 sleep were significantly reduced in OB adolescents during eucaloric compared to ad libitum sessions. Moreover, in accordance with the subjective sleep evaluation at baseline, poorer sleep quality was obtained in OB adolescents compared to NW during both sessions. This was marked by higher WASO, and Arousals (all $p < 0.05$). All these results are still valid without taking into account subjects with OSA as shown in Table S5.

Table 2. Sleep parameters among adolescents with obesity and their matched normal weight controls during 3rd night of ad libitum and eucaloric sessions.

	OB		NW		ANOVA					
	Ad libitum	**Eucaloric**	**Ad libitum**	**Eucaloric**	**Interaction (Weight Status × Condition)**		**Weight Status Effect**		**Condition Effect**	
	Mean (SD)	**Mean (SD)**	**Mean (SD)**	**Mean (SD)**	**F**	***p* Value**	**F**	***p* Value**	**F**	***p* Value**
SE (%)	77.0 (9.6)	78.0 (10.4)	90.5 (6.3)	91.1 (6.7)	0.367	0.550 ES = 0.24	17.310	**<0.001** ES = 1.64	5.591	**0.026** ES = 0.92
SL (min)	32.0 (24.8)	23.5 (21.6) **	19.7 (11.8)	16.1 (13.0)	3.699	0.065 ES = 0.76	2.012	0.168 ES = 0.56	22.576	**<0.001** ES = 1.86
WASO (min)	92.0 (46.3)	95.0 (49.8)	31.5 (23.8)	31.9 (25.3)	1.216	0.280 ES = 0.44	18.420	**<0.001** ES = 1.68	1.962	0.173 ES = 0.54
Wake (min)	124.0 (51.8)	118.6 (56.5)	51.2 (34.3)	48.0 (36.6)	0.367	0.550 ES = 0.24	17.310	**<0.001** ES = 1.64	5.591	**0.026** ES = 1.92
Stage REM (min)	74.1 (26.9)	76.0 (26.5)	103.5 (40.8)	95.8 (38.3)	3.149	0.088 ES = 0.70	3.873	0.060 ES = 0.78	1.199	0.280 ES = 0.42
Stage N1 (min)	36.6 (9.4)	30.6 (11.8) *	30.7 (16.3)	29.7 (17.8)	5.508	**0.027** ES = 0.92	0.421	0.522 ES = 0.24	10.797	**0.003** ES = 1.28
Stage N2 (min)	209.0 (29.1)	209.7 (33.1)	256.7 (46.1)	258.2 (48.6)	0.883	0.356 ES = 0.36	10.838	**0.003** ES = 1.3	0.044	0.836 ES = 0.08
Stage N3 (min)	96.0 (45.0)	105.0 (54.5)	97.7 (47.7)	108.0 (50.3)	0.034	0.856 ES = 0.06	0.002	0.961 ES < 0.01	6.434	**0.018** ES = 1.00
Arousals (events/h)	17.3 (2.3)	15.4 (3.7)	13.3 (5.4)	12.34 (3.68)	0.637	0.432 ES = 0.32	6.502	**0.017** ES = 1.00	7.099	**0.013** ES = 1.04

Eucaloric: dietary intake to energy requirement; NW: normal-weight controls; OB: adolescents with obesity; REM: rapid eye movement; SE: sleep efficiency; SL: sleep latency; SD: standard deviation; significant p values are bolded; *: significant difference in post hoc pairwise comparisons (ad libitum vs. eucaloric) with $p < 0.05$. **: significant difference in Post hoc pairwise comparisons (ad libitum vs. eucaloric) with $p < 0.01$. ES: effect size.

3.5. Correlation of Δ Energy Intake with Δ Sleep Outcomes between Sessions in OB Group

Figure 4 shows correlations of Δ energy intake with Δ sleep outcomes between sessions in OB group. Increased total energy intake was associated with reduced SE ($p < 0.05$), increased SL ($p < 0.01$), N2 stage ($p < 0.01$), and reduced N3 stage ($p < 0.01$). Increased energy intake during the second half of the day especially during the dinner might be responsible of this effect. Increased energy intake during dinner was associated with decreased SE ($p < 0.01$), N3 stage ($p < 0.01$), and increased N2 stage ($p < 0.05$).

	Total energy intake (kcal)	Breakfast energy intake (kcal)	Lunch energy intake (kcal)	Snack energy intake (kcal)	Dinner energy intake (kcal)	First half of the day energy intake (kcal)	Second half of the day energy intake (kcal)
SE (%)	-0,549 *	0,224	-0,113	-0,562 *	-0,54 *	-0,045	-0,605 *
SL (min)	0,612 *	-0,113	0,089	0,218	0,462	0,176	0,445
WASO (min)	0,183	-0,214	0,075	0,604 *	0,324	-0,114	0,436
N1 (%)	0,27	-0,159	0,059	-0,176	0,251	0,19	0,153
N2 (%)	0,638 *	-0,231	0,239	0,032	0,575 *	0,202	0,505
N3 (%)	-0,685 **	0,113	0,019	-0,176	-0,713 **	0,048	-0,676 **
REM (%)	0,185	0,215	-0,333	-0,046	0,3	-0,362	0,285
Arousals (events/h)	0,508	-0,335	0,267	0,478	0,333	0,184	0,389

-1 / 0 / 1

Figure 4. Heatmap representation of the correlations between Δ energy intake and Δ sleep outcomes between sessions (ad libitum—eucaloric) in OB group. Red indicates a negative relationship whereas blue indicates a positive relationship; the darker the color, the higher the Pearson coefficient; *: significant correlation with $p < 0.05$; **: significant correlation with $p < 0.01$.

4. Discussion

In view of reported poor sleep outcomes in adolescents and based on growing literature connecting sleep to obesity, there is an urgent need for effective non-pharmacological alternatives to improve sleep in this population. Several studies highlighted the adverse effect of poor sleep (duration/quality) on energy balance regulation resulting in increased dietary intakes [6,9]. However, a clearer understanding of the complex interrelationship between sleep and energy balance in the opposite direction is of significant importance particularly in adolescents with obesity since the latest recommendations strongly endorsed the consideration of sleep disturbances in the management of pediatric obesity [27]. In this context, the present study questioned the effect on sleep of adjusting dietary intake according to energy requirements (eucaloric) with a balanced macronutrient distribution compared to ad libitum intake condition designed to reflect habitual dietary patterns. Poorer sleep quality was observed in adolescents with obesity compared to their normal weight peers during both conditions (eucaloric and ad libitum). Only three days of a eucaloric diet were sufficient to allow a reduction of SL and N1 in adolescents with obesity. This result suggests that positive energy balance may exacerbate the effect of obesity on sleep. However, the results also suggest a strong weight status effect on sleep despite feeding condition which means that positive energy balance alone cannot justify worsened sleep among adolescents with obesity compared to normal weight counterparts. No effect was obtained on sleep in normal weight adolescents. However, this might be related to the smaller variation in dietary intake between eucaloric and ad libitum conditions in this group.

The effect of the present dietary interventions on sleep were not surprising. Despite the limited number of clinical trials, an early study by Phillips et al., (1975) reported acute sleep changes in response to dietary manipulation. In fact, lower time spent in slow wave sleep (SWS) was observed following a high-carbohydrate/low-fat diet and higher time spent in SWS upon a low-carbohydrate/high-fat diet compared to the control diet. Moreover, rapid eye movement sleep (REM) was longer with the high-carbohydrate/low-fat diet compared to the low-carbohydrate/high-fat and control diets; while non rapid eye movement sleep (NREM) was shorter with both high-carbohydrate/low-fat and low-carbohydrate/high-fat diets compared to the control diet [24]. Sleep quality changes measured by accelerometry were also notable after 4 days of intervention including isocaloric high-protein or high-fat or high-carbohydrate diets [34]. Others show that carbohydrate intake with high glycemic index was associated with reduced SL [35]. In another study, SWS decreased immediately during the first sleep cycle with a high carbohydrate diet compared to a high fat diet when macronutrient distribution was only manipulated at dinner [25]. However, we highlight that our results documented sleep changes in the absence of any substantial modification in macronutrient proportions. Therefore, the most tangible explanation here could be overall energy intake, or variation of energy consumption distribution on the four meals offered during the day. Correlations between changes in energy intake and changes in SL and N3 suggest that excess energy during the latter meals may be more related to difficulties initiating sleep as well as lighter sleep among adolescents with obesity, but it remains to be demonstrated in normal weight adolescents as the ad libitum condition resulted in a smaller non-significant increase in EI.

Despite the fact that the obtained effect was very modest, our results confirm that excess energy intake, mainly in the latter meals, could alter sleep quality. This is in line with previously published studies. In fact, St-Onge et al., (2016) showed an increase in SL and a decrease in slow wave sleep (Stage N3) following one day of ad libitum feeding compared to those obtained upon 3 days of adapted and balanced diet in 26 normal-weight adults [36]. In another study carried out among 45 adults with obesity and obstructive sleep apnea, a higher caloric intake at night was also associated with higher sleep latency [37]. As detailed above, most studies dealing with the effect of nutrition on sleep focused on macronutrient proportions and their "quality effect", on sleep. However, the current study highlights that sleep is also sensitive to energy intake variation underlining a "quantity effect". Excess energy especially in the latter meals appears to exert a negative effect on sleep initiation and structure with more light sleep and less deep sleep. Some clinical evidence suggests that consuming a large morning meal was not associated with obesity while a large evening meal was found to substantially increase the risk of obesity [38]. Moreover, following a 12-week weight loss intervention (isocaloric diet: 1400 kcal per day) among women with obesity, the group consuming a larger breakfast meal lost much more weight and showed better improvement in fasting glucose, insulin, and Homeostatic Model Assessment of Insulin Resistance (HOMA-IR) [39]. That is to say that in spite of the number of calories, early eaters lose more weight and enhance their metabolic profile in comparison with late eaters. The mechanisms behind this temporal effect on weight regulation remain to be further explored. However, it appears that excess energy intake during the latter meals could disrupt the circadian system by inducing a phase shift in the peripheral clock and not the master clock. This circadian misalignment appears to induce adverse effect on metabolism. Fewer studies focused on the effect of evening energy intake on sleep. Driver et al. (1999) reported increased body temperature and suggested that a high energy meal resulted in a long-lasting thermic effect of food that may interfere with sleep episodes. However, no effect of a high energy dinner on sleep compared to a control dinner containing half the amount of calories or even to a fasted condition were obtained. Elevated core body temperature is known to affect sleep onset latency and to decrease slow wave sleep (N3 stage) [40]. However, the aforementioned study was based on the acute effect of one meal and may have been underpowered to detect differences in sleep (n = 7). The effect of the digestive system on sleep remains poorly understood and should be further investigated. However, a decrease in the gastrointestinal tract activity from sleep stage N1 to N3 has been already reported [41]. Furthermore, previous studies showed that large meals expand the stomach and increase upward pressure against the lower esophageal sphincter [42]. Therefore, along with the thermic effect of food, mechanical stimulation of the gastrointestinal tract as well as heartburn symptoms near to bedtime could also be responsible for sleep alteration in response to larger energy intake during the latter meals. Finally, we emphasize that the timing of energy intake was restricted in this study. Thus, the effect of delayed meal timing on sleep in patients with obesity needs to be addressed in future studies.

Limitations, Strengths, and Future Research Directions

This study was strengthened by combining objective assessments of energy balance and sleep in youth under naturalistic free-living settings. However, as reported EI variations between ad libitum and eucaloric sessions were only significant in adolescents with obesity. Our initial hypothesis of better improvement of sleep quality in adolescents with obesity during eucaloric session was only based on celling effect of nutritional intervention on sleep. We actually expected a comparable increase (but lower than that observed in OB group) in EI in the two groups as previously reported in adults [43]. No study to our knowledge has compared the nutritional response between normal weight adolescents and those with obesity over 24 h. However, previous studies have reported similar nutritional response in both groups during ad libitum lunch and dinner [44]. Even if we failed to obtain a significant increase in EI in normal weight adolescents, this result may underline lower loss of control overeating in NW group compared to OB. The choice of this protocol was initially intended to reflect habitual dietary intake in comparison with eucaloric condition. However, eating behavior traits of participants, as well as the inclusion of subjects with obesity taking part in a multidisciplinary weight loss program may also have exaggerated the increase of EI in OB group. We underline though that the ad libitum sessions were organized identically with the same food items in both groups according to Thivel et al. (2016) [45]. On the other hand, although we carefully included adolescents without diagnosed OSA, it turned out that some participants from the OB group (n = 4) presented undiagnosed OSA. However, the study results were checked, and no major statistical differences resulted from the inclusion of these participants (Table S3–S5). This could be explained by the mildness of the disease (AHI < 7 episodes·h^{-1}).

The majority of previous studies exploring the effect of nutrition on sleep did not consider physical activity. We emphasize that future research should systematically assess physical activity during the intervention, giving the major impact that this behavior plays on metabolism and sleep regulation, even in the short term [46]. The current study opens the horizons for future research investigating the effect of energy balance (both EI and EE) on sleep. It would be interesting to examine the effect of positive, neutral, or negative energy balance on sleep and determine the effect of weight status per se on this relationship. Furthermore, determining if a differential effect on sleep would result from an energy deficit obtained by decreasing energy intake compared to that of increasing energy expenditure would also be interesting. Future studies should also examine if an interaction effect of nutrition and exercise on sleep exist. Finally, the effect of temporal variation in excess energy intake on sleep found in this study brings a new insight to the potential therapeutic effect of chrono-nutrition on sleep in adolescents.

5. Conclusions

Reduced energy intake during the eucaloric condition compared to ad libitum feeding for only three days improved sleep quality by decreasing SL and N1 stage. This effect was only obtained in adolescents with obesity given that normal weight adolescents didn't increase their energy intake between the two conditions.

Supplementary Materials: The following are available online at https://www.mdpi.com/article/10.3390/nu13103550/s1, Table S1: Food items included in the ad libitum buffet meals and the standardized eucaloric meals, Table S2: Energy derived from each macronutrient among adolescents with obesity and their matched normal weight controls during *ad libitum* and eucaloric sessions, Table S3: Energy derived from each macronutrient among adolescents with obesity (excluding the 4 participants with OSA) and their matched normal weight controls during ad libitum and eucaloric sessions, Table S4: Distribution of energy intake throughout the four meals among adolescents with obesity (excluding the 4 participants with OSA) and their matched normal weight controls during ad libitum and eucaloric sessions, Table S5: Sleep based on EEG-recording among adolescents with obesity (excluding the 4 participants with OSA) and their matched normal weight controls during 3rd night of ad libitum and eucaloric sessions. Figure S1: Heatmap representation of the correlations between Δ energy intake and Δ sleep outcomes between sessions (ad libitum–eucaloric) in NW group.

Author Contributions: Study conceptualization, O.S., P.D., and S.W., investigation, O.S., E.R., and É.D., formal analysis, G.D.S., O.S., and P.D., writing—original draft preparation, O.S. writing—review and editing, S.W., É.M., and P.D. All authors have read and agreed to the published version of the manuscript.

Funding: Financial support for this study was provided by the Auvergne-Rhône-Alpes region and the European Regional Development Fund (FEDER). [grant numbers DRV_SIP_2019-012_protmorpheus].

Institutional Review Board Statement: The study was conducted according to the guidelines of the Declaration of Helsinki, and based on preliminary assessments performed within the framework of the PROTMORPHEUS project which received approval by relevant Institutional Ethics Review (Clinical trial No. NCT04041934).

Informed Consent Statement: Informed consent was obtained from adolescents and their legal guardians prior to the study launch.

Data Availability Statement: The data presented in this study are available on request from the corresponding author and the permission of all parties involved in the study.

Acknowledgments: The authors thank all the adolescents, the staff of "Tza Nou" (Obesity Treatment Center for Children and Adolescents), in particular the director Mr Guettouche Twefik as well as the Pediatric Department, CHU (Clermont-Ferrand, France) for their time and effort in the completion of this study. A special thanks to Carole Brun and all implicated students in this work for their thoughtful and enthusiastic help.

Conflicts of Interest: The authors declare no conflict of interest. The funders had no role in the design of the study; in the collection, analyses, or interpretation of data; in the writing of the manuscript, or in the decision to publish the results.

References

1. Gradisar, M.; Gardner, G.; Dohnt, H. Recent Worldwide Sleep Patterns and Problems during Adolescence: A Review and Meta-Analysis of Age, Region, and Sleep. *Sleep Med.* **2011**, *12*, 110–118. [CrossRef] [PubMed]
2. Matricciani, L.; Olds, T.; Petkov, J. In Search of Lost Sleep: Secular Trends in the Sleep Time of School-Aged Children and Adolescents. *Sleep Med. Rev.* **2012**, *16*, 203–211. [CrossRef] [PubMed]
3. Carskadon, M.A. Sleep in Adolescents: The Perfect Storm. *Pediatric Clin.* **2011**, *58*, 637–647. [CrossRef] [PubMed]
4. Danisi, J.; Fernandez-Mendoza, J.; Vgontzas, A.N.; Bixler, E.O. Obesity and Sleep Disturbances. In *The Behavioral, Molecular, Pharmacological, and Clinical Basis of the Sleep-Wake Cycle*; Elsevier: Amsterdam, The Netherlands, 2019; pp. 123–142.
5. Chaput, J.-P. Sleep Patterns, Diet Quality and Energy Balance. *Physiol. Behav.* **2014**, *134*, 86–91. [CrossRef] [PubMed]
6. St-Onge, M.-P. The Role of Sleep Duration in the Regulation of Energy Balance: Effects on Energy Intakes and Expenditure. *J. Clin. Sleep Med.* **2013**, *9*, 73–80. [CrossRef] [PubMed]
7. Morselli, L.; Leproult, R.; Balbo, M.; Spiegel, K. Role of Sleep Duration in the Regulation of Glucose Metabolism and Appetite. *Best Pract. Res. Clin. Endocrinol. Metab.* **2010**, *24*, 687–702. [CrossRef]
8. Shechter, A.; Grandner, M.A.; St-Onge, M.-P. The Role of Sleep in the Control of Food Intake. *Am. J. Lifestyle Med.* **2014**, *8*, 371–374. [CrossRef]

9. St-Onge, M.-P. Sleep–Obesity Relation: Underlying Mechanisms and Consequences for Treatment. *Obes. Rev.* **2017**, *18*, 34–39. [CrossRef]
10. Beebe, D.W.; Simon, S.; Summer, S.; Hemmer, S.; Strotman, D.; Dolan, L.M. Dietary Intake Following Experimentally Restricted Sleep in Adolescents. *Sleep* **2013**, *36*, 827–834. [CrossRef]
11. Hart, C.N.; Carskadon, M.A.; Considine, R.V.; Fava, J.L.; Lawton, J.; Raynor, H.A.; Jelalian, E.; Owens, J.; Wing, R. Changes in Children's Sleep Duration on Food Intake, Weight, and Leptin. *Pediatrics* **2013**, *132*, e1473–e1480. [CrossRef]
12. Kredlow, M.A.; Capozzoli, M.C.; Hearon, B.A.; Calkins, A.W.; Otto, M.W. The Effects of Physical Activity on Sleep: A Meta-Analytic Review. *J. Behav. Med.* **2015**, *38*, 427–449. [CrossRef]
13. Lang, C.; Kalak, N.; Brand, S.; Holsboer-Trachsler, E.; Pühse, U.; Gerber, M. The Relationship between Physical Activity and Sleep from Mid Adolescence to Early Adulthood. A Systematic Review of Methodological Approaches and Meta-Analysis. *Sleep Med. Rev.* **2016**, *28*, 32–45. [CrossRef] [PubMed]
14. Binks, H.; E. Vincent, G.; Gupta, C.; Irwin, C.; Khalesi, S. Effects of Diet on Sleep: A Narrative Review. *Nutrients* **2020**, *12*, 936. [CrossRef] [PubMed]
15. Frank, S.; Gonzalez, K.; Lee-Ang, L.; Young, M.C.; Tamez, M.; Mattei, J. Diet and Sleep Physiology: Public Health and Clinical Implications. *Front. Neurol* **2017**, *8*, 393. [CrossRef] [PubMed]
16. Peuhkuri, K.; Sihvola, N.; Korpela, R. Diet Promotes Sleep Duration and Quality. *Nutr. Res.* **2012**, *32*, 309–319. [CrossRef] [PubMed]
17. St-Onge, M.-P.; Mikic, A.; Pietrolungo, C.E. Effects of Diet on Sleep Quality. *Adv. Nutr.* **2016**, *7*, 938–949. [CrossRef] [PubMed]
18. Dashti, H.S.; Zuurbier, L.A.; de Jonge, E.; Voortman, T.; Jacques, P.F.; Lamon-Fava, S.; Scheer, F.A.J.L.; Kiefte-De Jong, J.C.; Hofman, A.; Ordovás, J.M.; et al. Actigraphic Sleep Fragmentation, Efficiency and Duration Associate with Dietary Intake in the Rotterdam Study. *J. Sleep Res.* **2016**, *25*, 404–411. [CrossRef]
19. Grandner, M.A.; Kripke, D.F.; Naidoo, N.; Langer, R.D. Relationships among Dietary Nutrients and Subjective Sleep, Objective Sleep, and Napping in Women. *Sleep Med.* **2010**, *11*, 180–184. [CrossRef] [PubMed]
20. Tanaka, E.; Yatsuya, H.; Uemura, M.; Murata, C.; Otsuka, R.; Toyoshima, H.; Tamakoshi, K.; Sasaki, S.; Kawaguchi, L.; Aoyama, A. Associations of Protein, Fat, and Carbohydrate Intakes with Insomnia Symptoms among Middle-Aged Japanese Workers. *J. Epidemiol.* **2013**, *23*, 132–138. [CrossRef]
21. Spaeth, A.M.; Dinges, D.F.; Goel, N. Objective Measurements of Energy Balance Are Associated with Sleep Architecture in Healthy Adults. *Sleep* **2017**, *40*, zsw018. [CrossRef]
22. Crispim, C.A.; Zimberg, I.Z.; dos Reis, B.G.; Diniz, R.M.; Tufik, S.; de Mello, M.T. Relationship between Food Intake and Sleep Pattern in Healthy Individuals. *J. Clin. Sleep Med.* **2011**, *7*, 659–664. [CrossRef] [PubMed]
23. Afaghi, A.; O'Connor, H.; Chow, C.M. Acute Effects of the Very Low Carbohydrate Diet on Sleep Indices. *Nutr. Neurosci.* **2008**, *11*, 146–154. [CrossRef]
24. Phillips, F.; Crisp, A.H.; McGuinness, B.; Kalucy, E.C.; Chen, C.N.; Koval, J.; Kalucy, R.S.; Lacey, J.H. Isocaloric Diet Changes and Electroencephalographic Sleep. *Lancet* **1975**, *306*, 723–725. [CrossRef]
25. Yajima, K.; Seya, T.; Iwayama, K.; Hibi, M.; Hari, S.; NAkASHIMA, Y.; Ogata, H.; Omi, N.; Satoh, M.; TOkUYAMA, K. Effects of Nutrient Composition of Dinner on Sleep Architecture and Energy Metabolism during Sleep. *J. Nutr. Sci. Vitaminol.* **2014**, *60*, 114–121. [CrossRef] [PubMed]
26. Fatima, Y.; Doi, S.A.R.; Mamun, A.A. Sleep Quality and Obesity in Young Subjects: A Meta-Analysis. *Obes. Rev.* **2016**, *17*, 1154–1166. [CrossRef]
27. Styne, D.M.; Arslanian, S.A.; Connor, E.L.; Farooqi, I.S.; Murad, M.H.; Silverstein, J.H.; Yanovski, J.A. Pediatric Obesity—Assessment, Treatment, and Prevention: An Endocrine Society Clinical Practice Guideline. *J. Clin. Endocrinol. Metab.* **2017**, *102*, 709–757. [CrossRef]
28. Horne, J.A.; Östberg, O. A Self-Assessment Questionnaire to Determine Morningness-Eveningness in Human Circadian Rhythms. *Int. J. Chronobiol.* **1976**, *4*, 97–110. [PubMed]
29. Buysse, D.J.; Reynolds, C.F.; Monk, T.H.; Berman, S.R.; Kupfer, D.J. The Pittsburgh Sleep Quality Index: A New Instrument for Psychiatric Practice and Research. *Psychiatry Res.* **1989**, *28*, 193–213. [CrossRef]
30. Johns, M.W. A New Method for Measuring Daytime Sleepiness: The Epworth Sleepiness Scale. *Sleep* **1991**, *14*, 540–545. [CrossRef]
31. Trost, S.G.; Loprinzi, P.D.; Moore, R.; Pfeiffer, K.A. Comparison of Accelerometer Cut Points for Predicting Activity Intensity in Youth. *Med. Sci. Sports Exerc.* **2011**, *43*, 1360–1368. [CrossRef]
32. Levendowski, D.J.; Popovic, D.; Berka, C.; Westbrook, P.R. Retrospective Cross-Validation of Automated Sleep Staging Using Electroocular Recording in Patients with and without Sleep Disordered Breathing. *Int. Arch. Med.* **2012**, *5*, 21. [CrossRef]
33. Levendowski, D.J.; Ferini-Strambi, L.; Gamaldo, C.; Cetel, M.; Rosenberg, R.; Westbrook, P.R. The Accuracy, Night-to-Night Variability, and Stability of Frontopolar Sleep Electroencephalography Biomarkers. *J. Clin. Sleep Med.* **2017**, *13*, 791–803. [CrossRef] [PubMed]
34. Lindseth, G.; Lindseth, P.; Thompson, M. Nutritional Effects on Sleep. *West. J. Nurs. Res.* **2013**, *35*, 497–513. [CrossRef] [PubMed]
35. Afaghi, A.; O'Connor, H.; Chow, C.M. High-Glycemic-Index Carbohydrate Meals Shorten Sleep Onset. *Am. J. Clin. Nutr.* **2007**, *85*, 426–430. [CrossRef]
36. St-Onge, M.-P.; Roberts, A.; Shechter, A.; Choudhury, A.R. Fiber and Saturated Fat Are Associated with Sleep Arousals and Slow Wave Sleep. *J. Clin. Sleep Med.* **2016**, *12*, 19–24. [CrossRef] [PubMed]

37. de Melo, C.M.; Del Re, M.P.; dos Santos Quaresma, M.V.L.; Antunes, H.K.M.; Togeiro, S.M.; Ribeiro, S.M.L.; Tufik, S.; de Mello, M.T. Relationship of Evening Meal with Sleep Quality in Obese Individuals with Obstructive Sleep Apnea. *Clin. Nutr. ESPEN* **2019**, *29*, 231–236. [CrossRef]
38. Wang, J.B.; Patterson, R.E.; Ang, A.; Emond, J.A.; Shetty, N.; Arab, L. Timing of Energy Intake during the Day Is Associated with the Risk of Obesity in Adults. *J. Hum. Nutr. Diet.* **2014**, *27*, 255–262. [CrossRef]
39. Jakubowicz, D.; Barnea, M.; Wainstein, J.; Froy, O. High Caloric Intake at Breakfast vs. Dinner Differentially Influences Weight Loss of Overweight and Obese Women. *Obesity* **2013**, *21*, 2504–2512. [CrossRef]
40. Kräuchi, K.; Fattori, E.; Giordano, A.; Falbo, M.; Iadarola, A.; Aglì, F.; Tribolo, A.; Mutani, R.; Cicolin, A. Sleep on a High Heat Capacity Mattress Increases Conductive Body Heat Loss and Slow Wave Sleep. *Physiol. Behav.* **2018**, *185*, 23–30. [CrossRef] [PubMed]
41. Kanaly, T.; Shaheen, N.J.; Vaughn, B.V. Gastrointestinal Physiology and Digestive Disorders in Sleep. *Curr. Opin. Pulm. Med.* **2009**, *15*, 571–577. [CrossRef]
42. Jarosz, M.; Taraszewska, A. Risk Factors for Gastroesophageal Reflux Disease: The Role of Diet. *Prz. Gastroenterol.* **2014**, *9*, 297. [CrossRef] [PubMed]
43. Cornier, M.-A.; Grunwald, G.K.; Johnson, S.L.; Bessesen, D.H. Effects of Short-Term Overfeeding on Hunger, Satiety, and Energy Intake in Thin and Reduced-Obese Individuals. *Appetite* **2004**, *43*, 253–259. [CrossRef] [PubMed]
44. Thivel, D.; Metz, L.; Julien, A.; Morio, B.; Duché, P. Obese but Not Lean Adolescents Spontaneously Decrease Energy Intake after Intensive Exercise. *Physiol. Behav.* **2014**, *123*, 41–46. [CrossRef] [PubMed]
45. Thivel, D.; Genin, P.M.; Mathieu, M.-E.; Pereira, B.; Metz, L. Reproducibility of an In-Laboratory Test Meal to Assess Ad Libitum Energy Intake in Adolescents with Obesity. *Appetite* **2016**, *105*, 129–133. [CrossRef] [PubMed]
46. Saidi, O.; Rochette, E.; Bovet, M.; Merlin, E.; Duché, P. Acute Intense Exercise Improves Sleep and Decreases next Morning Consumption of Energy-Dense Food in Adolescent Girls with Obesity and Evening Chronotype. *Pediatr. Obes.* **2020**, *15*, e12613. [CrossRef] [PubMed]

nutrients

MDPI

Article

No Significant Effect of the Individual Chronotype on the Result of Moderate Calorie Restriction for Obesity—A Pilot Study

Zofia Strojny [1], Rafał Rutkowski [1,†], Alina Kanikowska [2,†], Agnieszka Zawada [2], Aldona Juchacz [2], Marian Grzymisławski [2], Maki Sato [3], Monika Litwinowicz [1], Katarzyna Korybalska [1], Andrzej Bręborowicz [1], Janusz Witowski [1] and Dominika Kanikowska [1,*]

1 Department of Pathophysiology, Poznan University of Medical Sciences, 61-701 Poznan, Poland; zstrojny@ump.edu.pl (Z.S.); rrutkowski@ump.edu.pl (R.R.); mlitwinowicz@ump.edu.pl (M.L.); koryb@ump.edu.pl (K.K.); abreb@ump.edu.pl (A.B.); jwitow@ump.edu.pl (J.W.)
2 Department of Gastroenterology, Dietetics and Internal Diseases, Poznan University of Medical Sciences, 60-355 Poznan, Poland; akanikowska@ump.edu.pl (A.K.); a.zawada@ump.edu.pl (A.Z.); ajuchacz@wcpit.org (A.J.); mariangrzym@ump.edu.pl (M.G.)
3 Department of Physiology, Institutional Research, School of Medicine, Aichi Medical University, Aichi 480-1195, Japan; msato@aichi-med-u.ac.jp
* Correspondence: dkanikowska@ump.edu.pl
† Authors contributed equally to this paper.

Citation: Strojny, Z.; Rutkowski, R.; Kanikowska, A.; Zawada, A.; Juchacz, A.; Grzymisławski, M.; Sato, M.; Litwinowicz, M.; Korybalska, K.; Bręborowicz, A.; et al. No Significant Effect of the Individual Chronotype on the Result of Moderate Calorie Restriction for Obesity—A Pilot Study. *Nutrients* **2021**, *13*, 4089. https://doi.org/10.3390/nu13114089

Academic Editor: Maria Luz Fernandez

Received: 13 September 2021
Accepted: 9 November 2021
Published: 15 November 2021

Publisher's Note: MDPI stays neutral with regard to jurisdictional claims in published maps and institutional affiliations.

Abstract: Background: Chronotype is the pattern of the circadian rhythm that allows an individual to optimize times of sleep and activity. It has been observed that chronotypes may associate with some conditions and diseases, including obesity. It is not known, however, whether chronotypes determine the effectiveness of weight loss regimens. Therefore, in the present study, we compared the outcomes of a 3-week moderate calorie restriction undertaken by individuals with obesity under the same controlled hospital conditions. Methods: A total of 131 participants with obesity (median BMI 40.0) were studied. The subjects underwent the same dietary intervention over 3 weeks, with a 30% reduction in daily caloric intake. The individual chronotypes were assessed by the morning and evening questionnaire (MEQ) according to Horne and Östberg. Anthropometric and biochemical parameters were assessed by routine methods. Results: Of all patients examined, 75% had the morning (lark) chronotype and 25% had the evening (owl) chronotype. These patient sub-groups did not differ in terms of demographic, anthropometric and biochemical characteristics at baseline. After 3 weeks of calorie restriction, both groups experienced a similar loss of weight and BMI (Body Mass Index) (3.4 ± 0.38% for larks vs. 4.1 ± 0.47% for owls, $p = 0.45$), with owls exhibiting a marginally greater loss of body fat (3.1 ± 0.79%) compared with larks (2.6 ± 0.64%), $p = 0.02$. On the other hand, the larks had a more discernable, but not statistically significant from owls, decrease in glycated haemoglobin and CRP (C Reactive Protein). Conclusions: The chronotype of individuals with obesity does not have a significant effect on the magnitude of the body weight loss, but there is a tendency observed towards the reduction in body fat content in owls through changing their meal and sleep timing to earlier hours, in response to moderate calorie restriction applied under the same controlled conditions.

Keywords: chronotype; calorie restriction; obesity

1. Introduction

The chronotype is a pattern of the circadian rhythm that determines the most optimal activity and rest time. It is divided into morning ("lark"), evening ("owl"), and mixed type.

It seems indisputable that there is a close correlation between the chronotype and health condition, including the incidence of specific diseases [1–3].

Based on the relationship between chronotype and obesity, people defined as evening type present a predisposition to an increased adipose tissue content [4]. Numerous studies suggest that this relationship may be determined by diet. Maukonen (2016) noted that people with the evening chronotype were much more likely to lead an unhealthy lifestyle while not showing a greater genetic predisposition to obesity compared to individuals from other chronotypes [5]. Similar conclusions were also reached by Lucassen et al., who determined that evening chronotype individuals were associated with the later consumption of meals, more meals per day, night snacking, and a higher level of stress hormone and the risk of cardiovascular events related to obesity [6].

Considering the scientific data, few reports exist on the relationship between chronotype and the effectiveness of a diet. Ross et al. focused their studies on the effectiveness of a reduction diet, taking into account the subjects' chronotype. Of the subjects studied, the majority (73%) were women. During the course of the study, the authors tried to determine whether the chronotype of the subject affected the results of the diet used. It was found that the "morning" chronotype individuals had a higher chance of weight loss. This phenomenon was determined by longer sleep and its better quality [7]. In the Ross experiment, participants were not subject to the same dietary restrictions but were recruited as members of the National Weight Control Registry (NWCR) who lost at least 13.6 kg and maintained that weight for at least 1 year.

Furthermore, we investigated a soluble form of the advanced glycation end products receptor (sRAGE) [8]. RAGE receptors show a strong relationship with obesity [9], and their expression within adipocytes is higher in the population with increased body weight [10]. The available literature on the subject provides ample evidence indicating a close relationship between the concentration of the RAGE receptor's soluble form and body weight [11,12]. Moreover, many authors suggest that sRAGE may be considered a universal biomarker used to treat obesity and its complications [9,13].

Although prior studies have focused on the eating and physical activity patterns of successful weight loss, there has been no research to date on the chronotype in control condition. As many individuals may have difficulties in adhering to prolonged and substantial caloric restriction resulting in a significant weight loss, here we have addressed this issue in a setting more likely to reflect a real-life situation but in hospital conditions, and assessed various biochemical parameters in obese volunteers undergoing only short-term (3 weeks) and rather moderate calorie restriction (25–30% energy deficit).

The aim of the study was to determine whether the individual chronotype is related to the weight loss.

2. Materials and Methods

The study protocol was approved by the Bioethics Committee of the Medical University of Poznań (Resolution No. 201/19) and was in line with the Helsinki Declaration's basic principles. Each subject signed a consent form to participate in the study and was informed in detail about its course.

2.1. Study Group

The study group consisted of 131 people (61 men and 70 women) treated for obesity at the Department and Clinic of Internal Diseases, Metabolic Diseases, and Dietetics of the Medical University of Poznań, outlined in Table 1. The subjects were individuals with obesity, falling into pre-defined criteria with BMI (Body Mass Index) >30 kg/m^2, age >18 years old, who were invited to initiate a weight loss regimen on a doctor's advice. The exclusion criteria for the study included: vegetarian (or another alternative) diet consumption; previous or current neoplastic disease (receiving radiotherapy, chemotherapy); cardiovascular, autoimmune, congenital metabolic or liver diseases; inflammatory bowel disease (Crohn's disease, ulcerative colitis); status after ischemic or hemorrhagic stroke (<6 months) and after STEMI (ST Elevation Myocardial Infarction) or NSTEMI (No ST

Elevation Myocardial Infarction) (<12 months); eating disorders (anorexia, bulimia); mental disorders; alcohol/drug abuse; and ongoing antibiotic therapy and steroid therapy.

Table 1. Anthropometric characteristics of all study participants by chronotype (n = 131); Kruskal–Wallis test. The values of the results in the study are given as a median (minimum–maximum). BMI (Body Mass Index).

Parameter	Morning Type (n = 98)	Evening Type (n = 33)	p
Age, years	47.0 (22.0–69.0)	38.0 (21.0–63.0)	0.0737
Body weight, kg	115.8 (75.0–205.0)	113.0 (89.2–194.0)	0.8011
BMI, kg/m^2	40.5 (30.1–63.3)	38.5 (30.5–73.0)	0.2233
Female, n(%)	55 (56)	15(45)	
Post-menopausal female, n (%)	26 (27)	7 (21)	
Diabetes mellitus patients, n (%)	16 (16)	6 (18)	

Individuals who qualified for the study were subjected to a three-week dietary treatment in a hospital setting. All patients were hospitalized and subjected to the same environmental factors, schedule of the day, amount of physical exercise, and catering diet including vegetables, meat, dairy, and grain products. They used an individually selected diet with a 25–30% reduction in the daily caloric supply (reduction of 500 to 1000 kcal) in relation to the total energy requirement, calculated according to the Harris and Benedict formula [14] and the physical activity index. All patients received the same type of diet provided by a catering company, with the same proportion of nutrients: 20% protein, 25–30% fat, and 50–55% carbohydrates.

The patients' daily schedule was as follows:

- 7.00 am—wake up;
- 7.30 am—breakfast;
- 8.30 am—breathing exercises 30 min;
- 9.30 am—strengthening exercises 30 min;
- 10.15 am—brunch;
- 11.00 am—stationary bike ride 40 min;
- 12.00 pm—meeting with a dietitian;
- 12.40 pm—lunch;
- 13.30 pm—breathing exercises 30 min;
- 17.00 pm—dinner;
- 23:00 pm—bedtime.

During their stay at the Department and Clinic of Internal Diseases, Metabolic Diseases and Dietetics of the Medical University of Poznań, the patients had two blood samples taken for biochemical tests, based on which the levels of glucose, insulin, C-reactive protein, HbA1c glycated hemoglobin in the blood serum, and insulin resistance were calculated. Sampling was performed at the same time of day (between 07:00 h and 08:00 h) and in a fasting state to avoid circadian influences.

Samples were centrifuged and stored at −80 °C until assay. Each sample was assayed for soluble receptor for advanced glycation end products (sRAGE) using enzyme-linked immunosorbent assays (ELISA) using the manufacturers' instructions (R&D system, Minneapolis, USA). The sensitivity of the sRAGE was 48 pg/mL. The routine biochemical investigations were carried out in the University Hospital's central laboratory, using standard commercial reagent kits. Bodyweight measurement was also conducted, including body composition analysis using the BIA electrical bioimpedance method, which allowed the estimation of both the percentage and weight of adipose tissue and muscle tissue in patients. The measurement was performed under standardized conditions (in the morning, in the fasting state, with an empty bladder, and with the body in a standardized spatial position).

Simultaneously, all participants completed the morning and evening questionnaire (MEQ) according to Horne and Östberg (1976) to assess individual daily preferences [15]. On this basis, they were divided according to the presented type of chronotype: lark (scoring $\geq$ 59), owl (scoring $\leq$ 41), and intermediate type (scoring 42–58).

However, ultimately it was decided to combine the last two types; therefore, owl and intermediate were classified as the evening type [4,16].

2.2. Statistical Analysis

The calculations were made using the Statsoft Statistica 12 program. The level of significance was set at $\alpha = 0.05$. The result was considered statistically significant when $p < \alpha$. The normality of the distribution of variables was checked using the Shapiro–Wilk test. The Mann–Whitney test was used to calculate the two groups of variables. To test the relationship between continuous variables, Spearman's rank correlation coefficient was calculated, and in the case of categorical variables, the chi2 test of independence. In order to analyze changes over time, for related samples, the Wilcoxon test was used. The sRAGE results were assessed in terms of outliers' values in the Grubbs' test. After removing two outliers above 2000 pg/mL, the statistical significance did not change significantly. Therefore, in the entire further study, the sRAGE results were presented with outliers included. The values of the results in the study are given as a median (minimum–maximum).

3. Results

Of 131 patients examined, 98 (75%) were found to have the morning chronotype, 7 (5%) had the evening chronotype, and 26 (20%) had the intermediate chronotype (Figure 1). Due to the small number of subjects with the pure evening chronotype, for the purpose of this analysis the participants with intermediate and evening chronotypes were grouped together and designated further as owls. These patients were compared to those with the morning chronotype, designated as larks. Such an approach was used previously in other studies [16,17].

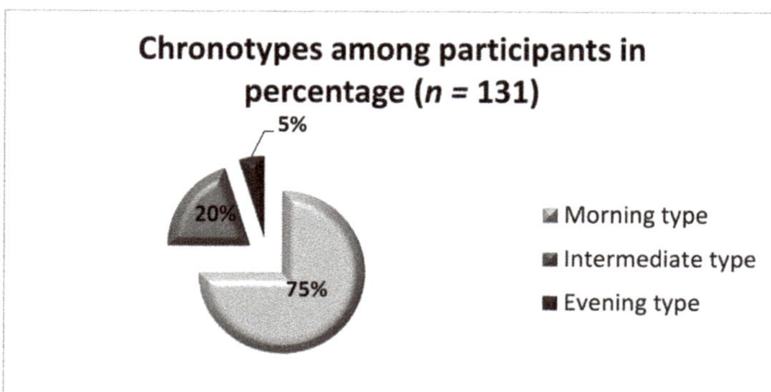

Figure 1. Distribution of chronotypes among all participants, $n = 131$.

The participants' characteristics are given in Table 1. At the beginning of the study, there were no differences between larks and owls in terms of basic demographic and anthropometric criteria. There were also no significant differences between the groups in biochemical parameters related to glycemic control and inflammation (blood glucose and insulin, insulin resistance, glycated haemoglobin, and soluble receptor for advanced glycation end-products) (Table 2).

Table 2. Anthropometric and metabolic characteristics of the study group (n = 131), taking into account measurement changes before and after caloric restriction, by chronotype; Wilcoxon signed-rank test. The values of the results in the study are given as a median (minimum–maximum). BMI (Body Mass Index), HOMA-IR (Homeostasis Model Assessment of Insulin Resistance), HbA1c (Glycated Haemoglobin), CRP (C Reactive Protein).

The Obese (n = 131) before and after Caloric Restriction, Broken Down by Chronotype									
Parameter	Morning Type before (M1) n = 98	Evening Type before (E1) n = 33	Morning Type after (M2) n = 98	Evening Type after (E2) n = 33	M1 vs. M2 p	E1 vs. E2 p	Δ (%) Morning Type	Δ (%) Evening Type	Δ (%) Morning Type vs. Δ (%) Evening Type p
Body weight, kg	115.8 (75.0–205.0)	113.0 (89.2–194.0)	115.2 (78.5–192.3)	110.9 (81.8–187.6)	<0.001	<0.001	3.4	4.1	0.4486
BMI, kg/m^2	40.5 (30.1–63.3)	38.5 (30.5–73.0)	39.9 (29.0–59.4)	38.4 (30.4–70.6)	<0.001	<0.001	3.25	4.0	0.3298
Fat tissue, %	41.7 (25.8–51.0)	43.6 (27.1–51.1)	40.9 (21.3–50.1)	41.8 (24.5–51.4)	<0.001	0.0019	2.6	3.1	0.0224
Visceral fat, kg	16.0 (5.0–43.0)	14.0 (10.0–45.0)	14.0 (4.0–39.0)	13.0 (10.0–44.0)	<0.001	0.0033	6.7	7.1	0.7605
sRAGE serum, pg/mL	371.8 (70.4–1259.0)	394.1 (194.3–1295.0)	389.7 (74.5–1011.0)	533.9 (82.0–1343.0)	0.8078	0.2652	0.85	5.7	0.6779
Glucose, mg/dl	101.0 (79.0–308.0)	103.0 (79.0–290.0)	101.0 (82.0–203.0)	105.0 (78.0–253.0)	0.3972	0.3203	0.0	2.1	0.7438
Insulin, μU/L	18.6 (4.7–79.6)	17.1 (7.6–45.3)	13.9 (1.3–78.0)	12.7 (7.3–43.8)	0.1249	0.2845	7.8	8.0	0.6749
HOMA-IR	4.5 (1.2–25.2)	4.3 (1.8–11.4)	3.0 (0.0–50.4)	2.5 (0.0–10.4)	0.0591	0.0107	5.9	9.7	0.4370
HbA1c, %	5.8 (4.4–9.80	5.7 (4.8–9.6)	5.4 (4.4–16.0)	5.5 (4.7–8.8)	0.0047	0.0712	4.3	5.7	0.8387
CRP, mg/L	5.0 (0.5–22.5)	3.7 (0.5–13.4)	3.1 (0.5–35.2)	3.9 (0.5–11.3)	0.0039	0.3507	30	24.3	0.9183

The 3-week calorie restriction resulted in a a modest, but significant reduction in body weight (Figure 2), and consistently in BMI, % of adipose tissue, and visceral fat. While there was no difference between larks and owls in the magnitude of the reduction in weight or BMI, there was a marginally greater loss of body fat in owls (3.1 ± 0.79%) compared with larks (2.6 ± 0.64%), p = 0.02 (Table 2). In addition, the larks had a clearer decrease in glycated haemoglobin and CRP (C Reactive Protein), but this did not differ significantly from that in owls (Table 2). Since it has previously been suggested that the concentration of sRAGE decreases in obesity [7,18], we assessed sRAGE levels in our patients. Indeed, we observed a weak, inverse correspondence between BMI and sRAGE concentration (r = −0.2327, p = 0.0075; Figure 3A), which, however, did not differ between larks and owls (Figure 3B,C). Moreover, a decrease in BMI following the dietary intervention was not associated with a significant change in sRAGE concentration in either group.

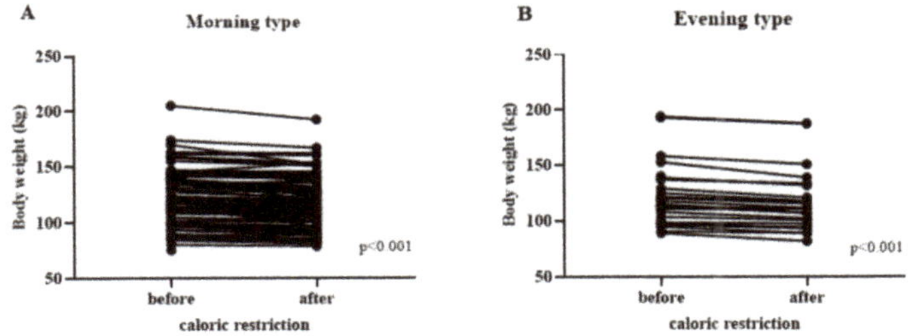

Figure 2. Body weight before and after caloric restriction. Wilcoxon signed-rank test. Individual datapoints linked by a line for: (**A**)—morning chronotype (n = 98), (**B**)—evening chronotype (n = 33).

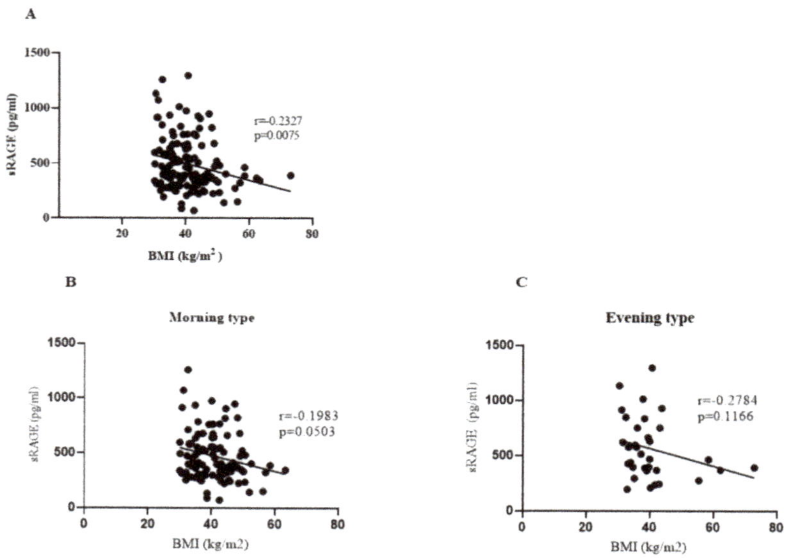

Figure 3. Correlation between serum sRAGE concentration (pg/mL) and BMI (kg/m^2) before caloric restriction; Spearman test. (**A**) In the study group, n = 131. (**B**,**C**) In each chronotype: (**B**)—morning type (n = 98), (**C**)—evening type (n = 33).

4. Discussion

To the best of our knowledge, this is the first study to assess the relationship between dietary restriction and chronotype. Due to its innovation and pioneering nature, this aspect seems particularly interesting to us in exploring obesity. Imposed dietary schedules allow us to observe if, according to chronotype, the same timing and caloric load of meals results in different weight reduction. This type of restricted conditions removes differences due to chronotype eating habits, which, according to the literature, appears to be less healthy in the owl chronotype, consuming more food in the evening.

Research provides ample evidence that the chronotype is associated with the occurrence of specific metabolic disorders [1,3]. Underlying these aberrations are disturbances in biological rhythms that control sleep and wakefulness [19].

When analyzing the study group, we found that 75% of obese patients were larks, while 25% were owls, a correlation that is opposite to that reported in the literature [20,21]. The prevalence of participants with a lark chronotype may suggest that this type of chronotype is more eager to take care of their health and search for medical help. According to Dashti et al., morning diurnal preference is connected with an increased intake of healthy foods and better adherence to recommendations [22].

Multiple authors note that people described as evening type present a predisposition to increased body fat content [4,23]. This dependence may be determined by the individuals' diet and unhealthy lifestyle [5,6].

The dietary restriction we carried out in our study allowed for a significant reduction in body weight, BMI, % of adipose tissue, and visceral fat both in the case of the lark type and representatives of the owl type. Comparing these two studied groups, owls reduced body fat more significantly than larks, possibly because of imposed earlier meal consuming—last meal at 5 p.m. and regular bedtime an hour before midnight. It suggests that, indeed, late eating and falling asleep are connected with negative metabolic shifts, predisposing patients to greater body fat accumulation. Circadian misalignment seems to be harmful only if it takes place during night hours, as it is observed in many studies of night-shift workers [18,24]. The shift to earlier diurnal hours in our study had a beneficial effect on the faster reduction in body fat content in participants with late hours activity before the study period.

It should be noted that the nutritional modification carried out was effective in relation to other parameters: a decrease in CRP, HOMA-IR (Homeostasis Model Assessment of Insulin Resistance), or HbA1c (Glycated Haemoglobin). The improvement in insulin resistance associated with HOMA-IR reduction was recorded in the evening type ($p = 0.0107$), and the glycosylated hemoglobin and CRP decreased in the morning chronotype ($p = 0.0047$ and $p = 0.0039$, respectively).

This result is similar to the Meydani study, which also reported a 40% reduction in CRP following a long-term moderate diet [25].

In turn, Reutrakul et al., in their work, emphasized the potential role of the evening chronotype as a prognostic factor for disturbances in glycemic control and promoting the subsequent development of insulin resistance [3].

Additionally, it is worth emphasizing that multiple studies reported a negative correlation between body weight, BMI, and serum sRAGE concentration [9,12,13]. Considering this dependence, the effect of diet therapy may be an appealing new topic for researchers.

In our study, weight loss did not correlate with the change in sRAGE; sRAGE concentration did not change among all subjects ($p = 0.7760$) and after taking into account the division by chronotype. The limitation of the study is its short duration but, on the other hand, all participants were in the same supervised environmental circumstances, which enabled the assessment of the influence of extrinsic factors on different types of chronotype. These small sample sizes and the short-term caloric restriction regimen were imposed by the hospital conditions and require further confirmation in a larger cohort and for a more extended period followed by a weight-stable period.

More research is needed to define the chronotype's exact role in the effectiveness of a diet.

5. Conclusions

Our research found that an individual's chronotype is not related to body weight loss or concentration of the soluble form of the receptor for advanced glycation products (sRAGE), but there is a tendency observed towards a faster reduction in body fat content in owls by changing their meal and sleep timing to earlier hours compared to larks. All other parameters such as caloric load, food composition, and physical exercise/activity levels were the same. The results of the study suggest that taking into consideration the chronotype of obese patients may possibly facilitate a particular body fat content reduction by changing the timing of the meals, but further research is needed to assess other correlations between chronotype and factors influencing weight changes.

Author Contributions: Z.S. conceived and designed the study, R.R. performed biochemical analyses, A.K., A.Z., A.J. and M.G. managed the patients and collected clinical samples, M.S., M.L., K.K., A.B. and J.W. critically reviewed the manuscript, D.K. supervised the study. All authors have read and agreed to the published version of the manuscript.

Funding: This research received no external funding.

Institutional Review Board Statement: The present study was approved by the Ethics Committee of the Medical University of Poznań (Resolution No. 201/19) and was performed in accordance with the Declaration of Helsinki.

Informed Consent Statement: All subjects gave their informed consent for inclusion before they participated in the study.

Data Availability Statement: The data supporting the conclusions of this article are included within the manuscript. The dataset is available from the corresponding author on request.

Acknowledgments: The authors wish to acknowledge the volunteers who participated in this study.

Conflicts of Interest: The authors declare no conflict of interest.

References

1. Anothaisintawee, T.; Lertrattananon, D.; Thamakaison, S.; Knutson, K.L.; Thakkinstian, A.; Reutrakul, S. Later chronotype is associated with higher hemoglobin A1c in prediabetes patients. *Chronobiol. Int.* **2017**, *34*, 393–402. [CrossRef] [PubMed]
2. Dumin, M.; O'Sullivan, K.; Deplewski, D.; Van Cauter, E.; Koren, D. Chronotype and type 2 diabetes risk in preadolescents. *ESPE Abstr.* **2016**, *86*, RFC9.7.
3. Reutrakul, S.; Hood, M.M.; Crowley, S.J.; Morgan, M.K.; Teodori, M.; Knutson, K.L.; Van Cauter, E. Chronotype is independently associated with glycemic control in type 2 diabetes. *Diabetes Care* **2013**, *36*, 2523–2529. [CrossRef] [PubMed]
4. Ruiz-Lozano, T.; Vidal, J.; de Hollanda, A.; Canteras, M.; Garaulet, M.; Izquierdo-Pulido, M. Evening-chronotype associates with obesity in severe obese subjects: Interaction with CLOCK 3111T/C. *Int. J. Obes.* **2016**, *40*, 1550–1557. [CrossRef] [PubMed]
5. Maukonen, M.; Kanerva, N.; Partonen, T.; Kronholm, E.; Konttinen, H.; Wennman, H.; Männistö, S. The associations between chronotype, a healthy diet and obesity. *Chronobiol. Int.* **2016**, *33*, 972–981. [CrossRef] [PubMed]
6. Lucassen, E.A.; Zhao, X.; Rother, K.I.; Mattingly, M.S.; Courville, A.B.; de Jonge, L.; Csako, G.; Cizza, G. Evening chronotype is associated with changes in eating behavior, more sleep apnea, and increased stress hormones in short sleeping obese individuals. *PLoS ONE* **2013**, *8*, e56519. [CrossRef]
7. Ross, K.M.; Thomas, J.G.; Wing, R.R. Successful weight loss maintenance associated with morning chronotype and better sleep quality. *J. Behav. Med.* **2016**, *39*, 465–471. [CrossRef] [PubMed]
8. Schmidt, A.M.; Vianna, M.; Gerlach, M.; Brett, J.; Ryan, J.; Kao, J.; Esposito, C.; Hegarty, H.; Hurley, W.; Clauss, M. Isolation and characterization of two binding proteins for advanced glycosylation end products from bovine lung, which are present on the endothelial cell surface. *J. Biol. Chem.* **1992**, *267*, 14987–14997. [CrossRef]
9. Miranda, E.; Fuller, K.; Perkins, R.; Kroeger, C.; Trepanowski, J.; Varady, K.; Haus, J. Endogenous secretory RAGE increases with improvements in body composition and is associated with markers of adipocyte health. *Nutr. Metab. Cardiovasc. Dis.* **2018**, *28*, 1155–1165. [CrossRef]
10. Gaens, K.H.; Goossens, G.H.; Niessen, P.M.; van Greeven, M.M.; van der Kallen, C.J.; Niessen, H.W.; Rensen, S.S.; Buurman, W.A.; Greve, J.W.M.; Blaak, E.E.; et al. Nε-(carboxymethyl)lysine- receptor for advanced glycation end product axis is a key modulator of obesity-induced dysregulation of adipokine expression and insulin resistance. *Arterioscler. Thromb. Vasc. Biol.* **2014**, *34*, 1199–1208. [CrossRef] [PubMed]
11. Eleazu, C.; Omar, N.; Lim, O.Z.; Yeoh, B.S.; Nik Hussain, N.H.; Mohamed, M. Obesity and comorbidity: Could simultaneous targeting of esRAGE and sRAGE be the panacea? *Front. Physiol.* **2019**, *10*, 787. [CrossRef]
12. Tavares, J.F.; Ribeiro, P.V.M.; Coelho, O.G.L.; Silva, L.E.D.; Alfenas, R.C.G. Can advanced glycation end-products and their receptors be affected by weight loss? A systematic review. *Obes. Rev.* **2020**, *21*, e13000. [CrossRef] [PubMed]
13. Hagen, I.; Schulte, D.M.; Müller, N.; Martinsen, J.; Tuerk, K.; Hedderich, J.; Schreiber, S.; Laudes, M. Soluble receptor for advanced glycation end products as a potential biomarker to predict weight loss and improvement of insulin sensitivity by a very low calorie diet of obese human subjects. *Cytokine* **2015**, *73*, 265–269. [CrossRef] [PubMed]
14. Harris, J.A.; Benedict, F.G. A biometric study of human basal metabolism. *Proc. Natl. Acad. Sci. USA* **1918**, *4*, 370–373. [CrossRef] [PubMed]
15. Horne, J.A.; Östberg, O. A self-assessment questionnaire to determine morningness- eveningness in human circadian rhythms. *Int. J. Chronobiol.* **1976**, *4*, 97–110. [PubMed]
16. Vera, B.; Dashti, H.S.; Gómez-Abellán, P.G.; Hernández-Martínez, A.M.; Esteban, A.; Scheer, F.A.; Saxena, R.; Garaulet, M. Modifiable lifestyle behaviors, but not a genetic risk score, associate with metabolic syndrome in evening chronotypes. *Sci. Rep.* **2018**, *8*, 945. [CrossRef] [PubMed]
17. Bhatti, P.; Mirick, D.K.; Davis, S. The impact of chronotype on melatonin levels among shift workers. *Occup. Environ. Med.* **2014**, *71*, 195–200. [CrossRef]
18. Hulsegge, G.H.; Picavet, S.J.; van der Beek, A.J.; Verschuren, W.M.; Twisk, J.W.; Proper, K.I. Shift work, chronotype and the risk of cardiometabolic risk factors. *Eur. J. Public Health* **2019**, *29*, 128–134. [CrossRef] [PubMed]
19. Gavrila, A.; Peng, C.K.; Chan, J.L.; Mietus, J.E.; Goldberger, A.L.; Mantzoros, C.S. Diurnal and ultradian dynamics of serum adiponectin in healthy men: Comparison with leptin, circulating soluble leptin receptor, and cortisol patterns. *J. Clin. Endocrinol. Metab.* **2003**, *88*, 2838–2843. [CrossRef]
20. Makarem, N.; Paul, J.; Giardina, E.V.; Liao, M.; Aggarwal, B. Evening chronotype is associated with poor cardiovascular health and adverse health behaviors in a diverse population of women. *Chronobiol. Int.* **2020**, *37*, 673–685. [CrossRef]
21. Sun, X.; Gustat, J.; Bertisch, S.M.; Redline, S.; Bazzano, L. The association between sleep chronotype and obesity among black and white participants of the Bogalusa Heart Study. *Chronobiol. Int.* **2020**, *37*, 123–134. [CrossRef] [PubMed]
22. Dashti, H.S.; Chen, A.; Daghlas, I.; Saxena, R. Morning diurnal preference and food intake: A Mendelian randomization study. *Am. J. Clin. Nutr.* **2020**, *112*, 1348–1357. [CrossRef] [PubMed]
23. Arora, T.; Taheri, S. Associations among late chronotype, body mass index and dietary behaviors in young adolescents. *Int. J. Obes.* **2015**, *39*, 39–44. [CrossRef] [PubMed]
24. Brum, M.C.B.; Filho, F.F.D.; Schnorr, C.C.; Bertoletti, O.A.; Bottega, G.B.; Rodrigues, T.D.C. Night shift work, short sleep and obesity. *Diabetol. Metab. Syndr.* **2020**, *12*, 13. [CrossRef]
25. Meydani, S.N.; Das, S.K.; Pieper, C.F.; Lewis, M.R.; Klein, S.; Dixit, V.D.; Gupta, A.K.; Villareal, D.T.; Bhapkar, M.; Huang, M.; et al. Long-term moderate calorie restriction inhibits inflammation without impairing cell-mediated immunity: A randomized controlled trial in non-obese humans. *Aging* **2016**, *8*, 1416–1426. [CrossRef] [PubMed]

nutrients

MDPI

Article

Breakfast Consumption Suppresses Appetite but Does Not Increase Daily Energy Intake or Physical Activity Energy Expenditure When Compared with Breakfast Omission in Adolescent Girls Who Habitually Skip Breakfast: A 7-Day Randomised Crossover Trial

Julia Kirstey Zakrzewski-Fruer [1,*], Claire Seall [1] and Keith Tolfrey [2]

[1] Institute for Sport and Physical Activity Research, School of Sport Science and Physical Activity, University of Bedfordshire, Bedford MK41 9EA, UK; claire.seall@bps.org.uk

[2] Paediatric Exercise Physiology Research Group, School of Sport, Exercise and Health Sciences, Loughborough University, Loughborough LE11 3TU, UK; K.Tolfrey@lboro.ac.uk

* Correspondence: Julia.Fruer@beds.ac.uk; Tel.: +44-(0)1234-793410

Abstract: With concerns that adolescent girls often skip breakfast, this study compared the effects of breakfast consumption versus breakfast omission on free-living physical activity (PA) energy expenditure (PAEE) and dietary intakes among adolescent girls classified as habitual breakfast skippers. The participants went through two 7-day conditions in a trial with a crossover design: daily standardised breakfast consumption (energy content: 25% of resting metabolic rate) before 09:00 (BC) and daily breakfast omission (no energy-providing nutrients consumed) until 10:30 (BO). Free-living PAEE, dietary intakes, and perceived appetite, tiredness, and energy levels were assessed. Analyses were linear mixed models. Breakfast manipulation did not affect PAEE or PA duration. Daily fibre intake was higher ($p = 0.005$; d = 1.31), daily protein intake tended to be higher ($p = 0.092$; d = 0.54), post-10:30 carbohydrate intake tended to be lower ($p = 0.096$; d = 0.41), and pre-10:30 hunger and fullness were lower and higher, respectively ($p \leq 0.065$; d = 0.33–1.01), in BC versus BO. No other between-condition differences were found. Breakfast-skipping adolescent girls do not compensate for an imbalance in energy intake caused by breakfast consumption versus omission through subsequent changes in PAEE but may increase their carbohydrate intakes later in the day to partially compensate for breakfast omission. Furthermore, breakfast can make substantial contributions to daily fibre intake among adolescent girls.

Keywords: adolescents; children; exercise; health; nutrient timing

Citation: Zakrzewski-Fruer, J.K.; Seall, C.; Tolfrey, K. Breakfast Consumption Suppresses Appetite but Does Not Increase Daily Energy Intake or Physical Activity Energy Expenditure When Compared with Breakfast Omission in Adolescent Girls Who Habitually Skip Breakfast: A 7-Day Randomised Crossover Trial. *Nutrients* **2021**, *13*, 4261. https://doi.org/10.3390/nu13124261

Academic Editors: Egeria Scoditti and Sergio Garbarino

Received: 28 October 2021
Accepted: 24 November 2021
Published: 26 November 2021

1. Introduction

Observational reports in children and adolescents suggest that more frequent breakfast consumption is associated with higher physical activity (PA) levels, which may contribute to reduced adiposity and cardiometabolic disease risk [1–3]. Indeed, a sustained energy deficit through increased physical activity energy expenditure (PAEE; i.e., bodily movements resulting in energy expenditures exceeding 1.5 metabolic equivalents (METs)) and/or reduced energy intake can reduce adiposity, and PA independently improves cardiometabolic health [4,5]. As such, the combination of consuming breakfast with the possible consequential improvements in PA could enhance an individual's health. However, only experimental study designs can determine whether breakfast consumption causes such improvements when compared with breakfast omission [1,6]. Randomised controlled trials in adults suggest that rather than reductions in daily energy intake [7], increased free-living PAEE occurs in response to breakfast consumption (typically carbohydrate-based) versus breakfast omission, possibly due to exogenous glucose availability being the primary fuel source for PA [8–10]. So far, evidence of a breakfast effect on PA has been inconclusive and has not necessarily targeted the most relevant populations [11–13]. Understanding such responses in adolescent girls has particular public health relevance because the adolescent decline in breakfast consumption [1] and PA [14] is more pronounced in girls than in boys, such that only ~20% of UK adolescent girls consume breakfast daily [2], and only ~9% meet the PA level recommendations [15].

Adolescent girls may respond differently to breakfast manipulations than adults because they have distinct hormonal, metabolic, and behavioural profiles [16], including pubertal insulin resistance [17], growth [18], fuel utilisation [19,20], and PA behaviours [15]. Furthermore, eating breakfast to aid the provision of a continuous carbohydrate supply may be more important to fuel daily PA in adolescents than adults due to their higher reliance on exogenous carbohydrate [19,20] and higher energy expenditures [21]. Nevertheless, protein-rich breakfasts may be more beneficial when the primary aim is to reduce appetite [22–24]. In young people, only two published experimental studies have examined the causal effects of breakfast manipulation on PA [25,26], and cross-sectional findings are unclear [2,3,27,28], as are experimental findings on energy intake [22–25,29]. In our 3-day crossover trial comparing breakfast omission versus consumption in adolescent girls, PA assessed via wrist-worn accelerometry was unaffected despite the incomplete (24%) energy intake compensation after breakfast omission [25]. Using combined heart rate and accelerometry devices to provide a more sensitive measure of PAEE, our follow-up 7-day crossover trial reported that adolescent girls spent more time in light PA before 10:30 and after school, and less time sedentary after school during daily versus intermittent breakfast consumption [26]. However, PAEE was unaffected, and energy intake was not assessed. It is logical that contrasting the extremes of complete daily breakfast omission across the week with breakfast consumption should result in more pronounced effects, which can be ethically achieved with girls who habitually skip breakfast. Moreover, data from girls who habitually skip breakfast ensure that the findings can be applied directly to those 'in need'. Thus, the primary aim of this cross-over study was to compare the effects of seven days of breakfast consumption with breakfast omission on free-living PAEE in adolescent girls classified as habitual breakfast skippers. The secondary aims were to examine the effects on dietary intakes and perceived appetite, energy, and tiredness. The key novel contributions of this study in extending previous breakfast–PA research were: (1) the targeting of adolescent girls who habitually skip breakfast specifically, (2) an extended breakfast omission experimental period of an entire week, and (3) an assessment of both PA and diet-related responses as outcomes.

2. Materials and Methods

2.1. Participants

This study was conducted in accordance with the ethical standards of the University of Bedfordshire Research Ethics Committee (ethical approval number: 2018SSPA002) and the Helsinki Declaration of 1975 as revised in 1983. The study was registered at clinicaltrials.gov with identifier NCT04481776. Data collection was completed between January 2018 and February 2019. Thirty-nine girls aged 11–14 years were recruited from schools located in Bedford, England. Parental informed consent and child assent were provided for all participants. Girls were excluded from the study if they had health-related issues identified from a health screening questionnaire (e.g., allergies to the breakfast meals, fitted with a pacemaker), were unable to walk or wear a combined heart rate and accelerometer on their chest, or were classified as habitual breakfast consumers during preliminary measures.

2.2. Sample Size Calculations

The sample size estimation was based on our primary outcome, PAEE. A positive energy balance of at least 628 kJ/d in excess of normal growth requirements [30] may explain the higher adiposity in infrequent breakfast consumers [1]. Post-breakfast energy intake compensation was expected to account for ~24% of breakfast energy intake in adolescent girls [25], which equates to ~312 kJ of the expected average breakfast of 1302 kJ (i.e., 25% of the resting metabolic rate (RMR) based on an RMR of 5207 kJ/d) used here. Therefore, energy intake was expected to be 989 kJ/d higher on the breakfast consumption days as compared with the breakfast omission days in the present study (i.e., the added energy consumed at breakfast minus the 24% energy intake compensation). Based on these figures, we deemed that the smallest worthwhile difference in the estimated PAEE between the conditions would be 1617 kJ/d (i.e., 989 kJ/d to achieve energy balance plus 628 kJ/d [30]). The expected SD for free-living PAEE in adolescents is ~1990 kJ/d [26,31,32]. Thus, a sample size of 15 participants was estimated to detect a significant difference in the estimated PAEE at 85% power, with an α of 0.05, a Cohen's f effect size of 0.43, and an assumed correlation between treatments of 0.5 in this two-treatment crossover design. Thirty-nine girls were recruited to allow for ineligible volunteers and dropouts.

2.3. Preliminary Measurements

Stature was measured to the nearest 0.01 m using a portable Leicester height measure (SECA Corporation, Hamburg, Germany). Body mass was measured, and percent body fat was estimated to the nearest 0.1 kg and 0.1%, respectively, using a Tanita Body Composition Analyser (BC-418 MA, Tanita Corporation, Tokyo, Japan). Body mass index (BMI) was calculated as body mass divided by stature squared ($kg{\cdot}m^{-2}$), with weight status subsequently defined according to the International Obesity Task Force age- and sex- specific cut-points [33]. Waist circumference was measured to the nearest millimetre in accordance with recommended procedures [34]. With the assistance of a primary home-based carer, the girls provided a validated [35,36] self-assessment of their physical maturation using secondary sexual characteristics [37]. A questionnaire was used to assess their breakfast habits on week and weekend days, including frequency (number of days per week), time and location of consumption, types of food and beverages consumed, and reasons for skipping breakfast. For study eligibility purposes, the girls were asked the following question: 'How often do you normally consume less than 50 kcal (e.g., less than a piece of fruit or a small glass of juice) before 10:30?'; the girls were given the opportunity to ask questions to clarify any aspects of this question that they did not understand and were asked to confirm their response verbally prior to the commencement of the study. Only girls who skipped breakfast (less than 50 kcal before 10:30) on at least four days/week were included in the study [8,9].

The participants then completed two tests required for the individual calibration of the combined heart rate and accelerometry devices used in the experimental conditions: (1) a submaximal treadmill exercise protocol consisting of 4 $\times$ 4 min stages to determine the relation between heart rate and estimated energy expenditure, and (2) a resting metabolic rate protocol where a 10 min resting expired air sample was collected after 20 min of quiet rest in the fasted state [38]. Expired air was sampled continuously during the treadmill and RMR tests using an online gas analysis (Metalyzer 3b, Cortex, Leipzig, Germany). Energy expenditure was estimated using the Weir equation [39,40].

2.4. Experimental Design

Using a cross-over design, each participant went through two 7-day conditions separated by a seven to ten-day washout: breakfast consumption (BC) and breakfast omission (BO). The conditions were completed in a counter-balanced order using block randomisation, with as close as possible to half of the participants randomised to each sequence (i.e., BC then BO or BO then BC), as determined using a computer-based number generator. The irregular menstrual cycles in this population coupled with the 7-day duration of each intervention, and the feasibility of allowing the girls to complete the study alongside a friend to help provide an enjoyable, exciting, and comfortable experience meant that it was not possible to align the experimental conditions to a specific menstrual cycle phase for each individual. On each of the seven days, the participants were asked to consume a standardised breakfast before 09:00 in BC and to abstain from all energy-providing nutrients before 10:30 in BO. These cut-off times were in line with proposed definitions of breakfast, which should be consumed within 2 to 3 h of waking, typically no later than 10:00 [41,42]. Due to the nature of the sample, it was also important that the breakfast omission cut-off time coincided with the participants' first opportunity to consume food or drink at school (i.e., break time at ~10:30). The participants could eat as and when they pleased from 10:30 onwards in both conditions. Throughout each 7-day condition, free-living PAEE was estimated using the combined heart rate and accelerometry devices (Actiheart, CamNtech, Cambridge, UK), as described elsewhere [22], and dietary intakes were recorded using a combined photographic and written food diary, also described elsewhere [21]. Additionally, the participants completed a Visual Analogue Scale (VAS) for hunger, fullness, tiredness, and energy levels on the dietary assessment days. The experimental conditions did not coincide with anticipated changes in PA or dietary habits (e.g., holidays, school sports days), as confirmed with the participants, their parents, and teachers.

The participants and their parents received telephone reminders during each condition to help maximise compliance to study procedures (i.e., adhering to the breakfast omission and omission protocols, wearing the Actiheart monitor, and recording dietary intakes). Compliance to the breakfast intervention was confirmed via photographs taken by the participants of all food and drink consumed before 10:30 using the digital camera provided (ViviCam 46, Vivitar Shenzhen, China), and through a written daily breakfast log that included the time breakfast was consumed for BC or the time that the first meal or snack was consumed after the breakfast omission cut-off (i.e., 10:30) for BO.

2.5. Breakfast Interventions

The quantity, composition, and time of consumption of the standardised breakfast was designed to align with proposed definitions of 'breakfast' [41,42]. The energy content was 25% of individual RMR, which is in line with recommendations that breakfast should contribute to ~15–25% of daily energy intake [41]. Based on our previous studies, RMR ranged from around 3348 to 7115 kJ/d in adolescent girls [26], which equated to a breakfast energy content of 837–1779 kJ/d in the present study. Prior to the experimental conditions, the participants selected one wholegrain, high-fibre, ready-to-eat cereal (with the option of adding raisins) and fruit juice from a limited selection that was based on the breakfast preferences of girls in our previous work [25,26]. The breakfast items that the girls chose are shown in Supplementary Table S1. The breakfast items selected were consumed on each day of BC; thus, breakfast composition was controlled within participants, but not between participants, in order to account for individual preferences. This CHO-based breakfast was chosen as higher exogenous glucose availability may act as a physiological mechanism that increases PA in response to breakfast consumption versus omission [8–10]. Furthermore, high fibre, wholegrain, cereal-based breakfasts may be particularly beneficial to health in adolescents [43,44]. The participants were instructed to consume the breakfast at home before 09:00. To ensure that the correct amount of each breakfast item was consumed, the food items were provided to the participants in pre-packaged containers, and the participants were provided with a marked beaker to measure their milk and juice each morning. The only exception was that parents were asked to provide the semi-skimmed (1.8%) milk. To be included in the final dataset, the participants were required to confirm verbally and using photographic evidence that they had consumed their breakfast on all seven days of BC, and to confirm verbally that they had consumed no energy-providing nutrients before 10:30 on all seven days of BO.

2.6. Physical Activity Energy Expenditure Assessment

Participants were fitted with a combined heart rate and accelerometer (Actiheart, CamNtech, Cambridge, UK) the day before each condition, which was removed after eight days. Combining heart rate and accelerometer data improves the validity of PAEE estimations in 12–13-year-olds as compared with accelerometry or heart rate monitoring alone [45]. The procedures used to fit the monitor and the instructions provided to the participants to ensure that only genuinely meaningful behavioural responses were recorded were in line with recommendations and are described in detail elsewhere [26]. Each participant's monitor was set to record data in 15-second epochs and was individually calibrated using the measured RMR and exercise energy expenditure values from preliminary testing. This calibration method accounted for individual differences in the heart rate–PAEE relationship, ensuring greater accuracy of the PAEE estimations when compared with group regression equations [45].

As only Actiheart data during waking hours were analysed, participants were required to record their answers to the following questions using a daily log: "what time did you wake up?", "what time did you get out of bed?", "what time did you turn off the light and go to bed?", and "what time did you fall asleep?". Using a standardised protocol, the self-reported wake and bedtimes were utilized to provide a region of interest for each 24 h Actiheart data file. Objective markers were then used to identify bed time (i.e., the beginning of prolonged minimal movement accompanied by a decline in heart rate) and wake time (i.e., the beginning of prolonged increased movement accompanied by an increase in heart rate) [26,32].

The procedures used to analyse the Actiheart data files are described in detail in our previous publication [26]. After excluding data classified as 'lost' and 'not worn', only datasets with four valid days (i.e., at least 10 h of useable data), including one weekend day for each condition, were included [2,26,28]. Branched equation modelling was used to estimate PAEE. Metabolic equivalent (MET) values were used to define sedentary (<1.5 METs), light (1.5–2.9 METs), moderate (3.0–5.9 METs), and vigorous (>5.9 METs) activity.

2.7. Dietary Assessment

Weighed food records were not considered to be suitable for the present study due to the high participant burden and poor compliance in adolescents [46,47], a population that has reported a preference for methods using technology such as a disposable camera [48]. Thus, the participants recorded their daily diet using a digital camera (ViviCam 46; Vivitar) and a written food diary during the final four consecutive days that included two weekdays and both weekend days during each condition. As dietary intakes vary significantly between Saturday and Sunday, including both weekend days and a selection of weekdays was recommended in order to obtain estimates that are representative of usual intakes [49]. Thus, all four days had to have been completed for a participant to be included in the final dataset for dietary analyses. The combined photographic and written food diary is described in detail elsewhere [25]. We have previously shown that the natural variation in free-living energy intake assessed using this method may be small enough to detect meaningful differences [25]. The mass of each food and beverage item consumed was estimated by comparing the digital photographs taken by the participants with the Young Person's Food Atlas [50,51], which has good agreement with weighed food diaries in children aged ≥11 years [52]. Energy and macronutrient intakes were estimated from the food diaries using the myfood24 online dietary analysis software (Nexus, Leeds, UK).

2.8. Perceived Appetite, Tiredness, and Energy Levels

Participants were asked to answer the following questions using a 100 mm visual analogue scale (VAS): "How hungry do you feel right now?", "How full do you feel right now?", "How tired or drowsy do you feel right now?", and "How energetic do you feel right now?". Responses were recorded on the four days that dietary intakes were assessed, at three time points: on waking (i.e., 'baseline'), at 09:00 (i.e., the BC cut-off), and at 10:30 (i.e., the BO cut-off); further assessments throughout the day were not taken due to the additional participant burden. VAS has been shown to be valid and reliable for assessing hunger, fullness, fatigue, and energy level in adults [53–55], and it has been successfully used in adolescent girls to assess perceptions of appetite and mood in free-living settings [56]. Based on the number of complete VAS available and the previous literature [21,56], participants were required to have had at least three days of complete

VAS data per condition in order to be included in the final dataset for VAS analyses; the mean of the three days was calculated.

2.9. Statistical Analyses

Statistical analyses were completed using the IBM SPSS statistics software for Windows version 26 (IBM Corporation, New York, NY, USA). One-way ANOVA was used to test for differences in participant characteristics and the nutrient content of the breakfast meal by outcome-specific analytical samples. As breakfast manipulation may affect PA and diet during specific times of the day [2,8–10,26], estimated PA and dietary variables were computed for three daily time segments: from wake time to before 10:30, from 10:30 to before 15:30, and from 15:30 until bed time [26]. These times coincided with the 10:30 breakfast omission cut-off and the end of the school day to account for potential differences in PA during and outside of school time [57]. The Shapiro–Wilks tests showed that the residuals were not normally distributed for the time spent in moderate and vigorous PA, for the PAEE from sedentary, light, moderate, and vigorous activity and total PAEE, and for post-10:30 CHO and fibre intakes ($p \leq 0.05$). Thus, for consistency, all PA and dietary outcomes were natural log-transformed (Ln) and presented as a geometric mean (95% confidence intervals (CI)), with analyses based on ratios of the geometric means and a 95% CI for the ratios. Linear mixed models were used to examine all outcome variables, with condition and time of day included as fixed factors. The linear mixed models included a random effect for each participant and were adjusted for period (order) effects [58]. Minutes of useable data (Ln) was included as a covariate for PA-related variables. Where significant condition or condition by time-of-day interactions were found, post hoc analysis was performed using the Holm–Bonferroni correction for multiple comparisons; data from each individual time segment were compared between the conditions for significant condition by time interactions [59]. Statistical significance was accepted as $p \leq 0.05$; p values with two decimal places are attributed to non-significant results or three to at least borderline significant results. Pooled p values are provided in the text where appropriate; the 95% CI for the differences are provided for all primary and secondary outcome variables in the tables. Absolute standardised effect sizes (Cohen's d) are provided to supplement important findings (i.e., potentially meaningful between-condition differences), with 0.2 considered the minimum important difference, 0.5 moderate, and 0.8 large [60]. Values are presented as means ± SDs for descriptive data; for the results of statistical analyses, values are presented as estimated marginal means or geometric means (95% confidence intervals (CIs)) unless stated otherwise. Figures of individual responses are provided in order to give a more detailed insight into inter-individual variability where appropriate.

3. Results

3.1. Participant Characteristics

The final sample included 15 participants for PA analyses, 11 participants for dietary analyses, and 11 participants for VAS analyses; the flow of participants from enrolment to analyses is shown in Figure 1. The physical characteristics of the participants are shown in Table 1. There were no significant differences in the physical characteristics or breakfast frequencies between the final samples for each outcome ($p \geq 0.85$ for all).

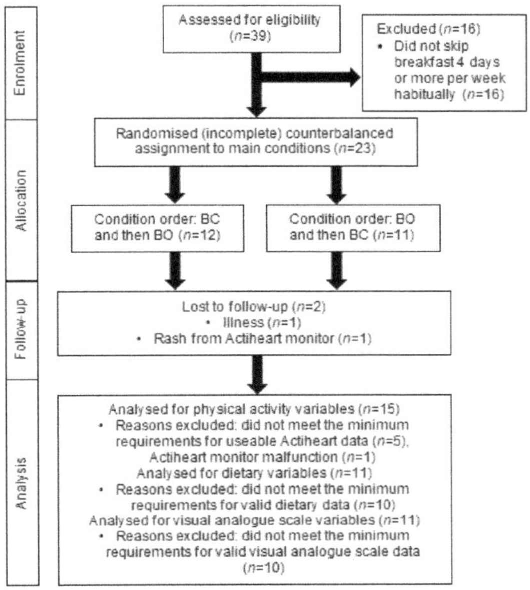

Figure 1. Schematic representation of recruitment, enrolment, and follow-up of adolescent girls who participated in the randomised crossover trial comparing seven days of daily breakfast consumption (BC) with seven days of breakfast omission (BO). BC was the consumption of a standardised breakfast with an energy content equivalent to 25% of individual resting metabolic rate before 09:00 for seven consecutive days; BO was the abstinence from all energy-providing nutrients until at least 10:30 for seven consecutive days.

Table 1. Characteristics of adolescent girls who participated in the randomised crossover trial comparing seven days of daily breakfast consumption (BC) with seven days of breakfast omission (BO) [1].

	PA Analyses Sample (*n* = 15)	**Diet Analyses Sample (*n* = 11)**	**VAS Analyses Sample (*n* = 11)**
Age (y)	13.3 ± 0.7	13.3 ± 0.8	13.3 ± 0.7
Stature (m)	1.58 ± 0.06	1.57 ± 0.06	1.60 ± 0.06
Body mass (kg)	52.9 ± 7.6	53.3 ± 11.0	54.5 ± 10.1
Body fat %	26.7 ± 5.5	27.0 ± 5.8	26.7 ± 5.9
Waist circumference (cm)	69.7 ± 8.5	68.3 ± 8.9	69.5 ± 9.1
BMI (kg·m^{-2})	21.3 ± 3.2	21.6 ± 3.8	21.3 ± 3.8
BMI classification (*n* NO, OW, OB) [2]	11, 3, 1	8, 2, 1	8, 2, 1
Breast development (stage) [3]	4 (0)	4 (0)	4 (0)
Pubic hair (stage) [3]	4 (0)	4 (0)	4 (0)
RMR (kJ/d)	6325 ± 1195	6172 ± 1420	6535 ± 1245
Weekdays skip breakfast habitually (d/week) [4]	4 ± 1	4 ± 1	4 ± 1
Weekend days skip breakfast habitually (d/week) [4]	1 ± 1	1 ± 0	1 ± 1

Table 1. *Cont.*

	PA Analyses Sample (n = 15)	Diet Analyses Sample (n = 11)	VAS Analyses Sample (n = 11)
Weekly days skip breakfast habitually (d/week) [4]	5 ± 1	5 ± 1	5 ± 1
Habitual weekday cereal-based breakfast consumption (n)	5	3	4
Habitual weekend cereal-based breakfast consumption (n)	5	4	5
Habitual weekday breakfast consumption time (h:min)	08:16 ± 01:08	08:09 ± 00:51	08:29 ± 01:25
Habitual weekend breakfast consumption time (h:min)	09:51 ± 01:40	09:37 ± 01:36	09:15 ± 01:50

[1.] Values are mean ± SDs or medians (IQRs). BC was the consumption of a standardised breakfast for seven consecutive days; BO was the abstinence from all energy-providing nutrients until at least 10:30 for seven consecutive days. PA, physical activity; VAS, visual analogue scale; NO, non-overweight; OW, overweight; OB, obese; RMR, resting metabolic rate. [2.] BMI classification according to the International Obesity Task Force [33]. [3.] Five stages of breast and pubic hair development described by Tanner [36].[4.] Less than 50 kcal consumed before 10:30.

3.2. Breakfast Meals

The standardised breakfast energy and macronutrient intakes of the girls who were included in the PA, dietary, and VAS analyses are shown in Table 2; there were no significant differences in these variables between the samples ($p \geq 0.43$). Data from the food diaries indicated that on average, the girls consumed their first meal or snack at 11:48 ± 00:52 in BO.

Table 2. Nutrient content of the breakfast providing 25% of individual resting metabolic rate to adolescent girls who participated in the randomised crossover trial that compared seven days of daily breakfast consumption (BC) with seven days of breakfast omission (BO) [1].

	PA Analyses Sample (n = 15)	Dietary Analyses Sample (n = 11)	VAS Analyses Sample (n = 11)
Energy (kJ)	1578 ± 303	1543 ± 355	1634 ± 311
Carbohydrate (g)	68.5 ± 18.4	63.7 ± 16.5	72.5 ± 19.9
Fat (g)	6.4 ± 4.0	4.9 ± 1.0	6.8 ± 4.5
Protein (g)	13.3 ± 2.6	13.2 ± 2.9	13.8 ± 2.8
Fibre (g)	8.1 ± 4.7	9.3 ± 5.0	7.9 ± 4.6

[1] Values are mean ± SDs. BC was the consumption of a standardised breakfast for seven consecutive days; BO was the abstinence from all energy-providing nutrients until at least 10:30 for seven consecutive days. PA, physical activity; VAS, visual analogue scale.

3.3. Wake Time and Useable Data

Useable data averaged at 910 min/d for BC and 942 min/d for BO. Minutes of useable data (Ln) were lower in BC versus BO across the three time segments (BC 305 (297–313) vs. BO 316 (308–324) min/d; condition main effect $p = 0.017$; condition by time-of-day interaction $p = 0.20$). For descriptive purposes, mean (SD) wake time did not differ significantly between the conditions (BC 07:06 ± 00:17 vs. BO 07:00 ± 00:22; $p = 0.26$), whereas bedtime was earlier in BC versus BO (BC 23:00 ± 00:32 vs. BO 23:13 ± 00:26; $p = 0.042$).

3.4. Physical Activity Energy Expenditure and Duration

Table 3 shows the estimated daily PAEE and the time spent in PA for each intensity, stratified by condition and time of day. Data for all PAEE variables and minutes of useable data were natural log-transformed. Adjusting for minutes of useable data, the main effects for the condition and the condition by time-of-day interaction were non-significant for the estimated PAEE from sedentary, light, and moderate activities ($p \geq 0.41$ for all). Total PAEE and vigorous PAEE both tended to be lower in BC versus BO, but the effect sizes were trivial-small ($p \leq 0.097$; d = 0.18 to 0.20) and the condition by time interactions were non-significant ($p \geq 0.21$ for all). Individual responses for total daily PAEE are shown in Figure 2. In 10 out of the 15 participants, PAEE was lower in BC than BO, and the difference exceeded 628 kJ/d in four girls. Consequently, for the remaining five girls, PAEE was higher in BC than in BO, and this difference exceeded 628 kJ/d for one of them. The main effect of time of day was significant for the PAEE for each intensity and for the total PAEE ($p < 0.0005$ for all); after adjusting for multiple comparisons, energy expenditure from sedentary activities tended to be lower at wake–10:30 as compared with 10:30–15:30 ($p = 0.083$), and light PAEE tended to be higher at 10:30–15:30 versus 15:30–bed ($p = 0.085$).

Data for the time spent sedentary and in light, moderate, and vigorous PA were log-transformed. Adjusting for minutes of useable data, sedentary time tended to be higher in BC versus BO, although the effect size was trivial ($p = 0.068$; d = 0.13), and the condition by time interaction was not significant ($p = 0.36$). The main effects of condition and the condition by time-of-day interaction were not significant for the time spent in light or moderate PA ($p \geq 0.34$ for all). Time spent in vigorous PA tended to be lower in BC versus BO, and the condition by time interaction tended to be significant ($p \leq 0.077$), but again, the between-condition effect size was trivial (d = 0.18). The main effect of the time of day was significant for all PA intensities ($p \leq 0.001$ for all), but there were no significant differences between the individual time segments after adjusting for multiple comparisons ($p \geq 0.14$ for all).

Table 3. Free-living physical activity energy expenditure and duration of adolescent girls who participated in the randomised crossover trial comparing seven days of breakfast consumption (BC) with seven days of breakfast omission (BO) [1].

	BC			BO			95% CI for BC vs. BO [2]
	Wake–10:30	10:30–15:30	15:30–bed	Wake–10:30	10:30–15:30	15:30–bed	
PAEE (kJ/d)							
Sedentary	54 (40–72)	72 (60–85)	60 (45—-78)	54 (41–72)	73 (61–87)	65 (48–87)	−12–5%
Light	321 (233–442)	342 (286–409)	260 (194–349)	321 (238–433)	369 (309–441)	264 (191–365)	−12–7%
Moderate	168 (84–338)	228 (142–366)	152 (79–292)	192 (99–372)	257 (160–411)	148 (73–300)	−24–13%
Vigorous	12 (5–35)	23 (11–47)	16 (6–41)	17 (6–44)	37 (18–76)	15 (5–41)	−40–4%
Total	601 (417–868)	712 (567–894)	547 (389–770)	666 (471–942)	857 (682–1076)	550 (379–797)	−18–1%
PA Duration (min/d)							
Sedentary	202 (170–241)	189 (172–207)	209 (178–245)	196 (167–231)	170 (155–187)	205 (172–245)	0–11%
Light	67 (49–90)	74 (64–86)	59 (45–78)	67 (50–89)	81 (70–95)	62 (46–84)	−13–5%
Moderate	15 (7–30)	19 (12–32)	13 (7–25)	17 (9–32)	22 (13–35)	13 (6–26)	−23–12%
Vigorous	1.44 (0.79–2.63)	2.14 (1.37–3.34)	1.93 (1.09–3.42)	1.73 (0.97–3.08)	2.95 (1.89–4.61)	1.78 (0.97–3.27)	−25–2%

[1] Based on natural log-transformed data with values presented as a geometric mean (95% confidence interval (CI)) adjusted for minutes of useable data, $n = 15$. [2] A total of 95% CI for the percentage difference of the geometric means between the experimental conditions using a condition by time-of-day linear mixed model adjusted for minutes of useable data and condition order. BC was the consumption of a standardised breakfast with an energy content equivalent to 25% of individual resting metabolic rate before 09:00 for seven consecutive days; BO was the abstinence from all energy-providing nutrients until at least 10:30 for seven consecutive days. PAEE, physical activity energy expenditure; PA, physical activity.

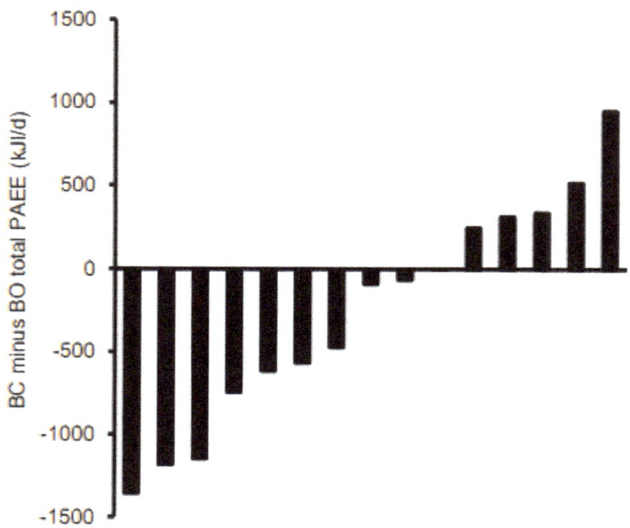

Figure 2. Individual total daily physical activity energy expenditure (PAEE) responses of adolescent girls who participated in the randomised crossover trial comparing seven days of breakfast consumption (BC) with seven days of breakfast omission (BO). BC was the consumption of a standardised breakfast with an energy content equivalent to 25% of individual resting metabolic rate before 09:00 for seven consecutive days; BO was the abstinence from all energy-providing nutrients until at least 10:30 for seven consecutive days.

3.5. Energy and Macronutrient Intakes

Table 4 shows the total and post-10:30 daily energy and macronutrient intakes stratified according to condition. For total daily intakes, there were no significant effects of condition for energy, CHO, and fat ($p \leq 0.63$ for all), whereas protein intakes tended to be higher ($p = 0.10$; d = 0.54) and fibre intakes were significantly higher ($p = 0.01$; d = 1.31) in BC versus BO. Given the seemingly very high accuracy of energy intake compensation at the group level shown in Table 4, Figure 3 shows the difference in total daily energy intake between the conditions at the individual level. Based on the 628 kJ/d cut-off (30), differences in total daily EI were accurate in four of the girls. There was one girl with a relatively extreme difference; removal of this outlier did not affect the statistical analyses but actually tightened the group means even further. The main effect of condition and condition by time interaction were non-significant for post-10:30 energy, fat, protein, and fibre ($p \geq 0.12$ for all). Post-10:30 CHO intake tended to be lower in BC versus BO ($p = 0.096$; d = 0.41), with no significant condition by time-of-day interaction ($p = 0.50$). Figure 4 shows individual responses for post-10:30 CHO intake; 8 of the 11 girls consumed less CHO after 10:30 in BC versus BO. The main effect of time was non-significant for all post-10:30 dietary variables ($p \leq 0.17$ for all) other than fibre intakes, which were higher at 10:30–15:30 than at 15:30–bed ($p = 0.048$).

Table 4. Dietary intakes of adolescent girls who participated in randomised crossover trial comparing seven days of breakfast consumption (BC) with seven days of breakfast omission (BO) [1].

	BC			BO			95% CI for Total in BC vs. BO [2]	95% CI for Post-10:30 in BC vs. BO [2]
	Total	10:30–15:30	15:30–Bed	Total	10:30–15:30	15:30–Bed		
Energy intake (kJ/d)	4206 (3157–5604)	1726 (1041–2863)	1181 (712–1959)	4078 (3063–5433)	2026 (1222–3360)	1628 (982–2700)	−19–31%	−50–23%
CHO (g/d)	118 (82–169)	42 (24–75)	37 (20–67)	127 (88–182)	69 (39–122)	46 (26–82)	−32–28%	−54–7%
Fat (g/d)	35 (25–49)	16 (9–28)	12 (7–21)	38 (27–53)	17 (10–29)	16 (9–27)	−35–33%	−46–38%
Protein (g/d)	42 (29–59)	19 (12–32)	13 (8–21)	32 (22–45)	13 (8–22)	15 (9–25)	−5–83%	−27–67%
Fibre (g/d)	10.0 (7.1–14.2)	2.1 (1.0–4.3)	1.1 (0.5–2.3)	4.7 (3.3–6.7	2.7 (1.4–5.1)	1.4 (0.7–2.7)	34–236%	−61–55%

[1.] Based on natural log-transformed data, with values presented as a geometric mean (95% confidence interval (CI)), $n = 11$. [2.] 95% CI for the percentage difference of geometric means between the experimental conditions using a between-condition (for total intakes) or condition by time-of-day (for post-10:30 intakes) linear mixed model adjusted for condition order; values in bold are significant. BC was the consumption of a standardised breakfast with an energy content equivalent to 25% of individual resting metabolic rate before 09:00 for seven consecutive days; BO was the abstinence from all energy-providing nutrients until at least 10:30 for seven consecutive days. CHO, carbohydrate.

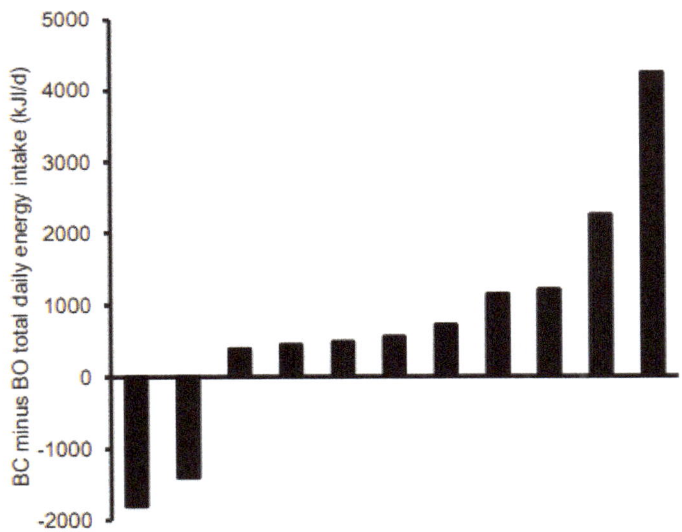

Figure 3. Individual total daily energy intakes of adolescent girls who participated in the randomised crossover trial comparing seven days of breakfast consumption (BC) with seven days of breakfast omission (BO). BC was the consumption of a standardised breakfast with an energy content equivalent to 25% of individual resting metabolic rate before 09:00 for seven consecutive days; BO was the abstinence from all energy-providing nutrients until at least 10:30 for seven consecutive days.

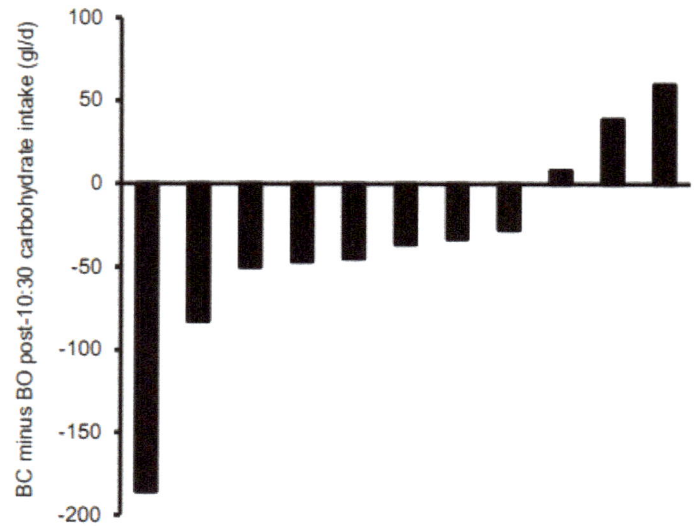

Figure 4. Individual post-10:30 carbohydrate intakes of adolescent girls who participated in the randomised crossover trial comparing seven days of breakfast consumption (BC) with seven days of breakfast omission (BO). BC was the consumption of a standardised breakfast with an energy content equivalent to 25% of individual resting metabolic rate before 09:00 for seven consecutive days; BO was the abstinence from all energy-providing nutrients until at least 10:30 for seven consecutive days.

3.6. Perceptions of Appetite, Tiredness, and Energy Levels

Table 5 shows the perceived appetite, tiredness, and energy levels in the morning, stratified by condition. The main effect for condition approached significance for perceived hunger, which tended to be lower in BC versus BO, with a small effect size ($p = 0.063$; d = 0.33). There was a significant main effect of time ($p = 0.003$) and condition by time interaction ($p = 0.001$) for hunger; whilst there were no significant differences over time in BC ($p \geq 0.97$ for all), hunger was lower at waking versus both 09:00 and 10:30 in BO ($p \leq 0.012$; d = 0.87–1.54) and tended to be lower at 09:00 versus 10:30 ($p = 0.074$; d = 0.67). Perceived fullness was higher in BC versus BO ($p = 0.045$; d = 0.35); the condition by time interaction ($p = 0.010$) was significant, whereas the main effect of time of day was not ($p = 0.23$). Fullness was higher in BC versus BO at 10:30 ($p = 0.004$; d = 1.01) and approached significance at 09:00 ($p = 0.065$; d = 0.58). Perceived tiredness was not different between the conditions, and the condition by time interaction was also non-significant ($p \leq 0.71$ for both); the main effect of time was significant ($p < 0.0005$), with tiredness being higher on waking versus 09:00 and 10:30 ($p \leq 0.006$). Similarly, perceived energy level was not different between the conditions, and the condition by time interaction was also non-significant ($p \leq 0.62$ for both); the main effect of time was significant ($p < 0.0005$), with energy levels being lower on waking versus 09:00 and 10:30, and lower at 09:00 versus 10:30 ($p \leq 0.017$ for both).

Table 5. Perceived appetite, tiredness, and energy levels (measured in mm out of 100) of adolescent girls who participated in the randomised crossover trial comparing seven days of breakfast consumption (BC) with seven days of breakfast omission (BO) [1].

	BC			BO			95% CI for BC vs. BO [2]
	Waking	09:00	10:30	Waking	09:00	10:30	
Hunger	35 (24–47)	30 (18–41)	34 (22–46)	24 (12–35)	41 (29–52)	54 (42–66)	−13.3–0.4
Fullness	46 (31–61)	56 (41–70)	53 (38–67)	54 (40–69)	43 (29–58)	33 (18–47)	0.2–15.9
Tiredness	48 (36–59)	38 (26–49)	34 (22–45)	46 (35–58)	36 (25–48)	33 (22–45)	−4.3–6.2
Energy	38 (30–46)	48 (40–56)	60 (52–68)	37 (29–45)	50 (42–58)	56 (48–64)	−4.5–6.0

[1] Values are estimated marginal mean (95% CIs), $n = 11$. [2] A total of 95% confidence interval (CI) of the mean absolute difference between the experimental conditions; values in bold are significant. BC was the consumption of a standardised breakfast with an energy content equivalent to 25% of individual resting metabolic rate before 09:00 for seven consecutive days; BC was the abstinence from all energy-providing nutrients until at least 10:30 for seven consecutive days.

4. Discussion

In adolescent girls who habitually skip breakfast, this 7-day crossover trial reported for the first time that PA duration and PAEE did not differ meaningfully between BC and BO. Total daily energy intake was almost identical between the conditions, which coincided with increased perceived morning appetite and a tendency for increased carbohydrate intake in response to BO (i.e., after 10:30). Thus, the girls were able to accurately compensate for the additional or missed energy at breakfast after 10:30.

Our finding that daily total PAEE and PA duration did not differ between BC as compared with BO agrees with some studies on adult samples, including female habitual breakfast skippers [12], obese mixed-sex habitual breakfast consumers and skippers [9], and mixed-sex habitual breakfast consumers and skippers varying in weight status [11,13], yet it contradicts research on adults showing that BC can increase PA in female habitual breakfast consumers [10] and lean mixed-sex habitual breakfast consumers and skippers [8]. The current findings also contrast with our previous crossover studies in adolescent girls, where the time spent in light PA was higher and after-school sedentary time was lower in daily BC versus intermittent BC and BO [26], although PA was unaffected by three days of BC versus BO when assessed using less sensitive measures [25]. It is possible that the girls in the current study responded differently because they had adapted metabolically and behaviourally to skipping breakfast habitually, whereas the girls in our previous research generally consumed breakfast habitually [25,26]. In the current sample, the 10:30 breakfast omission cut-off may have been too early to detrimentally affect PA, with studies on adults employing cut-offs at 12:00 [8–11] or 11:30 [12], which is difficult to justify ethically in adolescent girls. Additionally, MVPA averaged ~85 min/d in our sample, indicating high motivation for PA that may not have been overridden by breakfast manipulation. Interestingly, examination of the individual differences in total PAEE between the conditions provided insight into the high inter-individual variability in the responses to breakfast manipulation. Such inter- (and intra-) individual variability is common in the field of energy balance [61–63] and requires consideration in future breakfast-related work, along with addressing the potential modifying effect of breakfast and PA habits on the breakfast–PA relationship.

Under free-living conditions, low-intensity and sporadic PA may be most sensitive to the effects of breakfast consumption, whereas moderate or vigorous PA is typically planned and structured [8,26]. As such, the tendency for higher vigorous PAEE and total PAEE during BO is more likely to have resulted from a particularly active planned lesson at school or a sports club for some participants, for example, rather than whether or not they consumed breakfast. On this note, environmental constraints that are inherent to the lives of adolescent girls, including the school timetable, extra-curricular activities, family routines, and commitments, mean that it is difficult to truly assess 'free-living' PA responses in this population that would be amenable to any intervention effects. Thus, this may have contributed to the lack of difference in PA between the conditions, and future research may consider collecting data on such planned activities. It is also possible that the lower perceived energy levels and heightened tiredness early in the day across the conditions in our sample of female adolescent breakfast skippers may have overridden the potential effect of breakfast to increase morning PA, which is the time of day that is particularly sensitive to breakfast manipulation [8–10]. In support of this finding, adults whose activity peak is towards the later part of the day tend to skip breakfast [11]. As such, further research is warranted to examine the influence of chronotype on the breakfast–PA relationship in adolescent girls and to include measures of PA that may be particularly modifiable (e.g., PA during the school holidays or under laboratory conditions).

With total daily energy intake being almost identical between the conditions, post-10:30 energy intake was sufficient to compensate for the energy consumed or missed at breakfast. This finding is in accordance with previous research comparing carbohydrate-based BC versus BO in mixed-sex groups of so-called 'adolescents' who habitually skipped breakfast [22,24]. Nevertheless, the age range of 13–17 years [22] or the mean age of 19 years [24] in this past research indicates that the samples were either a mix of early- and late-pubertal adolescents [22] or a mix of late-pubertal adolescents and young adults [24]; furthermore, these findings were not consistent, with one study reporting increased daily energy intake during BC in 'breakfast skipping' overweight/obese females aged 15–20 years [23]. Moreover, total daily energy intake was higher during BC versus BO in adolescent girls [25] and in 8–10-year-old children [29] who tended to consume breakfast regularly. It is possible that the habit of skipping breakfast involves learning new eating patterns such that the habitual breakfast skippers in our study and previous work [22,24] were able to compensate by consuming extra energy later in the day. In adults, however, a recent systematic review of randomised controlled trials showed that the total daily energy intake was higher during periods of BC compared with BO regardless of breakfast habit [7]. The higher perceptions of hunger and lower fullness in response to breakfast omission in our study may have contributed to the accuracy of energy intake compensation between BC and BO, which complements previous findings on children [29] and adolescents classified as breakfast skippers [22] in controlled laboratory settings. However, no effect was found in women who habitually skipped breakfast, perhaps because they were older and thus more accustomed to skipping breakfast [12]. The energy intake compensation in response to BO was primarily in the form of carbohydrates, perhaps because the standardised breakfast was carbohydrate-rich, and dietary compensation may be macronutrient-specific [64], which aligns with previous research on adult women who habitually skipped breakfast [12]. In terms of diet quality, consuming breakfast also contributed to higher total daily fibre intakes, which can favourably affect long-term cardiometabolic health and thus has important health implications [43,44]. Overall, these findings warrant further research on the potential modifying effects of breakfast habits and age on the causal link between breakfast and dietary intakes, considering whether any compensatory behaviours are conscious and planned (i.e., to prepare for the next day) or unconscious and spontaneous.

Limitations of the present study include the 7-day intervention period; indeed, our sample of adolescents may respond differently once they are more accustomed to consuming breakfast and new PA, and when eating habits have had time to emerge. In addition, we did not assess diet-induced thermogenesis, which is another modifiable component of total energy expenditure and would be expected to negate at least part of the small increase in PAEE of ~258 kJ/d (62 kcal/d) during breakfast omission. Thus, any potential effects of breakfast manipulation on energy balance in our study and subsequent changes in body mass are unlikely to be meaningful [30]. Our drop-out rate and exclusion criteria based on fidelity and complete datasets also meant that our final sample was lower than expected for the dietary and VAS analyses (i.e., our secondary outcome variables), which may have affected our findings and are common issues when working with this population. Additionally, individual chronotype, circadian rhythms, PA habits, breakfast habits, and menstrual cycle phase could all influence the nature of the relationship between breakfast, PA, and diet, which we did not assess and may limit the generalisability of our conclusions. Furthermore, the potential interaction of our findings with the quantity, composition [22–24], and timing of the breakfast meal was beyond the scope of the present study and requires examination to inform an evidence-based 'definition' of breakfast. Finally, the increased risk of making a type I error from the testing of multiple secondary outcomes is a limitation of the statistical methods; however, as the sample size estimation was based on our primary outcome, there may been an increased risk of type II error for the secondary outcomes.

In conclusion, PA duration and PAEE were not meaningfully affected when adolescent girls who were habitual breakfast skippers consumed breakfast as compared with when they omitted breakfast over seven consecutive days. Total daily energy intake was almost identical during breakfast consumption and omission, with breakfast omission increasing perceptions of appetite during the early part of the morning and tending to increase carbohydrate intake after the 10:30 breakfast omission cut-off. Nevertheless, consuming breakfast can contribute to higher fibre intakes in 'breakfast skipping' adolescent girls. Considering the limited and inconclusive evidence to date, future research that accounts for the potential impact of breakfast habits will be important to understand the causal nature of associations between breakfast consumption and energy balance in adolescent girls.

Supplementary Materials: The following are available online at https://www.mdpi.com/article/10.3390/nu13124261/s1, Table S1: Breakfast choices of adolescent girls who participated in a randomised crossover trial comparing seven days of daily breakfast consumption (BC) with seven days of daily breakfast omission (BO).

Author Contributions: Conceptualization, J.K.Z.-F. and K.T.; methodology, J.K.Z.-F. and K.T.; formal analysis, C.S. and J.K.Z.-F.; investigation, J.K.Z.-F., K.T. and C.S.; resources, J.K.Z.-F. and K.T.; data curation, C.S. and J.K.Z.-F.; writing—original draft preparation, J.K.Z.-F.; writing—review and editing, J.K.Z.-F. and K.T.; visualization, J.K.Z.-F. and K.T.; supervision, J.K.Z.-F. and K.T.; project administration, J.K.Z.-F.; funding acquisition, J.K.Z.-F. and K.T. All authors have read and agreed to the published version of the manuscript.

Funding: This research was supported by the Nutrition Society Summer Studentship Scheme. The APC was funded by the National Institute for Health Research (NIHR) Leicester Biomedical Research Centre. The views expressed are those of the authors and not necessarily those of the NHS, the NIHR or the Department of Health.

Institutional Review Board Statement: This study was conducted in accordance with the ethical standards of the University of Bedfordshire Research Ethics Committee (ethical approval number: 2018SSPA002), in accordance with the Helsinki Declaration of 1975 as revised in 1983.

Informed Consent Statement: Parent/guardian informed consent and child (participant) assent were obtained from all participants involved in the study.

Data Availability Statement: The data presented in this study are available on request from the corresponding author.

Acknowledgments: The authors thank Katie Owen, Rahmeen Butt, Heidi Anukam, and Mia Chapman for assistance with data collection during the study.

Conflicts of Interest: The authors declare no conflict of interest. The funders had no role in the design of the study; in the collection, analyses, or interpretation of data; in the writing of the manuscript, or in the decision to publish the results.

References

1. Monzani, A.; Ricotti, R.; Caputo, M.; Solito, A.; Archero, F.; Bellone, S.; Prodam, F. A Systematic Review of the association of skipping breakfast with weight and cardiometabolic risk factors in children and adolescents. What should we better investigate in the future? *Nutrients* **2019**, *11*, 387. [CrossRef] [PubMed]
2. Corder, K.; van Sluijs, E.M.; Ridgway, C.L.; Steele, R.M.; Prynne, C.J.; Stephen, A.M.; Bamber, D.J.; Dunn, V.J.; Goodyer, I.M.; Ekelund, U. Breakfast consumption and physical activity in adolescents: Daily associations and hourly patterns. *Am. J. Clin. Nutr.* **2014**, *99*, 361–368. [CrossRef] [PubMed]
3. Corder, K.; van Sluijs, E.M.; Steele, R.M.; Stephen, A.M.; Dunn, V.; Bamber, D.; Goodyer, I.; Griffin, S.J.; Ekelund, U. Breakfast consumption and physical activity in British adolescents. *Br. J. Nutr.* **2011**, *105*, 316–321. [CrossRef]
4. Hill, J.O. Understanding and addressing the epidemic of obesity: An energy balance perspective. *Endocr. Rev.* **2006**, *27*, 750–761. [CrossRef]
5. Shaibi, G.Q.; Ryder, J.R.; Kim, J.Y.; Barraza, E. Exercise for obese youth: Refocusing attention from weight loss to health gains. *Exerc. Sport Sci. Rev.* **2015**, *43*, 41–47. [CrossRef]
6. Brown, A.W.; Bohan Brown, M.M.; Allison, D.B. Belief beyond the evidence: Using the proposed effect of breakfast on obesity to show 2 practices that distort scientific evidence. *Am. J. Clin. Nutr.* **2013**, *98*, 1298–1308. [CrossRef]
7. Sievert, K.; Hussain, S.M.; Page, M.J.; Wang, Y.; Hughes, H.J.; Malek, M.; Cicuttini, F.M. Effect of breakfast on weight and energy intake: Systematic review and meta-analysis of randomised controlled trials. *BMJ* **2019**, *364*, l42. [CrossRef] [PubMed]
8. Betts, J.A.; Richardson, J.D.; Chowdhury, E.A.; Holman, G.D.; Tsintzas, K.; Thompson, D. The causal role of breakfast in energy balance and health: A randomized controlled trial in lean adults. *Am. J. Clin. Nutr.* **2014**, *100*, 539–547. [CrossRef] [PubMed]
9. Chowdhury, E.A.; Richardson, J.D.; Holman, G.D.; Tsintzas, K.; Thompson, D.; Betts, J.A. The causal role of breakfast in energy balance and health: A randomized controlled trial in obese adults. *Am. J. Clin. Nutr.* **2016**, *103*, 747–756. [CrossRef]
10. Yoshimura, E.; Hatamoto, Y.; Yonekura, S.; Tanaka, H. Skipping breakfast reduces energy intake and physical activity in healthy women who are habitual breakfast eaters: A randomized crossover trial. *Physiol. Behav.* **2017**, *174*, 89–94. [CrossRef]
11. Halsey, L.G.; Huber, J.W.; Low, T.; Ibeawuchi, C.; Woodruff, P.; Reeves, S. Does consuming breakfast influence activity levels? An experiment into the effect of breakfast consumption on eating habits and energy expenditure. *Public Health Nutr.* **2012**, *15*, 238–245. [CrossRef] [PubMed]
12. LeCheminant, G.M.; LeCheminant, J.D.; Tucker, L.A.; Bailey, B.W. A randomized controlled trial to study the effects of breakfast on energy intake, physical activity, and body fat in women who are nonhabitual breakfast eaters. *Appetite* **2017**, *112*, 44–51. [CrossRef]
13. Reeves, S.; Huber, J.W.; Halsey, L.G.; Villegas-Montes, M.; Elgumati, J.; Smith, T. A cross-over experiment to investigate possible mechanisms for lower BMIs in people who habitually eat breakfast. *Eur. J. Clin. Nutr.* **2015**, *69*, 632–637. [CrossRef] [PubMed]
14. Metcalf, B.S.; Hosking, J.; Jeffery, A.N.; Henley, W.E.; Wilkin, T.J. Exploring the adolescent fall in physical activity: A 10-yr cohort study (EarlyBird 41]. *Med. Sci. Sports Exerc.* **2015**, *47*, 2084–2092. [CrossRef]
15. The Health and Social Care Information Centre. Health Survey for England—2016. Trend Tables. 2017. Available online: https://digital.nhs.uk/data-and-information/publications/statistical/health-survey-for-england/health-survey-for-england-2016 (accessed on 6 July 2021).
16. Boisseau, N.; Delamarche, P. Metabolic and hormonal responses to exercise in children and adolescents. *Sports Med.* **2000**, *30*, 405–422. [CrossRef] [PubMed]
17. Goran, M.I.; Gower, B.A. Longitudinal study on pubertal insulin resistance. *Diabetes* **2001**, *50*, 2444–2450. [CrossRef] [PubMed]
18. Rogol, A.D.; Clark, P.A.; Roemmich, J.N. Growth and pubertal development in children and adolescents: Effects of diet and physical activity. *Am. J. Clin. Nutr* **2000**, *72*, S521–S528. [CrossRef]
19. Timmons, B.W.; Bar-Or, O.; Riddell, M.C. Oxidation rate of exogenous carbohydrate during exercise is higher in boys than in men. *J. Appl. Physiol.* **2003**, *94*, 278–284. [CrossRef]
20. Timmons, B.W.; Bar-Or, O.; Riddell, M.C. Influence of age and pubertal status on substrate utilization during exercise with and without carbohydrate intake in healthy boys. *Appl. Physiol. Nutr. Metab.* **2007**, *32*, 416–425. [CrossRef]
21. Harrell, J.S.; McMurray, R.G.; Baggett, C.D.; Pennell, M.L.; Pearce, P.F.; Bangdiwala, S.I. Energy costs of physical activities in children and adolescents. *Med. Sci. Sports Exerc.* **2005**, *37*, 329–336. [CrossRef]
22. Leidy, H.J.; Racki, E.M. The addition of a protein-rich breakfast and its effects on acute appetite control and food intake in 'breakfast-skipping' adolescents. *Int. J. Obes.* **2010**, *34*, 1125–1133. [CrossRef]
23. Leidy, H.J.; Ortinau, L.C.; Douglas, S.M.; Hoertel, H.A. Beneficial effects of a higher-protein breakfast on the appetitive.; hormonal.; and neural signals controlling energy intake regulation in overweight/obese, "breakfast-skipping", late-adolescent girls. *Am. J. Clin. Nutr.* **2013**, *97*, 677–688. [CrossRef]
24. Leidy, H.J.; Hoertel, H.A.; Douglas, S.M.; Higgins, K.A.; Shafer, R.S. A high-protein breakfast prevents body fat gain.; through reductions in daily intake and hunger.; in 'breakfast skipping' adolescents. *Obesity* **2015**, *23*, 1761–1764. [CrossRef] [PubMed]
25. Zakrzewski-Fruer, J.K.; Plekhanova, T.; Mandila, D.; Lekatis, Y.; Tolfrey, K. Effect of breakfast omission and consumption on energy intake and physical activity in adolescent girls: A randomised controlled trial. *Br. J. Nutr.* **2017**, *118*, 392–400. [CrossRef]
26. Zakrzewski-Fruer, J.K.; Wells, E.K.; Crawford, N.S.G.; Afeef, S.M.O.; Tolfrey, K. Physical activity duration but not energy expenditure differs between daily compared with intermittent breakfast consumption in adolescent girls: A randomized crossover trial. *J. Nutr.* **2018**, *148*, 236–244. [CrossRef] [PubMed]

27. Vissers, P.A.; Jones, A.P.; Corder, K.; Jennings, A.; van Sluijs, E.M.; Welch, A.; Cassidy, A.; Griffin, S. Breakfast consumption and daily physical activity in 9–10-year-old British children. *Public Health Nutr.* **2013**, *16*, 1281–1290. [CrossRef] [PubMed]
28. Zakrzewski-Fruer, J.K.; Gillison, F.B.; Katzmarzyk, P.T.; Mire, E.F.; Broyles, S.T.; Champagne, C.M.; Chaput, J.P.; Denstel, K.D.; Fogelholm, M.; Hu, G.; et al. Association between breakfast frequency and physical activity and sedentary time: A cross-sectional study in children from 12 countries. *BMC Public Health* **2019**, *19*, 222. [CrossRef]
29. Kral, T.V.; Whiteford, L.M.; Heo, M.; Faith, M.S. Effects of eating breakfast compared with skipping breakfast on ratings of appetite and intake at subsequent meals in 8- to 10-y-old children. *Am. J. Clin. Nutr.* **2011**, *93*, 284–291. [CrossRef]
30. Wang, Y.C.; Gortmaker, S.L.; Sobol, A.M.; Kuntz, K.M. Estimating the energy gap among US children: A counterfactual approach. *Pediatrics* **2006**, *118*, e1721–e1733. [CrossRef]
31. Campbell, N.; Prapavessis, H.; Gray, C.; McGowan, E.; Rush, E.; Maddison, R. The Actiheart in adolescents: A doubly labelled water validation. *Pediatr. Exerc. Sci.* **2012**, *24*, 589–602. [CrossRef] [PubMed]
32. Collings, P.J.; Wijndaele, K.; Corder, K.; Westgate, K.; Ridgway, C.L.; Sharp, S.J.; Dunn, V.; Goodyer, I.; Ekelund, U.; Brage, S. Magnitude and determinants of change in objectively-measured physical activity.; sedentary time and sleep duration from ages 15 to 17.5y in UK adolescents: The ROOTS study. *Int. J. Behav. Nutr. Phys. Act.* **2015**, *12*, 61–70. [CrossRef] [PubMed]
33. Cole, T.J.; Bellizzi, M.C.; Flegal, K.M.; Dietz, W.H. Establishing a standard definition for child overweight and obesity worldwide: International survey. *BMJ* **2000**, *320*, 1240–1243. [CrossRef] [PubMed]
34. McCarthy, H.D.; Jarrett, K.V.; Emmett, P.M.; Rogers, I. Trends in waist circumferences in young British children: A comparative study. *Int. J. Obes.* **2005**, *29*, 157–162. [CrossRef]
35. Brooks-Gunn, J.; Warren, M.P.; Rosso, J.; Gargiulo, J. Validity of self-report measures of girls' pubertal status. *Child. Dev.* **1987**, *58*, 829–841. [CrossRef]
36. Morris, N.M.; Udry, J.R. Validation of a self-administered instrument to assess stage of adolescent development. *J. Youth Adolesc.* **1980**, *9*, 271–280. [CrossRef]
37. Tanner, J.M. *Growth at Adolescents*; Blackwell Scientific: Oxford, UK, 1962.
38. Mellecker, R.R.; McManus, A.M. Measurement of resting energy expenditure in healthy children. *J. Parenter. Enteral Nutr.* **2009**, *33*, 640–645. [CrossRef]
39. Weir, J.B. New methods for calculating metabolic rate with special reference to protein metabolism. *J. Physiol.* **1949**, *109*, 1–9. [CrossRef]
40. Kaiyala, K.J.; Wisse, B.E.; Lighton, J.R.B. Validation of an equation for energy expenditure that does not require the respiratory quotient. *PLoS ONE* **2019**, *14*, e0211585. [CrossRef]
41. Timlin, M.T.; Pereira, M.A. Breakfast frequency and quality in the etiology of adult obesity and chronic diseases. *Nutr. Rev.* **2007**, *65*, 268–281. [CrossRef]
42. O'Neil, C.E.; Byrd-Bredbenner, C.; Hayes, D.; Jana, L.; Klinger, S.E.; Stephenson-Martin, S. The role of breakfast in health: Definition and criteria for a quality breakfast. *J. Acad. Nutr. Diet* **2014**, *114*, S8–S26. [CrossRef] [PubMed]
43. Tolfrey, K.; Zakrzewski, J.K. Breakfast, glycaemic index and health in young people. *J. Sport Health Sci.* **2012**, *1*, 149–159. [CrossRef]
44. Smith, C.E.; Tucker, K.L. Health benefits of cereal fibre: A review of clinical trials. *Nutr. Res. Rev.* **2011**, *24*, 118–131. [CrossRef] [PubMed]
45. Corder, K.; Brage, S.; Wareham, N.J.; Ekelund, U. Comparison of PAEE from combined and separate heart rate and movement models in children. *Med. Sci. Sports Exerc.* **2005**, *37*, 1761–1767. [CrossRef] [PubMed]
46. Burrows, T.L.; Martin, R.J.; Collins, C.E. A systematic review of the validity of dietary assessment methods in children when compared with the method of doubly labeled water. *J. Am. Diet Assoc.* **2010**, *110*, 1501–1510. [CrossRef]
47. Livingstone, M.B.; Robson, P.J.; Wallace, J.M. Issues in dietary intake assessment of children and adolescents. *Br. J. Nutr.* **2004**, *92* (Suppl. S2), S213–S222. [CrossRef] [PubMed]
48. Boushey, C.J.; Kerr, D.A.; Wright, J.; Lutes, K.D.; Ebert, D.S.; Delp, E.J. Use of technology in children's dietary assessment. *Eur. J. Clin. Nutr.* **2009**, *63* (Suppl. S1), S50–S57. [CrossRef]
49. Thane, C.W.; Stephen, A.M. Day-to-day variation in food and nutrient intakes of British adults. *Public Health Nutr.* **2006**, *7A*, 102.
50. Foster, E.; Hawkins, A.; Adamson, A. *Young Person's Food Atlas—Secondary*; Food Standards Agency: London, UK, 2010.
51. Foster, E.; Adamson, A.J. Development and validation of the Young Person's Food Atlas. *Proc. Nutr. Soc.* **2012**, *71*, 195. [CrossRef]
52. Foster, E.; Hawkins, A.; Barton, K.L.; Stamp, E.; Matthews, J.N.S.; Adamson, A.J. Development of food photographs for use with children aged 18 months to 16 years: Comparison against weighed food diaries—The Young Person's Food Atlas (UK). *PLoS ONE* **2017**, *15*, 12. [CrossRef]
53. Lee, K.A.; Hicks, G.; Nino-Murcia, G. Validity and reliability of a scale to assess fatigue. *Psychiatry Res.* **1991**, *36*, 291–298. [CrossRef]
54. Flint, A.; Raben, A.; Blundell, J.E.; Astrup, A. Reproducibility, power and validity 496 of visual analogue scales in assessment of appetite sensations in single test meal studies. *Int. J. Obes. Relat. Metab. Disord.* **2000**, *24*, 38–48. [CrossRef] [PubMed]
55. Parker, B.A.; Sturm, K.; MacIntosh, C.; Feinle, C.; Horowitz, M.; Chapman, I.M. Relation between food intake and visual analogue scale ratings of appetite and other sensations in healthy older and young subjects. *Eur. J. Clin. Nutr.* **2004**, *58*, 212–218. [CrossRef]
56. Rumbold, P.L.; St Clair Gibson, A.; Allsop, S.; Stevenson, E.; Dodd-Reynolds, C.J. Energy intake and appetite following netball exercise over 5 days in trained 13–15 year old girls. *Appetite* **2011**, *56*, 621–628. [CrossRef]

57. Harrington, D.M.; Dowd, K.P.; Bourke, A.K.; Donnelly, A.E. Cross-sectional analysis of levels and patterns of objectively measured sedentary time in adolescent females. *Int. J. Behav. Nutr. Phys. Act.* **2011**, *8*, 120–126. [CrossRef]
58. Senn, S. *Cross-Over Trials in Clinical Research*; Wiley: Chichester, UK, 1993.
59. Atkinson, G. Analysis of repeated measurements in physical therapy research: Multiple comparisons amongst level means and multifactorial designs. *Phys. Ther. Sport* **2002**, *3*, 191–203. [CrossRef]
60. Cohen, J. *Statistical Power Analysis for the Behavioral Sciences*, 2nd ed.; Lawrence Erlbaum Associates: Hillsdale, NJ, USA, 1988.
61. Hopkins, M.; Blundell, J.E. Energy balance, body composition, sedentariness and appetite regulation: Pathways to obesity. *Clin. Sci.* **2006**, *130*, 1615–1628. [CrossRef]
62. Goltz, F.R.; Thackray, A.E.; Atkinson, G.; Lolli, L.; King, J.A.; Dorling, J.L.; Dowejko, M.; Mastana, S.; Stensel, D.J. True interindividual variability exists in postprandial appetite responses in healthy men but is not moderated by the FTO genotype. *J. Nutr.* **2019**, *149*, 1159–1169. [CrossRef] [PubMed]
63. Thivel, D.; Chaput, J.P. Are post-exercise appetite sensations and energy intake coupled in children and adolescents? *Sports Med.* **2014**, *44*, 735–741. [CrossRef]
64. De Castro, J.M. Prior day's intake has macronutrient-specific delayed negative feedback effects on the spontaneous food intake of free-living humans. *J. Nutr.* **1998**, *128*, 61–67. [CrossRef]

nutrients

MDPI

Article

Too Jittery to Sleep? Temporal Associations of Actigraphic Sleep and Caffeine in Adolescents

Gina Marie Mathew [1,*], David A. Reichenberger [2], Lindsay Master [2], Orfeu M. Buxton [2], Anne-Marie Chang [2] and Lauren Hale [1]

1 Program in Public Health, Department of Family, Population, and Preventive Medicine, Renaissance School of Medicine, Stony Brook University, Stony Brook, NY 11794, USA; lauren.hale@stonybrookmedicine.edu

2 Department of Biobehavioral Health, College of Health and Human Development, The Pennsylvania State University, University Park, PA 16802, USA; reichenberger@psu.edu (D.A.R.); lmm5346@psu.edu (L.M.); orfeu@psu.edu (O.M.B.); amchang@psu.edu (A.-M.C.)

* Correspondence: gina.mathew@stonybrookmedicine.edu

Abstract: Caffeine consumption has been linked to poor sleep health in adolescents, but it is unknown whether poor sleep predicts caffeine consumption, and/or whether caffeine consumption predicts poor sleep, particularly when sleep is measured objectively. Data were collected from a micro-longitudinal sub-study of the age 15 wave of the Fragile Families and Child Wellbeing Study (n = 589). Adolescents wore an actigraphy device and completed daily surveys for ~1 week. Daily surveys assessed subjective sleep quality and caffeinated beverage consumption (0 = no caffeine, 1 = any caffeine). Separate mixed models assessed whether actigraphy-measured sleep duration, timing, maintenance efficiency, and subjective quality predicted next-day caffeinated beverage consumption within and between adolescents. Variability (standard deviation) of sleep duration and timing, sleep regularity index, and social jetlag were tested as additional between-person predictors. Lagged models tested whether daily caffeinated beverage consumption predicted sleep that night (n = 458). Adolescents with more variable sleep duration and midpoint had higher average odds of consuming caffeinated beverages compared to others. After adolescents consumed ≥1 caffeinated beverage, they had later sleep onset that night and wake time the next morning than usual versus when they did not consume caffeine. Curbing caffeinated beverage consumption may aid in the maintenance of regular sleep schedules and advance sleep timing in adolescents.

Keywords: sleep duration; sleep timing; sleep maintenance efficiency; subjective sleep quality; sleep variability; social jetlag; caffeine; adolescence; actigraphy; diary

Citation: Mathew, G.M.; Reichenberger, D.A.; Master, L.; Buxton, O.M.; Chang, A.-M.; Hale, L. Too Jittery to Sleep? Temporal Associations of Actigraphic Sleep and Caffeine in Adolescents. *Nutrients* **2022**, *14*, 31. https://doi.org/10.3390/nu14010031

Academic Editors: Egeria Scoditti and Sergio Garbarino

Received: 20 November 2021
Accepted: 20 December 2021
Published: 23 December 2021

1. Introduction

Caffeine is a stimulant drug readily available in the United States and present in coffee, black and green tea, some sodas, and energy drinks [1]. Consumption of caffeine is high among American adolescents, with 75% reporting consuming a caffeinated beverage on a typical day [2,3]. Adolescents who consume caffeine may become caffeine dependent [4], and high caffeine consumption is associated with cigarette use, psychosocial problems [5], and lower grade point average [6] in adolescents. The high prevalence of regular caffeine consumption in adolescents and the potential negative consequences on health and wellbeing indicate the need for more research into the antecedents and consequences of caffeine consumption in this population.

The potential negative effects of adolescent caffeine consumption include short sleep duration [2,6–19]. Over 70% of American high schoolers report obtaining fewer than the recommended minimum of 8 h per night [20]. Cross-sectional studies measuring sleep through self-report generally find that adolescents who consume more caffeine have shorter sleep duration [2,6–19]. Some studies with similar designs, however, have found no association between caffeine and sleep duration in adolescents [21–23]. The few studies examining the link between objectively measured sleep (i.e., through wrist actigraphy) and caffeine consumption in adolescents have also demonstrated no association [24,25], warranting further research with objective measures of sleep.

Later sleep timing has also been linked to caffeine consumption in adolescents [6,8,12, 22–24,26–30]. Adolescents tend to have a later chronotype, or preference for timing of sleep and other behaviors, than children or adults [31]. Studies employing a cross-sectional design and measuring sleep through self-report have found that greater caffeine consumption is associated with later sleep timing [6,8,12,22–24,27,28] and evening preference [26,29,30] in adolescents, with one finding no association between caffeine consumption and evening preference [32]. There is a lack of research on the association between caffeine consumption and objectively measured sleep timing in adolescents.

Few studies have examined the association between caffeine consumption and other dimensions of sleep, such as quality (whether measured objectively or through self-report) and variability, in adolescents. One study demonstrated that higher caffeine intake was associated with greater self-reported wake after sleep onset (a measure of poor sleep quality) [19], and adolescents who consumed at least two servings of caffeine per day demonstrated less slow-wave activity (indicating poorer sleep quality) as measured through polysomnography (PSG), the gold standard for sleep monitoring [33], compared to adolescents who did not consume caffeine [24]. However, two studies that measured sleep through actigraphy found no association between caffeine consumption and wake after sleep onset in adolescents [24,25], and another found a null association between caffeine consumption and self-reported sleep quality [23]. These mixed findings indicate the need for more research on the associations of objective and subjective sleep quality with caffeine consumption in adolescents. There is also a lack of research on the associations of variability in sleep duration and timing, important dimensions of sleep that suggest poorer sleep health [34], with caffeine consumption in adolescents.

The association between sleep and caffeine may be bidirectional. Some adolescents report consuming caffeinated beverages to counteract daytime sleepiness and fatigue [35], and caffeine consumption may reduce that night's sleep duration and quality [36]. However, most studies that examined the association between sleep and caffeine consumption in adolescents employed a cross-sectional design [2,5–16,18,20–29,31], which precludes examination of temporal associations. Whereas cross-sectional studies are only able to describe between-person associations [37] (e.g., whether adolescents who sleep longer on average tend to consume caffeine more often than other adolescents), longitudinal studies can establish temporal precedence [38] (e.g., whether adolescents are more likely to consume caffeine the next morning following nights when they sleep shorter than usual). Only a few studies have examined longitudinal associations between caffeine consumption and dimensions of sleep among adolescents. Two studies found no within-person associations when measuring sleep through self-report [18,39]. One micro-longitudinal study measuring sleep through PSG found that increased caffeine consumption predicted less total sleep time and sleep efficiency that night, and reduced sleep efficiency predicted more afternoon caffeine consumption the next day [40], but the study did not assess sleep timing, subjective quality, or variability and included fewer than 100 adolescents. Determining the direction of the sleep—caffeine relationship while assessing multiple dimensions of sleep may allow for future targeted interventions to improve health behaviors, including sleep and dietary habits.

There is minimal research examining temporal within-person associations between caffeine consumption and objectively measured sleep among adolescents. Self-reported sleep can deviate considerably from sleep assessed through objective measures such as actigraphy [41], and between-person effects assessed in cross-sectional research do not necessarily translate to within-person effects in direction or magnitude [37]. It may be particularly important to examine associations between caffeine consumption and dimensions of sleep health given that adolescents begin to gain autonomy over their own sleep behaviors [42] and diet [43] versus childhood.

The present study examined bidirectional associations of objectively measured sleep dimensions and subjective sleep quality with caffeinated beverage consumption in adolescents. We additionally examined whether sleep variability was associated with greater caffeinated beverage consumption between adolescents. We hypothesized that poor sleep health (shorter or longer sleep duration, later sleep timing, lower sleep maintenance efficiency, and lower self-reported sleep quality) would predict caffeinated beverage consumption the next day, and that caffeinated beverage consumption would predict poor sleep health that night. We further hypothesized that greater variability in sleep duration and timing would be associated with more caffeinated beverage consumption between adolescents.

2. Materials and Methods

2.1. Participants

Data for the current analyses come from the Fragile Families and Child Wellbeing Study (FFCWS; www.fragilefamilies.princeton.edu, accessed on 11 August 2020), a longitudinal birth cohort oversampled for nonmarital births, which resulted in a greater proportion of racial/ethnic minority mothers and those of lower socioeconomic status and education level compared to the national population. More details regarding the sample and design may be found elsewhere [44]. Survey data from the Fragile Families and Child Wellbeing study (https://fragilefamilies.princeton.edu/documentation, accessed on 22 March 2017) are publicly available from Princeton University's Office of Population Research (OPR) data archive: https://opr.princeton.edu/archive/restricted/Default.aspx, accessed on 22 March 2017. The sleep actigraphy and daily diary datasets generated and analyzed during the current study are not publicly available yet, but will be available through an application process at the above link. This study was conducted according to the guidelines laid down in the Declaration of Helsinki of 1975 (revised 2013), and all procedures involving human participants were approved by the Princeton University and Stony Brook University (CORIHS B) (FWA #00000125) Institutional Review Boards. Written (for in-home interviews) or recorded verbal (for phone interviews) informed consent was obtained from primary caregivers, and assent was obtained from adolescents.

The original FFCWS birth cohort consists of 4898 children born from 1998–2000 in 20 large cities in the United States [45]. Families were recruited from local hospitals at the time of the child's birth. The study staff maintained records about the participants and their families for follow up at subsequent waves, when participants were approximately ages 1, 3, 5, 9, and 15 years of age. Families were eligible for inclusion in the age 15 follow-up wave if the child was alive and not legally adopted (95% of the birth sample were eligible). Data in the current analyses were collected from February 2014 to March 2016. During the year 15 wave of the FFCWS (wave 6), 3444 adolescents and their primary caregivers completed separate surveys querying household and demographic characteristics, administered either over the phone or in person at the participant's place of residence. A randomly selected subsample (n = 1090) was asked to participate in a micro-longitudinal FFCWS sub-study. Adolescents who agreed to participate (n = 1049) were asked to wear a wrist-worn accelerometer at all times and answer a daily diary for seven consecutive days in the evening.

The design of the study was such that a night's sleep predicted the next day's diary report, including assessment of caffeinated beverage consumption. Therefore, lagging analyses to examine if caffeinated beverage consumption predicted that night's sleep required dropping the first sleep recording and the last diary report. In the originally structured non-lagged dataset (where nighttime sleep predicted next-day caffeinated beverage consumption), of n = 1049 assenting adolescents, n = 414 were excluded due to not providing at least 3 valid days of actigraphy recordings (dataset has average of 5.6 $\pm$ 1.4 recordings per adolescent; range 3–10; IQR 5–7) and next-day diary reports (dataset has average of 5.5 $\pm$ 1.4 reports per adolescent; range 3–9; IQR 4–7), and n = 46 were excluded due to not reporting whether they consumed a caffeinated beverage on at least three days, leaving a total sample of n = 589 adolescents (56.1% of the subsample). A further n = 219 adolescents were excluded from social jetlag analyses due to not providing at least one school night and one non-school night, resulting in $n = 370$ included adolescents.

For analysis of caffeinated beverage consumption predicting that night's sleep (in the lagged dataset), of $n = 1049$ assenting adolescents, $n = 492$ adolescents were excluded due to not providing at least 3 valid lagged days of actigraphy recordings (lagged dataset has average of 5.2 ± 1.1 recordings per adolescent; range 3–9; IQR 4–6) and next-day diary reports (lagged dataset has average of 4.7 ± 1.2 reports per adolescent; range 3–8; IQR 4–6), $n = 98$ adolescents were excluded due to not reporting whether they consumed a caffeinated beverage on at least three lagged days, and $n = 1$ adolescent was excluded due to missing body mass index data, leaving $n = 458$ adolescents included in the lagged analyses (43.7% of the subsample). Supplemental Figure S1 (for sleep predicting caffeinated beverage consumption) and S2 (for caffeinated beverage consumption predicting sleep in lagged analyses) depict participant flow charts, and a "Strengthening the Reporting of Observational Studies in Epidemiology—Nutritional Epidemiology" (STROBE-nut) checklist is included as Supplemental Table S1 [46].

Separate logistic regression analyses were conducted to compare sex, race/ethnicity, and income between the adolescents included in the analyses and those who assented to participate in the sub-study but were ultimately excluded due to missing data (assented $n = 1049$). In non-lagged analyses (where sleep predicted next-day caffeinated beverage consumption; analytical sample $n = 589$, excluded $n = 460$), male sex marginally predicted missingness (odds ratio, OR = 1.13, $p = 0.052$), and Black/African American race/ethnicity (vs. White/Caucasian; OR = 1.38, $p < 0.001$) and lower household income (in thousands of dollars; OR = 0.997, $p = 0.012$) predicted higher odds of data missingness. In lagged analyses (where caffeinated beverage consumption predicted that night's sleep; analytical sample $n = 458$, excluded $n = 591$), Black/African American race/ethnicity (OR = 1.31, $p = 0.005$) and lower household income (OR = 0.998, $p = 0.020$) predicted higher odds of data missingness. All analyses were adjusted for sex, race/ethnicity, and household income.

2.2. Actigraphy Device and Scoring

Sleep measures were collected with a wrist-worn accelerometer with off-wrist detection (Actiwatch Spectrum; Philips-Respironics, Murrysville, PA, USA) and study participants were asked to wear the watch on their non-dominant wrist for one week. Data from the Spectrum device were downloaded with Actiware software (Version 6.0.4, Philips-Respironics, 2017). At least two trained independent scorers (blinded to each other) scored the data using an algorithm validated against polysomnography [33]. Sleep diaries were not used for actigraphy scoring given the frequent inaccuracy, uncertainty, and potential for systematic bias caused by discrepant diary and actigraphy reports [47]. The two scorers looked for disagreements in the number of valid days, cut-time (i.e., start and end time that defines a 24 h day), presence of all-nighters (i.e., no sleep interval within a 24 h day), the presence of false on-wrist detection, presence of naps, number of sleep intervals, and each sleep onset or offset that differed by >15 min. The scorers met to resolve any discrepancies related to these aspects of the data, and a final dataset was established [48]. The scorers determined sleep intervals using a decrease in activity levels and the aid of light levels for sleep onset and sleep offset [49], and a nighttime sleep interval was split into two intervals (main sleep and nap) if there was an awakening $\geq$1 h during this interval. A sleep actigraphy day was determined invalid and no sleep interval was set if there were $\geq$4 total hours of off-wrist time, with the exception of the first and last day (device should have been worn at least 2 h on the first day). Other invalidation criteria were constant false activity due to battery failure, data unable to be retrieved or recovered, or an off-wrist period of $\geq$60 min within 10 min of the scored beginning or end of the main sleep period for that day.

2.3. Daily Alignment of Sleep and Diary and Data Lagging

Participants were asked to complete a diary each evening after 7:00 PM (19:00) and before going to sleep. Sleep and daily diary data were merged by participant identification number and date/time. A previous night's sleep and the following day's diary answers were aligned on the same observation record in the originally structured merged data. Nights were excluded from analyses if the adolescent had an all-nighter and received no sleep.

To investigate whether caffeinated beverage consumption predicted that night's sleep, diary entries were each lagged by one row (i.e., day) through the "lag1" function in SAS 9.4. Lagging data resulted in one row of data loss per adolescent and reduced the sample size, due to adolescents requiring 4 non-lagged days to produce 3 lagged days.

2.4. Variables

2.4.1. Nightly Actigraphic Sleep Measures

Sleep onset and sleep offset were the start and end of the main nighttime sleep interval calculated in hours from midnight, respectively. Sleep midpoint (also midnight-centered) was calculated as midway between sleep and sleep offset. Sleep duration was calculated as the number of hours between sleep onset and sleep offset of the main nighttime sleep interval. Nighttime sleep maintenance efficiency (considered an objective measure of sleep quality) represents the percent of sleep duration that the individual spent asleep and was calculated as 1 − (wake after sleep onset in hours/sleep duration in hours) and multiplied by 100 to produce a percentage [50].

2.4.2. Actigraphic Sleep Measures Calculated per Adolescent

We calculated measures of sleep variability per adolescent across the monitoring days in the non-lagged dataset ($n = 589$). Sleep duration variability, sleep onset variability, sleep midpoint variability, and sleep offset variability were calculated as the standard deviation (*SD*, per adolescent) of each measure. Sleep regularity index (SRI) was calculated based on the formula from Phillips et al. [51] and ranges from 0 (low regularity) to 100 (high regularity). The score represents the percentage probability that an individual is in the same state (sleeping or awake) at any two time points 24 h apart, averaged across the interval of measurement. Social jetlag, a misalignment of sleep midpoint between school and free days, was calculated in hours through the following formula: |sleep midpoint on free nights−sleep midpoint on school nights| [52]. Only adolescents with at least one school night and one free night were included in the social jetlag measure; therefore, some adolescents did not have social jetlag values ($n = 370$ of the 589 total adolescents had social jetlag values).

2.4.3. Subjective Sleep Quality

Adolescents rated their last-night sleep quality on the daily diary by answering, "How would you rate your sleep quality?" with possible answers of "very good" (1), "fairly good" (2) "fairly bad" (3) or "very bad" (4). This variable was reverse coded such that "very bad" became 0, and "very good" became 3.

2.5. *Caffeinated Beverage Consumption*

Adolescents were asked on the daily diary, "How many caffeinated beverages (such as coffee, soda, energy drinks) did you have? One beverage is about 8 ounces" with response options of 0 (0), 1 (1), 2 (2), 3 (3), 4 (4), or 5 or more (5). Items were "coffee or tea (iced or hot)," "caffeinated soda (such as Coca-cola, Pepsi, Mountain Dew)," and "energy drinks (such as Red Bull, Monster, 5-h Energy, RockStar, Full Throttle, Amp, etc.)." A "number of caffeinated beverages that day" variable was constructed as the sum of coffee/tea, caffeinated soda, and energy drinks consumed each day. Due to the considerable number of days on which adolescents reported not consuming any caffeinated beverages (39.7% of the 3215 observations), the caffeinated beverages variable was dichotomized into 0 = none; 1 = at least one 8 oz beverage that day.

2.6. Covariates

Adolescents also reported whether they went to school (0 = no school, 1 = school) on the daily diary.

Other covariates were assessed through surveys administered once to youth and their primary caregivers during the year 15 wave. Race/ethnicity was reported on the youth survey and grouped into exclusive categories of White/Caucasian (not Hispanic or Latino), Black/African (not Hispanic or Latino), Hispanic and/or Latino (any race), or a category with other (including Asian, Central American/Caribbean, Native American/Alaska Native, and/or Native Hawaiian/Pacific Islander), mixed, and no race/ethnicity reported. The primary caregiver's education level ("some high school," "completed high school," "some college," or "college graduate"), annual income (in USD), and whether the youth lived with two biological parents were assessed on a survey administered to the youth's primary caregiver.

Information about biological sex was collected at birth. Weight and height were assessed objectively during in-person interviews by trained research assistants. Body mass index (BMI) percentile (range 0–100) was calculated based on the 2000 Centers for Disease Control and Prevention (CDC) growth charts [53], which matches BMI (weight in kg/(height in m^2)) [54] for the adolescent's sex and age.

2.7. Statistical Analyses

Analyses were conducted in SAS 9.4 software (SAS Institute, Cary, NC, USA). Most variables met standards for normality (skew < |3| and kurtosis < |10|) [55]. Variability (*SD*s) of sleep onset, midpoint, and offset and social jetlag were positively skewed (skew $\geq$ 3) and/or leptokurtotic (kurtosis $\geq$ 10) and were winsorized (i.e., values beyond the 99th percentile were replaced with the 99th percentile value). Of 589 adolescents, 5 sleep onset *SD* values, 5 sleep midpoint *SD* values, 6 sleep offset *SD* values, and 2 social jetlag values were winsorized. After winsorization, these variables also met criteria for normality.

Within-person reliability for each repeated measure (i.e., assessed through actigraphy or diary) was analyzed. Intraclass correlation coefficients (ICCs) were obtained for continuous measures (random effect variance/total variance [56]), which included all nightly sleep measures and number of caffeinated beverages. Pseudo-ICCs were obtained for binary measures (generated by the generalized estimating equation procedure [57,58]), which included the dichotomized caffeinated beverage consumption and school attendance. Higher ICCs indicate less variation within adolescents.

2.7.1. Main Analyses

In non-lagged analyses where sleep predicted next-day caffeinated beverage consumption, multilevel models (PROC GLIMMIX, with a binary distribution for the outcome caffeine consumption) tested whether dimensions of nightly sleep (sleep duration, sleep onset, midpoint, and offset, and sleep maintenance efficiency, measured through actigraphy; subjective sleep quality, measured through diary) predicted the odds of consuming $\geq$1 caffeinated beverage the next day, within adolescents. A quadratic association between sleep duration and odds of caffeinated beverage consumption was assessed using the predictor sleep duration^2 (sleep duration $\times$ sleep duration). The between-person predictor (within the same model as the within-person predictor) tested whether average sleep per adolescent across the monitoring period was associated with average odds of consuming $\geq$1 caffeinated beverage.

For the sleep measures that varied within-person (sleep duration, onset, midpoint, offset, maintenance efficiency, and subjective sleep quality), two-level models examined 3215 total daily observations that were clustered within 589 adolescents. Variances for nightly sleep measures were decomposed into within-person (level-1) and between-person (level-2) levels [38]. Within-person predictor variables were centered around the person mean, such that positive values indicated that value was higher than the person's own cross-day average. Between-person predictor variables were calculated as the mean per person across all time points. We specified the denominator degrees of freedom to be computed by dividing the residual degrees of freedom into between-subject and within-subject portions (DDFM = BETWITHIN) [59]. For the sleep variability measures that were between-person only (*SD* of duration, onset, midpoint, and offset, SRI, and social jetlag), the analyses were conducted with the sleep variability measure predicting the odds of consuming $\geq$1 caffeinated beverage (on average) per adolescent. Covariates that were significantly associated with odds of caffeinated beverage consumption were also included in models: school attendance and primary caregiver's highest education level. Analyses with predictors other than sleep duration were further adjusted for sleep duration and sleep duration^2.

Lagged analyses with caffeinated beverage consumption predicting sleep that night were conducted similarly as analyses that predicted next-day caffeinated beverage consumption from sleep, except using PROC MIXED for the continuous sleep outcomes. There were 2128 observations nested within 458 adolescents. Covariates that were significantly associated with any sleep outcome were also included in models: school attendance, BMI percentile, primary caregiver's education level, and whether the adolescent was living with two biological parents.

All models used autoregressive (AR) (1) covariance structure, included a random intercept for participant variation, and adjusted for sex, race/ethnicity, and household income. Alpha < 0.05 (two-sided) was deemed statistically significant.

2.7.2. Sensitivity Analyses by Adolescent's Average Caffeinated Beverage Consumption

To examine whether the associations between sleep and caffeinated beverage consumption varied as a function of the adolescent's average consumption, we conducted two sets of sensitivity analyses. One set of analyses examined whether the within-person effect of sleep on next-day caffeinated beverage consumption varied as a function of the proportion of days each adolescent consumed caffeine (a between-person moderator). These analyses included the interaction term within-person dimension of sleep x proportion of days the adolescent consumed caffeine predicting next-day caffeinated beverage consumption. The second set of analyses examined whether the within-person effect of caffeine on that night's sleep varied as a function of the proportion of days each adolescent consumed caffeine (lagged analyses). These analyses included the interaction term within-person caffeinated beverage consumption $\times$ proportion of days the adolescent consumed caffeine predicting sleep that night.

3. Results

3.1. Demographic Information

For the non-lagged analyses where sleep predicted caffeinated beverage consumption, 589 adolescents provided at least 3 nights of actigraphy, next-day diary, and complete covariate data (53% female, n = 311; mean age $\pm$ *SD* = 15.4 $\pm$ 0.5 years, range 14.7–17.7), with an average of 5.6 $\pm$ 1.4 actigraphy nights per adolescent (range 3–10 days; IQR 5–7) and 5.5 $\pm$ 1.4 reports of caffeinated beverage consumption (range 3–9 days; IQR 4–7). Ethno-racial composition of the sample was as follows: 41% Black/African American (n = 240), 25% Hispanic or Latino (n = 149), 19% White/Caucasian (n = 112), and 15% other, mixed, or none (n = 88). The mean percent of days that adolescents reported consuming $\geq$1 caffeinated beverage was 61% $\pm$ 34%. Demographic information and caffeinated beverage consumption were similar for lagged analyses (where caffeinated beverage consumption predicted sleep, n = 458 adolescents); the average number of actigraphy nights provided by each adolescent was 5.2 $\pm$ 1.1 (range 3–9 days; IQR 4–6) and the average number of reports of caffeinated beverage consumption was 4.7 $\pm$ 1.2 (range 3–8; IQR 4–6). Other sample information for the non-lagged dataset (n = 589), including descriptive statistics for sleep variables of interest and covariates, is in Table 1.

Table 1. Average descriptive statistics for analytical sample (n = 589).

Variable	*M* or %	*SD* or *n*
Demographic and household		
Age	15.39	0.52
Sex [a]		
Female	53%	311
Male	47%	278
Race/ethnicity		
Black/African American	41%	240
Hispanic and/or Latino	25%	149
White/Caucasian	19%	112
Other, [b] mixed, or none	15%	88
Body mass index percentile [c]	73.87	25.21
Annual household income (USD)	$64,906	$57,879
Primary caregiver's highest education level		
Did not graduate high school	14%	85
High school graduate	18%	106
Completed some college	47%	276
College graduate	21%	122
Youth living arrangements		
Lives with 2 married/cohabiting biological parents	32%	187
Lives with <2 biological parents	68%	402
School attendance		
Attended school (proportion of days)	0.44	0.34
Nightly sleep measures		
Sleep duration (h)	7.79	1.08
Sleep onset (clock time)	0:28	1:44
Sleep midpoint (clock time)	4:21	1:42
Sleep offset (clock time)	8:20	1:47
Sleep maintenance efficiency (%)	90.70	3.43
Subjective sleep quality [d]	2.36	0.50
Sleep variability measures [e]		
Variability (*SD*) of sleep duration (h)	1.57	0.80
Variability (*SD*) of sleep onset (h)	1.30	0.74
Variability (*SD*) of sleep midpoint (h)	1.22	0.65
Variability (*SD*) of sleep offset (h)	1.56	0.90
SRI [f]	48.35	13.38
Social jetlag (h) [g]	1.79	1.15
Dietary intake		
Consumed ≥1 cup caffeinated beverage (proportion of days) [h]	0.61	0.34
Consumed ≥1 caffeinated beverage 0–24% of the days	19%	112
Consumed ≥1 caffeinated beverage 25–49% of the days	15%	88
Consumed ≥1 caffeinated beverage 50–74% of the days	22%	129
Consumed ≥1 caffeinated beverage 75–100% of the days	44%	260

Notes. The mean number of actigraphy recordings per adolescent was 5.6 ± 1.4 (range 3–10; IQR 5–7) and the mean number of reports of caffeinated beverage consumption was 5.5 ± 1.4 (range 3–9; IQR 4–7). [a] Data collected at birth. [b] Other category includes Asian, Central American/Caribbean, Native American/Alaska Native, and/or Native Hawaiian/Pacific Islander. [c] Calculated based on 2000 Centers for Disease Control and Prevention (CDC) growth charts, matched for age and sex [53]. [d] Ranges from 0 (very bad)–3 (very good). [e] Higher value means greater variability, except the reverse for the sleep regularity index. [f] Calculated based on formula from Phillips et al. [51]; ranges from 0 (low)–100 (high). [g] Calculated based on formula from Wittmann et al. [52]. n = 370 (adolescent included only if provided at least one school night and one free night of actigraphy). [h] Includes coffee or tea, caffeinated soda, and energy drinks. H, hours; *M*, mean; *n*, number; *SD*, standard deviation; SRI, sleep regularity index; USD, United States dollar.

The ICCs for sleep measures ranged from 0.15 to 0.53, indicating poor to moderate reliability and therefore considerable within-person variation [56]. The ICC for the continuous caffeinated beverages variable (number of caffeinated beverages on a given day) was 0.58, and the pseudo-ICC for the dichotomized caffeinated beverages variable (caffeinated beverage consumption) was 0.39, demonstrating that the dichotomized measure had higher within-person variability. ICC and pseudo-ICC values for each nightly sleep measure, caffeinated beverages (continuous and dichotomized), and school attendance are in Supplemental Table S2.

3.2. Nightly Sleep Measures Predicting Next-Day and Average Caffeinated Beverage Consumption (Within- and Between-Person Associations)

No dimension of sleep measured nightly (sleep duration, onset, midpoint, offset, maintenance efficiency, or subjective quality) predicted next-day caffeinated beverage consumption within adolescents, nor were there any significant between-person associations for any of these sleep measures with odds of consuming ≥1 caffeinated beverage (all $p \geq 0.10$).

We additionally examined whether the within-person effects of sleep on next-day caffeinated beverage consumption varied depending on adolescent's average caffeinated beverage consumption (i.e., the proportion of days on which an adolescent reported consuming ≥1 caffeinated beverage). There were no significant interactions (all $p > 0.26$).

3.3. Associations of Sleep Variability with Average Caffeinated Beverage Consumption (Between-Person Associations)

There were significant associations between variability of sleep duration ($p = 0.042$) and variability of sleep midpoint ($p = 0.045$) with average odds of caffeinated beverage consumption, such that for every one *SD*-hour increase in variability, the average odds of an adolescent consuming one or more caffeinated beverages on a given day increased by 21% and 27%, respectively (see Figure 1 and Table 2). Variability of sleep onset ($p = 0.093$) and variability of sleep offset ($p = 0.058$) were marginally associated with average odds of caffeinated beverage consumption, such that for every one *SD*-hour increase in variability, the average odds of an adolescent consuming one or more caffeinated beverages on a given day increased by 19% and 17%, respectively. There were no significant associations of SRI ($p = 0.214$) or social jetlag ($p = 0.420$) with caffeinated beverage consumption.

Table 2. Between-person associations of sleep variability per youth across monitoring days with average odds of caffeinated beverage consumption ($n = 589$).

Model Predictor	OR	95%CI OR	
Sleep duration (*SD*, h)	1.21 *	1.01	1.45
Sleep onset (*SD*, h)	1.19 †	0.97	1.46
Sleep midpoint (*SD*, h)	1.27 *	1.01	1.59
Sleep offset (*SD*, h)	1.17 †	1.00	1.38
SRI [a]	0.99	0.98	1.00
Social jetlag (h) [b]	1.07	0.91	1.25

Notes. Each row represents a separate multilevel model that adjusts for mean sleep duration (linear and quadratic, sleep duration × sleep duration) and demographic/household covariates: birth sex, race/ethnicity, household income, and primary caregiver's highest education level. Caffeinated beverage consumption (the outcome) includes coffee or tea, caffeinated soda, and energy drinks and was coded as 0 = none; 1 = at least one 8 oz beverage that day. The between-person effect for sleep variability measures is represented by *SD* or SRI [51] (across all time points) or social jetlag (average midpoint on free nights−average sleep midpoint on school nights [52]) The mean number of valid actigraphy nights per youth included in present analyses was 5.6 ± 1.4 (range 3–10; IQR 5–7) and the mean number of reports of caffeinated beverage consumption was 5.5 ± 1.4 (range 3–9; IQR 4–7). Higher value means greater variability, except the reverse for the SRI. [a] Calculated based on formula from Phillips et al. [51]; ranges from 0 (low)–100 (high). [b] Calculated based on formula from Wittmann et al. [52]. $n = 370$ (adolescent included only if provided one school night and one free night of actigraphy). † $p < 0.10$, * $p < 0.05$, two-tailed. CI, confidence interval; h, hours; OR, odds ratio; *SD*, standard deviation; SRI, sleep regularity index.

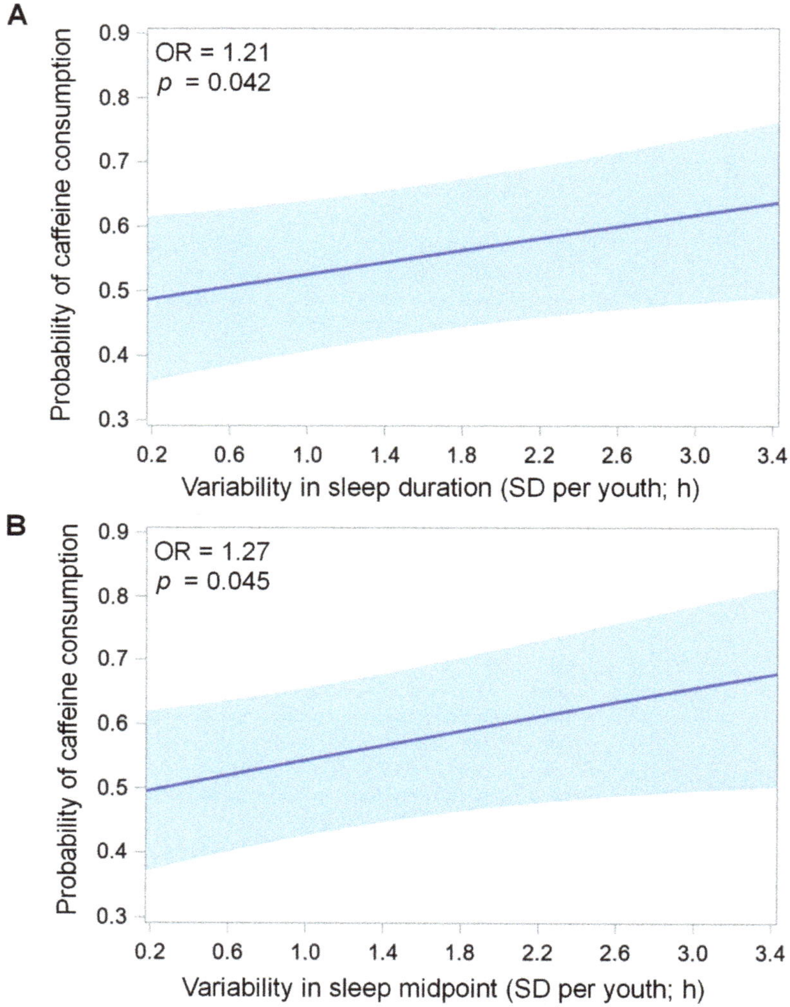

Figure 1. Associations of variability in sleep duration (**A**) and midpoint (**B**) (each in standard deviation, *SD* hours) per youth with average probability of caffeinated beverage consumption (0 = none; 1 = at least one 8 oz beverage that day), which includes coffee or tea, caffeinated soda, and energy drinks, across monitoring days in two separate mixed models. The mean number of valid actigraphy nights per youth included in present analyses was 5.6 ± 1.4 (range 3–10; interquartile range, IQR 5–7) and the mean number of reports of caffeinated beverage consumption was 5.5 ± 1.4 (range 3–9; IQR 4–7). Both models adjust for mean sleep duration (linear and quadratic, sleep duration x sleep duration) and demographic/household covariates: birth sex, race/ethnicity, household income, and primary caregiver's education level. Shaded light blue bands depict 95% confidence interval of mean probability of caffeinated beverage consumption per youth predicted from each sleep measure. H, hours; OR, odds ratio; SD, standard deviation.

3.4. Caffeinated Beverage Consumption Predicting Sleep Measures That Night and on Average (Within- and Between-Person Associations)

There were significant within-person associations between caffeinated beverage consumption and later sleep onset (p = 0.003), midpoint (p = 0.002), and offset (p = 0.011) (see Figure 2 and Table 3). On nights following consumption of ≥1 caffeinated beverage, adolescents' sleep onset and midpoint were delayed by 0.28 h (17 min) and their sleep offset the next morning was delayed by 0.31 h (19 min), compared to nights following no caffeinated beverage consumption. There were no significant within-person associations of sleep duration (p = 0.968), sleep maintenance efficiency (p = 0.810), or subjective sleep quality (p = 0.703) and no significant between-person associations for caffeinated beverage consumption with any of these sleep measures (all $p > 0.20$).

We additionally examined whether the within-person effects of caffeinated beverage consumption on that night's sleep varied depending on an adolescent's average caffeinated beverage consumption. There were no significant interactions (all $p > 0.17$).

Table 3. Caffeinated beverage consumption predicting sleep within and between adolescents (n = 458).

Model Outcome	Within-Person			Between-Person		
	b	95% CI		*b*	95% CI	
Sleep duration (h)	<0.01	−0.22	0.23	0.04	−0.24	0.32
Sleep onset (h)	0.28 **	0.10	0.47	−0.08	−0.46	0.30
Sleep midpoint (h)	0.28 **	0.11	0.46	−0.06	−0.41	0.30
Sleep offset (h)	0.31 *	0.07	0.55	−0.05	−0.41	0.32
Sleep maintenance efficiency (%)	−0.05	−0.45	0.35	−0.13	−0.98	0.72
Subjective sleep quality [a]	0.01	−0.05	0.08	−0.07	−0.20	0.06

Notes. Each row represents a separate multilevel model that adjusts for school attendance and demographic/household covariates: birth sex, race/ethnicity, body mass index (BMI) percentile, household income, primary caregiver's education level, and whether the adolescent was living with two biological parents. Caffeinated beverage consumption (predictor) includes coffee or tea, caffeinated soda, and energy drinks and was coded as 0 = none; 1 = at least one 8 oz beverage that day. Sleep timing measures (onset, midpoint, and offset) were centered around midnight (0:00). The mean number of valid actigraphy nights per youth included in present analyses was 5.2 ± 1.1 (range 3–9; IQR 4–6) and the mean number of reports of caffeinated beverage consumption was 4.7 ± 1.2 (range 3–8; IQR 4–6). [a] Ranges from 0 (very bad)–3 (very good). * $p < 0.05$, ** $p < 0.01$, two-tailed. *b*, unstandardized beta coefficient; CI, confidence interval; h, hours.

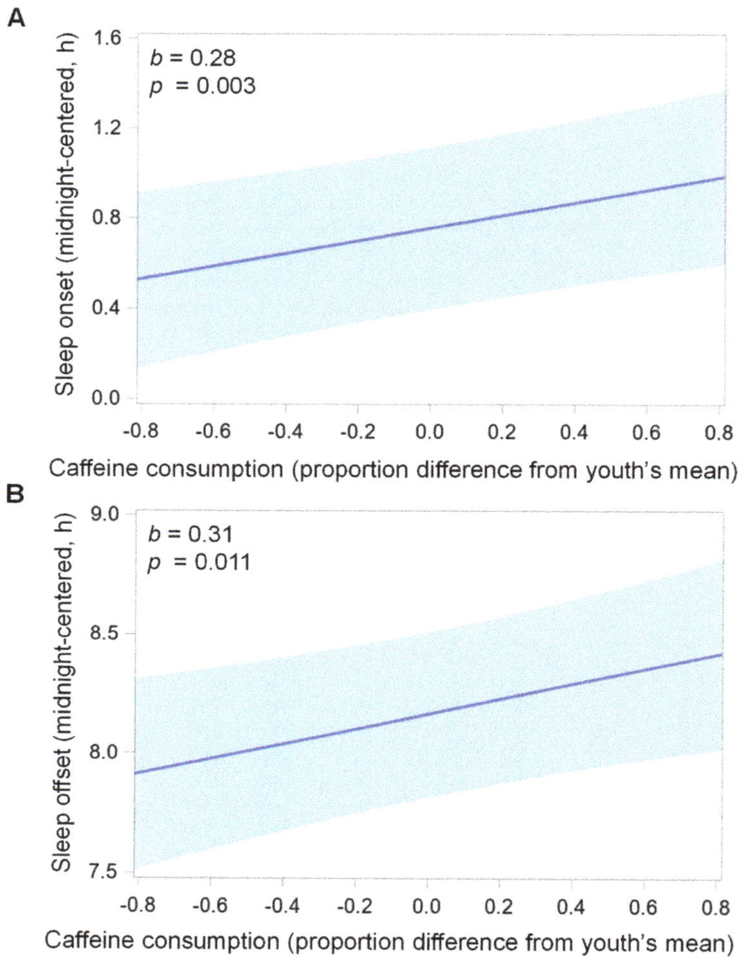

Figure 2. Caffeinated beverage consumption (0 = none; 1 = at least one 8 oz beverage that day), which includes coffee or tea, caffeinated soda, and energy drinks, predicting sleep (**A**) onset and (**B**) offset (each in hours from midnight) that night within each adolescent in two separate mixed models. Positive *x*-axis values indicate the adolescent consumed $\geq$1 caffeinated beverage that day; negative values indicate the adolescent did not consume a caffeinated beverage that day. The mean number of valid actigraphy nights per youth included in present analyses was 5.2 $\pm$ 1.1 (range 3–9; interquartile range, IQR 4–6) and the mean number of reports of caffeinated beverage consumption was 4.7 $\pm$ 1.2 (range 3–8; IQR 4–6). Both models adjust for school day and demographic/household covariates: birth sex, race/ethnicity, body mass index (BMI) percentile, household income, primary caregiver's education level, and whether the adolescent was living with two biological parents. Shaded light blue bands depict 95% confidence interval of caffeinated beverage consumption predicting each sleep measure. *b*, unstandardized beta (in hours); h, hours.

4. Discussion

The current study assessed bidirectional associations of caffeinated beverage consumption with multiple dimensions of sleep measured through actigraphy and subjective sleep quality in adolescents. Variability in sleep duration and midpoint across days was associated with higher average odds of an adolescent consuming ≥1 caffeinated beverage on a given day. Consuming ≥1 caffeinated beverage predicted later sleep timing that night compared to days on which the adolescent did not consume caffeine, but sleep timing did not predict next-day caffeinated beverage consumption. There were no within- or between-person associations of caffeinated beverage consumption with sleep duration or sleep quality (measured either through sleep efficiency or subjective ratings). These findings demonstrate that consuming caffeine may increase the likelihood that an adolescent shifts their sleep later and has more variable sleep duration and timing, potentially predisposing the adolescent to poorer psychological and physical health.

This study is the among the first to find that adolescents in a real-world study who consume caffeinated beverages more often are more likely to have variable sleep duration and timing. Specifically, for every extra *SD*-hour of variability in sleep duration and midpoint, adolescents were 17% and 21% more likely to consume at least one caffeinated beverage on average across monitoring days. We also measured sleep through objective measures, unlike previous studies that measured sleep through self-report [2,6–19,21–23,26–30,32,39]. Variable sleep schedules may contribute to daytime sleepiness due to circadian misalignment [60], which could lead to increased caffeinated beverage consumption to maintain alertness. Alternatively, later sleep timing, specifically on days when adolescents consume caffeine (as found in the current study), could drive greater variability across the week, particularly in those who consume more caffeine. Sleep variability has been associated with poor psychological and physical health, including mood instability [61] and increased odds of metabolic syndrome, greater adiposity, and poorer glycemic control [62]. More studies are needed to probe the link between caffeine consumption and sleep variability, particularly with objective measures of sleep.

We found that on days when adolescents consumed ≥1 caffeinated beverage, they fell asleep about 17 min later that night and woke up about 19 min later the next morning. The current study is one of the first to assess the link between caffeine consumption and objectively measured sleep timing in adolescents. The present findings suggest that adolescents may have more difficulty going to bed and rising early in time for morning activities such as school or extracurriculars after consuming caffeine. Similar to sleep variability, later sleep timing is a risk factor for poorer psychological [63] and metabolic health, including higher adiposity and BMI [64]. One study found that young adults with a 22 min earlier sleep onset and lower sleep variability following a light intervention reported higher positive affect compared to those with more delayed sleep onset [65], suggesting that even a minor advance in sleep timing and reduction of sleep variability may be associated with better psychological health. Furthermore, those with later sleep timing tend to perform poorly in the morning [66], which poses issues for adolescents who are expected to perform early due to school start times.

As both sleep onset and sleep offset were shifted later following caffeinated beverage consumption in the current study, there was no net effect on sleep duration. Previous studies examining the link between caffeine consumption and self-reported sleep duration in adolescents have been mixed [2,6–19,21–23], while two studies that measured sleep through actigraphy as in the current study found no cross-sectional association between caffeine consumption and sleep [21–23]. It is possible that self-reported and objectively measured sleep duration capture different constructs, and that adolescents who consume more caffeine perceive their sleep to be shorter without an actual difference in sleep duration.

We found no within- or between-person associations of caffeinated beverage consumption with either sleep efficiency or subjective sleep quality. These findings align with previous studies that found no association between caffeine consumption and wake after sleep onset [24,25] or self-reported [23] sleep quality. It should be noted that caffeine may have effects on nighttime sleep that are not captured by actigraphy, such as sleep onset latency or sleep architecture. For example, adolescents who consumed 80 mg caffeine 4 h before bedtime experienced unchanged total sleep time and sleep efficiency, yet they experienced increased non-rapid eye movement (REM) stage 1 ("light") sleep and decreased non-REM stage 3 ("deep") sleep [67], and sleep staging is not captured by actigraphy [33]. Such stage shifting may not translate into changes in the perception of sleep quality. Further research into bidirectional associations between caffeine consumption and both subjective and objective measures of sleep quality is warranted in adolescents.

We found that adolescents who consumed caffeinated beverages more often did not differ in average sleep duration, timing, efficiency, or subjective quality from other adolescents; that is, there were no between-person associations. It is possible that individuals develop a tolerance to caffeine consumption, such that caffeine no longer affects the sleep of habitual users. Tolerance to the effects of caffeine on sleep efficiency, for example, may develop after four days of caffeine consumption [68]. In addition, there is considerable interindividual variability in the effects of caffeine on sleep, with half-lives of caffeine consumption ranging from 2 to 10 h among individuals [69]. Adolescents in the current sample who were aware that caffeine did not affect their sleep may have been more likely to consume caffeinated beverages than more sensitive individuals. Tolerance to the effects of caffeine may explain why only the within-person association between caffeinated beverage consumption and sleep timing was significant in the current study, warranting further within-person research.

The current study has some limitations and certain strengths. Caffeinated beverage consumption was measured through daily self-report, which may not represent the adolescent's routine consumption outside of the monitoring period. We also did not measure the timing of caffeinated beverage consumption, which is an important factor to consider given that caffeine consumed close to the nighttime sleep episode may have a stronger impact on that night's sleep than caffeine consumed earlier in the day [70]. However, one study found that the impact of caffeine consumed 0, 3, or 6 h before bedtime on sleep was similar [36]. Furthermore, we did not account for caffeine content, which may vary among beverages such as coffee, tea, and energy drinks [71]. Future research may examine bidirectional associations of sleep and caffeine amount in adolescents. In addition, we cannot establish causal relationships within this observational study. A strength of the study is the large, diverse sample of adolescents across several areas of the United States. Other strengths of the current research are the objective measurement of multiple dimensions of sleep with actigraphy and the assessment of within-person effects, allowing for examination of temporal precedence.

5. Conclusions

The current study demonstrated that adolescents who consumed caffeinated beverages more frequently had greater sleep variability, and sleep was delayed on nights following daytime caffeinated beverage consumption. The findings suggest that reducing caffeinated beverage consumption in adolescents may assist in preventing negative consequences associated with sleep variability and delayed sleep timing, such as poor psychological [61,63] and metabolic [62,64] health.

Supplementary Materials: The following are available online at https://www.mdpi.com/article/10.3390/nu14010031/s1, Figure S1: Participant flow chart for sample included in analyses where sleep predicts caffeinated beverage consumption and sleep variability analyses (n = 589 except for social jetlag, n = 370); Figure S2: Participant flow chart for sample included in analyses where caffeinated beverage consumption predicts sleep (n = 458). Table S1: STROBE-nut: An extension of the STROBE statement for nutritional epidemiology; Table S2: Intraindividual consistency of repeated-measures variables.

Author Contributions: Conceptualization, G.M.M.; methodology, G.M.M., D.A.R. and L.M.; validation, D.A.R.; formal analysis, G.M.M.; resources, O.M.B., A.-M.C. and L.H.; data curation, G.M.M. and L.M.; writing—original draft preparation, G.M.M.; writing—review and editing, G.M.M., D.A.R., L.M., O.M.B., A.-M.C. and L.H.; visualization, G.M.M.; supervision, O.M.B., A.-M.C. and L.H.; project administration, O.M.B., A.-M.C. and L.H.; funding acquisition, L.H. All authors have read and agreed to the published version of the manuscript.

Funding: Research reported in this publication was supported by the Eunice Kennedy Shriver National Institute of Child Health and Human Development (NICHD) of the National Institutes of Health under award numbers R01HD073352 (to LH), R01HD36916, R01HD39135, and R01HD40421, as well as a consortium of private foundations. NICHD had no role in the design, analysis, or writing of this article.

Institutional Review Board Statement: The study was conducted according to the guidelines of the Declaration of Helsinki of 1975 (revised 2013) and approved by the Institutional Review Board of both Princeton University and Stony Brook University (IRB 404049, initial approval 23 April 2013).

Informed Consent Statement: Written (for in-home interviews) or recorded verbal (for phone interviews) informed consent was obtained from primary caregivers, and assent was obtained from adolescents.

Data Availability Statement: Survey data from the Fragile Families and Child Wellbeing study (https://fragilefamilies.princeton.edu/documentation), accessed on 22 March 2017. are publicly available from Princeton University's Office of Population Research (OPR) data archive: https://opr.princeton.edu/archive/restricted/Default.aspx (accessed on 22 March 2017). The sleep actigraphy and daily diary datasets generated and analyzed during the current study (accessed on 11 August 2020) are not publicly available yet but will be available through an application process at the above link.

Acknowledgments: We would like to thank the participants, their families, and the members of the Actigraphy Data Coordinating Center (ADCC) at the Pennsylvania State University for scoring the actigraphy data.

Conflicts of Interest: None of the authors have conflicts of interests related to the material presented. Outside of the current work, O.M.B. received subcontract grants to Pennsylvania State University from Proactive Life (formerly Mobile Sleep Technologies) doing business as SleepSpace (National Science Foundation grant #1622766 and National Institutes of Health/National Institute on Aging Small Business Innovation Research Program R43AG056250, R44 AG056250), honoraria/travel support for lectures from Boston University, Boston College, Tufts School of Dental Medicine, Harvard Chan School of Public Health, New York University, and Allstate, consulting fees from Sleep Number, and an honorarium for his role as the Editor-in-Chief of Sleep Health (sleephealthjournal.org). The funders had no role in the design of the study; in the collection, analyses, or interpretation of data; in the writing of the manuscript, or in the decision to publish the results.

References

1. Heckman, M.A.; Weil, J.; de Mejia, E.G. Caffeine (1, 3, 7-trimethylxanthine) in foods: A comprehensive review on consumption, functionality, safety, and regulatory matters. *J. Food Sci.* **2010**, *75*, R77–R87. [CrossRef]
2. National Sleep Foundation. 2006 Sleep in America Poll—Teens and Sleep. *Sleep Health* **2015**, *1*, e5. [CrossRef]
3. Temple, J.L. Review: Trends, safety, and recommendations for caffeine use in children and adolescents. *J. Am. Acad. Child. Adolesc. Psychiatry* **2019**, *58*, 36–45. [CrossRef]
4. Bernstein, G.A.; Carroll, M.E.; Thuras, P.D.; Cosgrove, K.P.; Roth, M.E. Caffeine dependence in teenagers. *Drug Alcohol Depend.* **2002**, *66*, 1–6. [CrossRef]
5. Martin, C.A.; Cook, C.; Woodring, J.H.; Burkhardt, G.; Guenthner, G.; Omar, H.A.; Kelly, T.H. Caffeine use: Association with nicotine use, aggression, and other psychopathology in psychiatric and pediatric outpatient adolescents. *Sci. World J.* **2008**, *8*, 512–516. [CrossRef]
6. Dimitriou, D.; Le Cornu, K.F.; Milton, P. The role of environmental factors on sleep patterns and school performance in adolescents. *Front. Psychol.* **2015**, *6*, 1–9. [CrossRef] [PubMed]
7. Sampasa-Kanyinga, H.; Hamilton, H.A.; Chaput, J.P. Sleep duration and consumption of sugar-sweetened beverages and energy drinks among adolescents. *Nutrition* **2018**, *48*, 77–81. [CrossRef] [PubMed]
8. Troxel, W.M.; Tucker, J.S.; Ewing, B.; Miles, J.N.V.; D'Amico, E.J. Sleepy teens and energy drink use: Results from an ethnically diverse sample of youth. *Behav. Sleep Med.* **2018**, *16*, 223–234. [CrossRef]
9. Veselska, Z.D.; Husarova, D.; Kosticova, M. Energy drinks consumption associated with emotional and behavioural problems via lack of sleep and skipped breakfast among adolescents. *Int. J. Environ. Res. Public Health* **2021**, *18*, 6055. [CrossRef]
10. Yen, C.F.; Ko, C.H.; Yen, J.Y.; Cheng, C.P. The multidimensional correlates associated with short nocturnal sleep duration and subjective insomnia among Taiwanese adolescents. *Sleep* **2008**, *31*, 1515–1525. [CrossRef] [PubMed]
11. Yilmaz, K.; Kilinçaslan, A.; Aydin, N.; Kul, S. Understanding sleep habits and associated factors can help to improve sleep in high school adolescents. *Turk. J. Pediatr.* **2011**, *53*, 430–436.
12. Chung, K.F.; Cheung, M.M. Sleep-wake patterns and sleep disturbance among Hong Kong Chinese adolescents. *Sleep* **2008**, *31*, 185–194. [CrossRef] [PubMed]

13. Drescher, A.A.; Goodwin, J.L.; Silva, G.E.; Quan, S.F. Caffeine and screen time in adolescence: Associations with short sleep and obesity. *J. Clin. Sleep Med.* **2011**, *7*, 337–342. [CrossRef] [PubMed]
14. Halldorsson, T.I.; Kristjansson, A.L.; Thorisdottir, I.; Oddsdóttir, C.; Sveinbjörnsson, J.; Benediktsson, R.; Sigfúsdóttir, I.D.; Jörundsdóttir, H.; Gunnlaugsdóttir, H. Caffeine exposure from beverages and its association with self-reported sleep duration and quality in a large sample of Icelandic adolescents. *Food Chem. Toxicol.* **2021**, *157*, 112549. [CrossRef]
15. Kim, S.Y.; Sim, S.; Choi, H.G. High stress, lack of sleep, low school performance, and suicide attempts are associated with high energy drink intake in adolescents. *PLoS ONE* **2017**, *12*, 1–12. [CrossRef] [PubMed]
16. Lodato, F.; Araújo, J.; Barros, H.; Lopes, C.; Agodi, A.; Barchitta, M.; Ramos, E. Caffeine intake reduces sleep duration in adolescents. *Nutr. Res.* **2013**, *33*, 726–732. [CrossRef] [PubMed]
17. Nuss, T.; Morley, B.; Scully, M.; Wakefield, M. Energy drink consumption among Australian adolescents associated with a cluster of unhealthy dietary behaviours and short sleep duration. *Nutr. J.* **2021**, *20*, 1–10. [CrossRef]
18. Patte, K.A.; Qian, W.; Leatherdale, S.T. Modifiable predictors of insufficient sleep durations: A longitudinal analysis of youth in the COMPASS study. *Prev. Med.* **2018**, *106*, 164–170. [CrossRef]
19. Pollak, C.P.; Bright, D. Caffeine consumption and weekly sleep patterns in US seventh-, eighth-, and ninth-graders. *Pediatrics* **2003**, *111*, 42–46. [CrossRef]
20. Wheaton, A.G.; Jones, S.E.; Cooper, A.C.; Croft, J.B. Short sleep duration among middle school and high school students—United States, 2015. *Morb. Mortal. Wkly. Rep.* **2018**, *67*, 85–90. [CrossRef]
21. Calamaro, C.J.; Mason, T.B.A.; Ratcliffe, S.J. Adolescents living the 24/7 lifestyle: Effects of caffeine and technology on sleep duration and daytime functioning. *Pediatrics* **2009**, *123*, 4–11. [CrossRef]
22. Merdad, R.A.; Merdad, L.A.; Nassif, R.A.; El-Derwi, D.; Wali, S.O. Sleep habits in adolescents of Saudi Arabia; distinct patterns and extreme sleep schedules. *Sleep Med.* **2014**, *15*, 1370–1378. [CrossRef]
23. Reid, A.; Maldonado, C.; Baker, F. Sleep behavior of South African adolescents. *Sleep* **2002**, *25*, 423–427. [CrossRef] [PubMed]
24. Aepli, A.; Kurth, S.; Tesler, N.; Jenni, O.; Huber, R. Caffeine consuming children and adolescents show altered sleep behavior and deep sleep. *Brain Sci.* **2015**, *5*, 441–455. [CrossRef]
25. Cusick, C.N.; Langberg, J.M.; Breaux, R.; Green, C.D.; Becker, S.P. Caffeine use and associations with sleep in adolescents with and without adhd. *J. Pediatr. Psychol.* **2020**, *45*, 643–653. [CrossRef]
26. Yim, S.H.; Yang, K.I.; Kim, J.H.; Hwangbo, Y.; Kim, D.; Hong, S.B. Association between eveningness preference, socio-behavioral factors, and insomnia symptoms in Korean adolescents. *Sleep Med.* **2021**, *82*, 144–150. [CrossRef] [PubMed]
27. Fleig, D.; Randler, C. Association between chronotype and diet in adolescents based on food logs. *Eat. Behav.* **2009**, *10*, 115–118. [CrossRef]
28. Gariépy, G.; Doré, I.; Whitehead, R.D.; Elgar, F.J. More than just sleeping in: A late timing of sleep is associated with health problems and unhealthy behaviours in adolescents. *Sleep Med.* **2019**, *56*, 66–72. [CrossRef]
29. Giannotti, F.; Cortesi, F.; Sebastiani, T.; Ottaviano, S. Circadian preference, sleep and daytime behaviour in adolescence. *J. Sleep Res.* **2002**, *11*, 191–199. [CrossRef] [PubMed]
30. Saxvig, I.W.; Evanger, L.N.; Pallesen, S.; Hysing, M.; Sivertsen, B.; Gradisar, M.; Bjorvatn, B. Circadian typology and implications for adolescent sleep health. Results from a large, cross-sectional, school-based study. *Sleep Med.* **2021**, *83*, 63–70. [CrossRef]
31. Roenneberg, T.; Kuehnle, T.; Pramstaller, P.P.; Ricken, J.; Havel, M.; Guth, A.; Merrow, M. A marker for the end of adolescence. *Curr. Biol.* **2004**, *14*, R1038–R1039. [CrossRef]
32. Shimura, A.; Hideo, S.; Takaesu, Y.; Nomura, R.; Komada, Y.; Inoue, T. Comprehensive assessment of the impact of life habits on sleep disturbance, chronotype, and daytime sleepiness among high-school students. *Sleep Med.* **2018**, *44*, 12–18. [CrossRef]
33. Marino, M.; Li, Y.; Rueschman, M.N.; Winkelman, J.W.; Ellenbogen, J.M.; Solet, J.M.; Dulin, H.; Berkman, L.F.; Buxton, O.M. Measuring sleep: Accuracy, sensitivity, and specificity of wrist actigraphy compared to polysomnography. *Sleep* **2013**, *36*, 1747–1755. [CrossRef]
34. Buysse, D.J. Sleep health: Can we define It? Does it matter? *Sleep* **2014**, *37*, 9–17. [CrossRef] [PubMed]
35. O'Dea, J.A. Consumption of nutritional supplements among adolescents: Usage and perceived benefits. *Health Educ. Res.* **2003**, *18*, 98–107. [CrossRef] [PubMed]
36. Drake, C.; Roehrs, T.; Shambroom, J.; Roth, T. Caffeine effects on sleep taken 0, 3, or 6 hours before going to bed. *J. Clin. Sleep Med.* **2013**, *9*, 1195–1200. [CrossRef] [PubMed]
37. Hoffman, L.; Stawski, R.S. Persons as contexts: Evaluating between-person and within-person effects in longitudinal analysis. *Res. Hum. Dev.* **2009**, *6*, 97–120. [CrossRef]
38. Curran, P.J.; Bauer, D.J. Effects in longitudinal models of change. *Annu. Rev. Psychol.* **2011**, *62*, 583–619. [CrossRef] [PubMed]
39. Whalen, D.J.; Silk, J.S.; Semel, M.; Forbes, E.E.; Ryan, N.D.; Axelson, D.A.; Birmaher, B.; Dahl, R.E. Caffeine consumption, sleep, and affect in the natural environments of depressed youth and healthy controls. *J. Pediatr. Psychol.* **2008**, *33*, 358–367. [CrossRef]
40. Lunsford-Avery, J.R.; Kollins, S.H.; Kansagra, S.; Wang, K.W.; Engelhard, M.M. Impact of daily caffeine intake and timing on electroencephalogram-measured sleep in adolescents. *J. Clin. Sleep Med.* **2021**, jcsm-9736. [CrossRef]
41. Short, M.A.; Gradisar, M.; Lack, L.C.; Wright, H.; Carskadon, M.A. The discrepancy between actigraphic and sleep diary measures of sleep in adolescents. *Sleep Med.* **2012**, *13*, 378–384. [CrossRef]

42. Crowley, S.J.; van Reen, E.; LeBourgeois, M.K.; Acebo, C.; Tarokh, L.; Seifer, R.; Barker, D.H.; Carskadon, M.A. A longitudinal assessment of sleep timing, circadian phase, and phase angle of entrainment across human adolescence. *PLoS ONE* **2014**, *9*, e112199. [CrossRef]
43. Ziegler, A.M.; Kasprzak, C.M.; Mansouri, T.H.; Gregory, A.M.; Barich, R.A.; Hatzinger, L.A.; Leone, L.A.; Temple, J.L. An ecological perspective of food choice and eating autonomy among adolescents. *Front. Psychol.* **2021**, *12*, 654139. [CrossRef] [PubMed]
44. Reichman, N.E.; Teitler, J. Fragile Families: Sample and design. *Child. Youth Serv. Rev.* **2001**, *23*, 303–326. [CrossRef]
45. Wagmiller, R.L.J. *How Representative are the Fragile Study Families? A Comparison of the Early Childhood Longitudinal Study-Birth Cohort and Fragile Families Samples*; Working Papers 1216; Princeton University: Princeton, NJ, USA, 2010; Volume 716, pp. 1–28.
46. Lachat, C.; Hawwash, D.; Ocké, M.C.; Berg, C.; Forsum, E.; Hörnell, A.; Larsson, C.L.; Sonestedt, E.; Wirfält, E.; Åkesson, A.; et al. Strengthening the reporting of observational studies in epidemiology-nutritional epidemiology (STROBE-nut): An extension of the STROBE statement. *PLoS Med.* **2016**, *13*, e1002036. [CrossRef] [PubMed]
47. Lauderdale, D.S.; Knutson, K.L.; Yan, L.L.; Rathouz, P.J.; Hulley, S.B.; Sidney, S.; Liu, K. Objectively measured sleep characteristics among early-middle-aged adults: The CARDIA study. *Am. J. Epidemiol.* **2006**, *164*, 5–16. [CrossRef]
48. Master, L.; Nye, R.T.; Lee, S.; Nahmod, N.G.; Mariani, S.; Hale, L.; Buxton, O.M. Bidirectional, daily temporal associations between sleep and physical activity in adolescents. *Sci. Rep.* **2019**, *9*, 1–14.
49. Buxton, O.M.; Lee, S.; Beverly, C.; Berkman, L.F.; Moen, P.; Kelly, E.L.; Hammer, L.B.; Almeida, D.M. Work-family conflict and employee sleep: Evidence from IT workers in the Work, Family and Health Study. *Sleep* **2016**, *39*, 1911–1918. [CrossRef]
50. Chen, X.; Wang, R.; Zee, P.; Lutsey, P.L.; Javaheri, S.; Alcántara, C.; Jackson, C.L.; Williams, M.A.; Redline, S. Racial/ethnic differences in sleep disturbances: The Multi-Ethnic Study of Atherosclerosis (MESA). *Sleep* **2015**, *38*, 877–888. [CrossRef]
51. Phillips, A.J.K.; Clerx, W.M.; O'Brien, C.S.; Sano, A.; Barger, L.K.; Picard, R.W.; Lockley, S.W.; Klerman, E.B.; Czeisler, C.A. Irregular sleep/wake patterns are associated with poorer academic performance and delayed circadian and sleep/wake timing. *Sci. Rep.* **2017**, *7*, 1–13. [CrossRef]
52. Wittmann, M.; Dinich, J.; Merrow, M.; Roenneberg, T. Social jetlag: Misalignment of biological and social time. *Chronobiol. Int.* **2006**, *23*, 497–509. [CrossRef]
53. Kuczmarski, R.; Ogden, C.; Guo, S.; Grummer-Strawn, L.; Flegal, K.; Mei, Z.; Wei, R.; Curtin, L.R.; Roche, A.F.; Johnson, C.L. 2000 CDC growth charts for the United States: Methods and development. *Vital Health Stat.* **2002**, *11*, 1–190.
54. Keys, A.; Fidanza, F.; Karvonen, M.J.; Kimura, N.; Taylor, H.L. Indices of relative weight and obesity. *J. Chronic Dis.* **1972**, *25*, 329–343. [CrossRef]
55. Kline, R.B. *Principles and Practice of Structural Equation Modeling, 4th ed*; Guilford Publications: New York, NY, USA, 2016.
56. Koo, T.K.; Li, M.Y. A guideline of selecting and reporting intraclass correlation coefficients for reliability research. *J. Chiropr. Med.* **2016**, *15*, 155–163. [CrossRef]
57. Horton, N.J.; Lipsitz, S.R. Review of software to fit generalized estimating equation regression models. *Am. Stat.* **1999**, *53*, 160–169.
58. Liang, K.Y.; Zeger, S.L. Longitudinal data analysis using generalized linear models. *Biometrika* **1986**, *73*, 13–22. [CrossRef]
59. Howard, A.L. Leveraging time-varying covariates to test within- and between-person effects and interactions in the multilevel linear model. *Emerg. Adulthood* **2015**, *3*, 400–412. [CrossRef]
60. Van Dyk, T.R.; Zhang, N.; Combs, A.; Howarth, T.; Whitacre, C.; McAlister, S.; Beebe, D.W. Feasibility and impact on daytime sleepiness of an experimental protocol inducing variable sleep duration in adolescents. *PLoS ONE* **2019**, *14*, e0218894. [CrossRef]
61. Fuligni, A.J.; Arruda, E.H.; Krull, J.L.; Gonzales, N.A. Adolescent sleep duration, variability, and peak levels of achievement and mental health. *Child. Dev.* **2018**, *89*, e18–e28. [CrossRef] [PubMed]
62. Zuraikat, F.M.; Makarem, N.; Redline, S.; Aggarwal, B.; Jelic, S.; St-Onge, M.-P. Sleep regularity and cardiometabolic heath: Is variability in sleep patterns a risk factor for excess adiposity and glycemic dysregulation? *Curr. Diab. Rep.* **2020**, *20*, 38. [CrossRef]
63. Asarnow, L.D.; McGlinchey, E.; Harvey, A.G. The effects of bedtime and sleep duration on academic and emotional outcomes in a nationally representative sample of adolescents. *J. Adolesc. Health* **2014**, *54*, 350–356. [CrossRef]
64. Collings, P.J. Independent associations of sleep timing, duration and quality with adiposity and weight status in a national sample of adolescents: The UK Millennium Cohort Study. *J. Sleep Res.* **2021**, *22*, e13436. [CrossRef]
65. Nagare, R.; Woo, M.; Macnaughton, P.; Plitnick, B.; Tinianov, B.; Figueiro, M. Article access to daylight at home improves circadian alignment, sleep, and mental health in healthy adults: A crossover study. *Int. J. Environ. Res. Public Health* **2021**, *18*, 9980. [CrossRef]
66. Lara, T.; Madrid, J.A.; Correa, Á. The vigilance decrement in executive function is attenuated when individual chronotypes perform at their optimal time of day. *PLoS ONE* **2014**, *9*, e88820. [CrossRef] [PubMed]
67. Reichert, C.F.; Veitz, S.; Bühler, M.; Gruber, G.; Deuring, G.; Rehm, S.S.; Rentsch, K.; Garbazza, C.; Meyer, M.; Slawik, H.; et al. Wide awake at bedtime? Effects of caffeine on sleep and circadian timing in male adolescents—A randomized crossover trial. *Biochem. Pharmacol.* **2021**, *191*, 114283. [CrossRef] [PubMed]
68. Zwyghuizen-Doorenbos, A.; Roehrs, T.A.; Lipschutz, L.; Timms, V.; Roth, T. Effects of caffeine on alertness. *Psychopharmacology* **1990**, *100*, 36–39. [CrossRef] [PubMed]
69. Blanchard, J.; Sawers, S.J.A. The absolute bioavailability of caffeine in man. *Eur. J. Clin. Pharmacol.* **1983**, *24*, 93–98. [CrossRef]

70. Clark, I.; Landolt, H.P. Coffee, caffeine, and sleep: A systematic review of epidemiological studies and randomized controlled trials. *Sleep Med. Rev.* **2017**, *31*, 70–78. [CrossRef]
71. U.S. Department of Health and Human Services and U.S. Department of Agriculture. *2015–2020 Dietary Guidelines for Americans*, 8th ed.; U.S. Department of Health and Human Services and U.S. Department of Agriculture: Washington, DC, USA, 2015; pp. 1–125.

nutrients

MDPI

Review

Sleep Quality: A Narrative Review on Nutrition, Stimulants, and Physical Activity as Important Factors

Monika Sejbuk *, Iwona Mirończuk-Chodakowska and Anna Maria Witkowska

Department of Food Biotechnology, Medical University of Bialystok, Szpitalna 37, 15-295 Bialystok, Poland; iwona.mironczuk-chodakowska@umb.edu.pl (I.M.-C.); anna.witkowska@umb.edu.pl (A.M.W.)

* Correspondence: sejbuk.monika99@gmail.com; Tel.: +48-85-686-5088; Fax: +48-85-686-5089

Abstract: Sleep is a cyclically occurring, transient, and functional state that is controlled primarily by neurobiological processes. Sleep disorders and insomnia are increasingly being diagnosed at all ages. These are risk factors for depression, mental disorders, coronary heart disease, metabolic syndrome, and/or high blood pressure. A number of factors can negatively affect sleep quality, including the use of stimulants, stress, anxiety, and the use of electronic devices before sleep. A growing body of evidence suggests that nutrition, physical activity, and sleep hygiene can significantly affect the quality of sleep. The aim of this review was to discuss the factors that can affect sleep quality, such as nutrition, stimulants, and physical activity.

Keywords: sleep; nutrition; physical activity; stimulants

Citation: Sejbuk, M.; Mirończuk-Chodakowska, I.; Witkowska, A.M. Sleep Quality: A Narrative Review on Nutrition, Stimulants, and Physical Activity as Important Factors. *Nutrients* **2022**, *14*, 1912. https://doi.org/10.3390/nu14091912

Academic Editors: Egeria Scoditti and Sergio Garbarino

Received: 24 March 2022
Accepted: 29 April 2022
Published: 2 May 2022

1. Introduction

Sleep is a natural and reversible condition that is controlled primarily by neurobiological processes, and it is a physiological part of human life that is necessary to the maintenance of health and wellbeing [1,2]. Sleep is associated with a reduction in the perception of external stimuli and the cessation of motor activity [1]. The quality of sleep is influenced by many factors, such as diet [3], physical activity [4], and genetic [5] and environmental factors [6]. Sleep has a multifactorial effect on the body: it reduces energy consumption and increases the recovery of the energy storage in the brain, it regulates the adaptive and innate immune response, and it contributes to memory consolidation (the fixing of acquired information in the brain) [2,7,8]. Sleep disorders are associated with the onset and progression of many different diseases, which include cardiovascular disease, depression, and cancer [9–11]. Sleep disorders also increase the risk of infectious diseases [12–15].

In modern times, with a significant increase in the occurrence of both noncommunicable diseases and sleep disorders, our understanding of the factors that are involved in improving the quality of sleep is of great importance. The purpose of this narrative review is to discuss the factors that can affect sleep quality, such as nutrition, stimulants, and physical activity.

1.1. Sleep Phases and Duration

Sleep continuity is assessed by the total time of sleep, the delay in falling asleep (i.e., the time between switching off the lights and falling asleep), and by the type and amount of sleep throughout the duration of sleep [14,16]. Physiological sleep consists of two main phases—the REM (rapid eye movements) phase and the NREM (non-REM) phase—which are repeated during sleep. The REM phase is associated with the activation of the sympathetic nervous system, and it leads to an increase in temperature and blood pressure and to an accelerated heart rate [17]. During the REM sleep phase, there is also a decrease in muscle tone [17,18], and activation in the limbic regions, which suggests that REM plays a role in emotional regulation [19]. The NREM phases are longer and are associated with the function of the parasympathetic nervous system and, in contrast to the REM phase, with decreases in body temperature, blood pressure, and pulse. The NREM sleep phase also supports memory consolidation, metabolic regulation, and brain regeneration [17,19].

Sleep disorders that are associated with an insufficient or missing NREM phase are an increasing public health problem that affects the overall functioning of the body [17]. Adults generally spend about 20–25% of their total sleep time in the REM phase, 75–80% in the NREM phase, and they have between four and five NREM cycles [17].

There are large inter- and intraindividual differences in the duration of sleep between people. A study in monozygotic and bizygotic twins indicates the inheritance of sleep duration [19]. The duration of sleep is influenced between 31 and 55%, which shows a significant influence of genetics on the duration of sleep [19]. This study also shows that, not only can the duration be inherited, but also insomnia, habitual sleep time, midday sleep, and the subjective quality of sleep between identical and fraternal twins [19].

In addition to genetic factors, environmental factors, such as the duration and type of work, the distance between home and work, commuting, professional and family responsibilities, and social relationships, also influence sleep needs. In order to enable the body to sleep "healthily", it is necessary to have adequate sleep duration, regularity, and quality, and the absence of sleep disturbances [20].

The length of sleep of healthy people decreases with aging: a newborn needs 14–17 h of sleep per day, while adults sleep 7–9 h, and older people sleep for 7–8 h [21]. Less than 7 h of sleep is associated with poorer wellbeing and poorer health. In addition, people who sleep less have a higher risk of illness (e.g., depression, mental disorders, coronary heart disease, metabolic syndrome, high blood pressure) when compared with people who sleep a sufficient number of hours (7–8 h) per day [10,12,20].

1.2. Insomnia and Its Risk Factors

Insomnia is a clinical condition that is characterized by difficulty in maintaining sleep or falling asleep, and tiredness and irritability during the day [22]. It cannot be determined only by the number of hours of sleep per day. According to the current state of the knowledge, insomnia disorders are found in about 10–20% of the adult population. They are influenced by factors such as a longer duration of sleep, being awake after sleep, respiratory disturbances during sleep, a shortened sleep duration, and sleep fragmentation [22]. Insomnia leads to arousal in the waking state, sleep with an increased metabolic rate, and increased levels of the adrenocorticotropic hormone and cortisol during the early sleep phase [23].

The most common symptoms that are associated with insomnia are sleep disturbance (which occurs in about 50–70% of patients), problems falling asleep, and insufficient regeneration during sleep [23]. Sleep disorders can be classified as secondary (e.g., medicines) or primary (e.g., mental disorders) [23]. Sleep disorders can also manifest as getting up early in the morning, regardless of when one goes to bed, which results in reduced productivity and concentration, irritability, a risk of mistakes and accidents, and lower quality of life [24].

Insomnia in children leads to poor concentration, which can impair learning performance. This may depend on certain situations and behaviors, such as weighing the baby at bedtime or bottle feeding. For example, if there is no stimulus before bedtime, the child may have difficulty falling asleep [24].

Insomnia is a factor that increases the risk of various diseases, such as asthma, gastroesophageal reflux, hypertension, cardiovascular diseases, and type 2 diabetes [24]. The malposition of the hypothalamic pituitary axis in chronic insomnia influences the fluctuations of the thyroid hormones by increasing the concentrations of cortisol, corticotropin-releasing hormone, and thyrotropin [24].

Careful diagnosis is required to detect or rule out insomnia, as there are many sleep-related disorders that patients may mistake for insomnia, such as sleep apnea, restless legs syndrome, or nocturnal cramps [24]. Restless legs syndrome (RLS), which is also known as

Willis Ekbom disease, is characterized by uncontrolled movement that is due to discomfort and pain in the legs. The symptoms decrease with movement [25]. Often, other diseases (such as obstructive sleep apnea and nocturnal cramps) are accompanied by insomnia, which is why a thorough diagnosis is so important. The aim of the treatment of insomnia is to improve the quality and quantity of sleep, which thereby improves the wellbeing and quality of life of the patient [23].

In the diagnosis of insomnia, the effectiveness of sleep plays a decisive role (i.e., the ratio between the total amount of sleep and the amount of time spent in bed). Spending too long in bed and in trying to fall asleep is one of the main problems of people who struggle with insomnia. Sleep disorders influence the development of anxiety during and before falling asleep, which also affects the development of insomnia [26]. Cognitive behavioral therapy can be helpful, which aims to treat insomnia by changing thoughts, attitudes, and beliefs about sleep. This therapy aims to reduce negative sleep thoughts, to improve sleep hygiene, and to reduce the hours spent in bed, which result in an improved sleep performance [26].

Every year, more and more people complain about a deterioration in their quality of sleep, and about insomnia, constant sleep problems, waking up at night, and prolonged dreams. The deterioration in the quality of sleep and the incidence of insomnia occur in the general population, but they more often affect women and older people (65 years and over) [12]. There are many factors that influence the development and occurrence of insomnia, such as caffeine abuse, stress at work, the loss of a loved one, divorce, domestic violence, and shiftwork. People who have perfectionism, neuroses, a suppressed personality, or an increased susceptibility to anxiety are more susceptible to sleep disorders [25].

Sleep disorders may also occur in children. They can be influenced, for example, by delayed milestones (in child development, there are certain skills that a child learns at certain points in time), separation anxiety, and hyperactivity [25]. Children's sleep problems can also be caused by a lack of certain items (such as favorite cuddly toys), stimulation (telling stories before bedtime, reading books, or swinging), and the parents' absence from the room. These factors can lead to increased anxiety in children, which may cause insomnia [25].

The sleep–wake cycle may be affected by stimulants (such as alcohol, caffeine, and tobacco) and the use of electronic devices. The use of electronics in the bedroom reduces the sleep time and leads to permanent exposure to external stimuli (the sound of a ringing telephone) and a reduction in melatonin release (bright screen light) [27].

1.3. Functions of Sleep

Two effector systems are responsible for regulating the immune response (inborn and adaptive): the sympathetic nervous system and the hypothalamic–pituitary–adrenal axis, which are both influenced by sleep. When you sleep too little, the immune system produces a reduced number of antibodies, which are involved in the body's defensive reactions [28,29].

During sleep, there is a decrease in the release of cortisol, norepinephrine, and adrenaline. The concentration of hormones that affect cell growth, such as growth hormones, melatonin, and prolactin, increases. Prolactin and growth hormone influence the differentiation and formation of new T cells and stimulate the function of type 1 cytokines that control the antigenic response of lymphocytes [29].

Sleep reduces energy consumption (the basic metabolic rate decreases) because, among other things, it reduces the body temperature. The glucose that is consumed by the brain is also reduced: in slow sleep, twice as much glucose is consumed as in the waking state (brain cells have a lower glucose demand). This decrease is not due to an excessively low blood-sugar level, as it is at the same level as when awake [2]. During the REM phase, the metabolic rate increases, which results in increased glucose consumption compared to during the NREM sleep phase [2].

The glymphatic system is a macroscopic system that uses perivascular canal systems to remove certain substances from the central nervous system [30]. The function of the glymphatic system is to remove toxins from the brain that are produced during cellular respiration. During sleep, there is an increased mass flow in which toxins are excreted from the brain [2]. The glymphatic system also contributes to the distribution of glucose, amino acids, lipids, and certain neurotransmitters [29].

As the body becomes older, or if it does not procure enough sleep, the toxin removal may be reduced, which leads to the formation of amyloid plaques, which can be seen in many neurodegenerative diseases, such as Alzheimer's disease [31].

Sleep and insomnia influence the different connections of the brain. During sleep, there is a spontaneous fusion of the glia and the neurons by the synapses, which leads to the formation of cell networks. The properties of the network are altered by synapses and signal molecules. During sleep, old superfluous memories are erased, new ones are strengthened, and the neuromuscular cycles are strengthened [2].

High blood pressure is a disease of civilization that is the main risk factor for the development of other cardiovascular diseases. High blood pressure is affected by the length of sleep. Studies in humans show that sleep deprivation ($\leq$5 h/day) and insomnia increase the risk of high blood pressure by a factor of five. The risk of high blood pressure is also higher in people who wake up early in the morning (e.g., going to bed late) and who have difficulty maintaining sleep [10,14].

There is an obvious link between insufficient sleep, cardiovascular diseases, and the development of inflammation in the body. The greatest risk of developing cardiovascular disease is in people who sleep for less than 5 h per day. People who sleep for less than 7 h are also at an increased risk of cardiovascular diseases and mortality that are caused by a disturbance of the functioning of this system. The duration of sleep should, thus, be at least 7 h [13,14].

Insufficient sleep leads to increased concentrations of inflammatory markers, such as C-reactive protein (CRP), while inflammation is thought to be related to the incidence of breast cancer and lung tumors [14]. People who work long hours on night shifts have an increased risk of cancers, such as breast cancer, colon cancer, and non-Hodgkin's lymphoma [32].

Sleep disorders and insomnia may occur as one of the symptoms during depression. Studies show a doubly increased risk of depression in people with sleep disorders [33]. Increased markers of inflammation, which can be caused by insufficient sleep time, are often high in people who suffer from depression. Experimental studies in which inflammation has been activated in the body have shown that the symptoms of depression are accompanied by the increased activation of the brain, and especially the areas that are responsible for regulating the negative and positive effects [9,11,14,33].

2. Search Strategy

In order to present the scientific reports of the last decade (2012–March 2022), the electronic databases Medline and the Web of Science were investigated. Publications on sleep quality in relation to nutrition, stimulants, and physical activity in observational studies, experimental studies, and meta-analyses were collected. This search included human studies and animal experiments. In some cases, studies that are important for this review but that are outside of the search period are described. The search terms were multiple, with "sleep" as the main search term, in combination with "insomnia", "nutrition", "protein", "carbohydrate", "fat", "vitamins", "vitamin D", "minerals", "tryptophan", "melatonin", "gamma amino butyric acid", "GABA", "caffeine", "nicotine", "alcohol", "marijuana", "cannabis", and "physical activity" as the most critical for this review. The full-text articles published in English that were relevant for this review were selected. To qualify the publications for further evaluation, the titles and abstracts were initially screened according to the search criteria. Studies that did not meet the search criteria were excluded.

3. Review

3.1. Diet and Sleep Quality

Proper nutrition involves providing all of the necessary nutrients in order to maintain health and wellbeing. The foods that people consume can not only influence their wakefulness during the day, but also their quality of sleep. Sleep is not only influenced by the energy efficiency of the diet, but also by the content of macronutrients, such as proteins, carbohydrates, and fats [34]. Insufficient protein intake may impair sleep quality, while too much protein intake may lead to difficulties in maintaining sleep [34,35].

An important relationship has been identified between the quality of carbohydrates ingested (fiber content of the products and the degree of processing of the food) and the quality of sleep [35]. The glycemic index and the frequency and time of meals are influenced not only by the carbohydrate intake, but also by the quality of the carbohydrates that are consumed [35]. The high consumption of noodles, sweets, and sugary drinks, as well as the omission of breakfast and irregular meals, are associated with poor sleep, while a diet that is rich in fish, seafood, and vegetables contributes to good sleep. The inadequate intake of macronutrients, excessive calorie intake, and late meals contribute to a reduction in sleep quality and may influence the development of insomnia [35].

Eating foods that are rich in tryptophan, melatonin, and serotonin improves sleep quality. In adults, after consuming foods rich in tryptophan, a longer downtime, increased performance, and an increased total sleep time have been observed [36]. Vitamins and minerals (e.g., B vitamins, zinc) influence sleep quality, and when a deficiency was compensated for, an improvement in the sleep rate and the overall sleep quality was observed [37].

Increasingly, a tendency to sleep for less than 6 h per night is being observed in the general population, which is influencing the increased consumption of coffee, which contains caffeine. The half-life of caffeine averages 2 to 10 h, but it can be up to 20 h. Caffeine obviously increases performance, but it also has side effects: it affects the quality of sleep. People who consume large amounts of caffeine are more likely to be drowsy in the morning than those who consume moderate amounts [37,38].

Adults often drink alcohol, and some of them believe that alcohol even helps them to fall asleep. However, alcohol has a negative effect on sleep, and it impairs the electrophysiological structure of sleep, affects biorhythms, and increases insomnia. In studies that used moderate doses of alcohol (<1 g/kg body weight), there was a shortening of the REM sleep phase, which was mainly in the second half of sleep [39].

Nutrition also has a significant influence on the wellbeing of sleep. However, the nutritional mechanisms that influence sleep regulation are complex [35]. Individual ingredients in the diet can directly influence sleep, such as caffeine, which prolongs the duration of sleep induction but reduces the overall duration and quality of sleep [40]. Many food metabolites may be important in the regulation of sleep through the regulation of other related factors. Foods may also influence the commensal microbiota, which may lead to the formation of metabolites [41]. Inadequate nutrition in the long term may contribute to inflammation, which is closely related to insomnia [42]. Adequate nutrition that is rich in fruits, vegetables, and whole grains has a positive effect on sleep [43].

3.1.1. Sleep and Energy Intake

Overweight and obesity are a growing problem in developed and developing countries. Overweight and obesity influence many concomitant diseases, such as type II diabetes, cancer, and cardiovascular diseases. Inadequate sleep habits and poor sleep hygiene may correlate with overweight and obesity [44]. The available data show that up to 95% of pupils in upper secondary education do not meet the requirements for adequate sleep time [45]. In the last hundred years, sleep duration has decreased by one hour in all age groups [46]. The available data also show a decrease in the sleep duration in recent years of 10–15 min, and an increase in the number of people who sleep less than 6 h [47].

Short sleep has been shown to increase the risk of developing obesity. Insufficient sleep leads to increased food intake, which leads to an excessive calorie diet. Studies have shown a link between insufficient sleep duration and biological changes in hunger [44].

Insufficient sleep is associated with hormonal changes in the body, including the release of leptin, ghrelin, cortisol, and growth hormone. Hormonal changes may result in reduced tissue insulin sensitivity. These changes have an impact on inappropriate food selection, changes in energy regulation, excessive food intake, and reduced physical activity [48,49].

People who sleep less have a shorter REM period, which probably plays a role in the link between weight gain and insufficient sleep. A study with 335 participants showed significant differences between the sleep phases in overweight children and those of children of adequate body weight [50]. The subjects with excessive body weight had reduced sleep function, a longer delay in the first REM phase, reduced REM time, and reduced REM activity.

Leptin and ghrelin are hormones that are involved in the regulation of appetite. Ghrelin is responsible for the feeling of hunger, while leptin is responsible for the feeling of satiety [51]. The level of leptin in persons with insufficient sleep durations decreases, while the level of ghrelin, which leads to a subjective feeling of hunger, is increased [52,53]. Poor sleep hygiene, which goes hand in hand with insufficient sleep time, has an impact on poor dietary choices, such as increased portions, increased calorie intake, an increased feeling of hunger, and increased consumption of sugary drinks and foods [49].

Our current knowledge shows a connection between sleep quality and obesity. Overweight and obese individuals have been shown to have lower quality of sleep than those of normal body weight, regardless of the length of sleep [49,54], which involves the number of wakes within 5 min, the sleep performance, delayed sleep, and waking up after sleep.

High energy and fat consumption, binge eating, and nighttime snacking lead to sleep disorders, which can subsequently lead to disturbances in the feelings of satiety and hunger. Short sleepers prefer high-calorie foods and frequent snacks, and they skip meals more frequently [51].

The relationship between poor sleep quality and the development of overweight/obesity is illustrated in Figure 1.

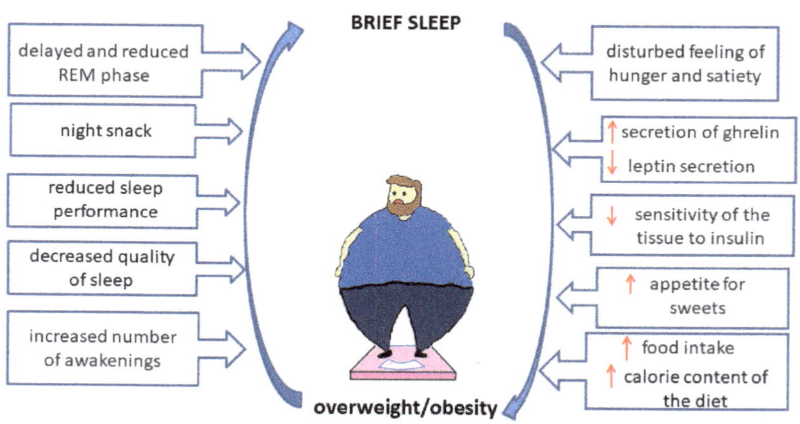

Figure 1. The relationship between inadequate sleep duration and calorie intake.

Sleep disorders have a significant impact on people's quality of life. Proper nutrition can significantly improve the quality of sleep. A balanced diet should contain all of the necessary minerals, vitamins, and amino acids. Poor nutrition can influence the onset of insomnia, which, in turn, is a factor in the development of many serious diseases, such as high blood pressure, type 2 diabetes, and cardiovascular diseases. Food-borne substances may impair sleep quality (e.g., by causing inflammation or alterations in hormone regulation) [41].

3.1.2. Dietary Fat and Sleep Quality

Nuts, vegetable oils, and olive oil are characterized by high contents of unsaturated fatty acids, but low contents of saturated fatty acids. The consumption of these products is lower than recommended for the majority of the population that is in favor of saturated fats. The excessive consumption of foods that are rich in saturated fatty acids contributes to the development of noncommunicable diseases (NCDs) [55–57]. Studies also show that people with insomnia have a higher consumption of high-fat foods than people without sleep disorders [35].

Eating fatty fish (more than 5% fat, such as salmon, mackerel, and trout) has a positive effect on sleep regulation. Fatty fish are a good source of omega-3 and omega-6 fatty acids, as well as of vitamin D. These nutrients may influence the regulation of serotonin secretion and, thus, the regulation of sleep [58]. Eating fatty fish leads to an increased feeling of drowsiness, which leads to better sleep and a more efficient performance during the day. Current evidence suggests that the consumption of fatty fish may have a positive impact on daily functioning and sleep [58].

Polyunsaturated omega-3 fatty acids are an important component of the diet. Diets that are low in omega-3 acids may impair sleep at night because of an endogenous disturbance of the daily clock and a reduction in melatonin secretion. Studies in hamsters with omega-3 deficiency have shown a disturbance in the melatonin-secretion rhythm and chronic locomotor hyperactivity [59,60].

Animal fats contain, almost exclusively, saturated fatty acids. Foods that are fried in hydrogenated oil are also a rich source of saturated fatty acids [59]. Studies on the effect of saturated fatty acids on sleep have shown that the consumption of saturated fatty acids leads to a greater number of wakes at night and shortens the duration of slow-wave sleep, which is the stage of sleep during which the body can recover [59]. The regular consumption of saturated fatty acids contributes to the development of diabetes, which is often associated with sleep problems [61].

The effect of dietary fat on sleep quality is shown in Figure 2.

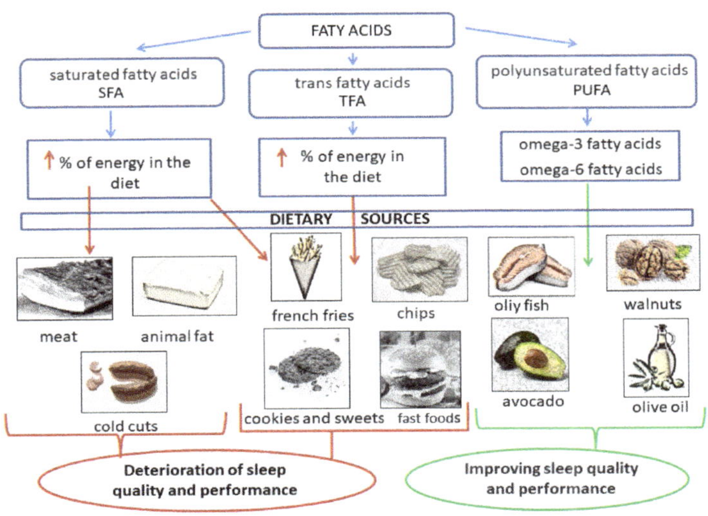

Figure 2. Dietary sources of fatty acids and their association with sleep quality.

Of the unsaturated fatty acids, arachidonic acid, which is a precursor to the production of the prostaglandin PGD2, which is a sleep-promoting eicosanoid, is of crucial importance for sleep quality [62].

3.1.3. Dietary Protein and Sleep Quality

Protein is one of the three main nutrients that meets the body's energy needs. Between 1999 and 2016, there was an increase in the estimated consumption of protein-source products, which was associated with an increased consumption of poultry, eggs, and soy [63]. Protein requirements change with age and are dependent on the condition of the body. For example, people who are ill or who have extensive burns will have an increased need for protein. Proteins have many functions in the body, such as transport, building, and structural functions [64].

A study of 4435 nonshift workers showed that protein intake can influence the symptoms of insomnia [65]. Low protein intake (<16% of total energy) was associated with a difficulty falling asleep and poor sleep quality, whereas high protein intake (≥19% of total energy) was linked to difficulty maintaining sleep. On this basis, it is recommended that protein accounts for between 16 and 19% of the energy efficiency of one's food intake [65].

Protein may consist of the amino acid tryptophan, which is a precursor of cerebral serotonin, which acts as a sleeping pill. Too little protein intake can lead to a deficiency of tryptophan, which can lead to sleep disturbances [66]. However, an excess of protein in the diet may lead to a decrease in the level of tryptophan in the brain, since the protein also contains other large neutral amino acids (LNAAs) (wide neutral amino acids) that affect the transport of tryptophan by the blood–brain barrier of tryptophan [66].

Protein that is taken in the evening has a positive effect on muscle protein synthesis during sleep. The availability of amino acids overnight is limited, and the rate of muscle protein synthesis is limited; however, by absorbing protein before bedtime, it is possible to digest and absorb it effectively. During sustained resistance training, the consumption of protein before bedtime may additionally influence muscle buildup and muscle strength [67].

The effect of proteins on sleep quality is presented in Figure 3.

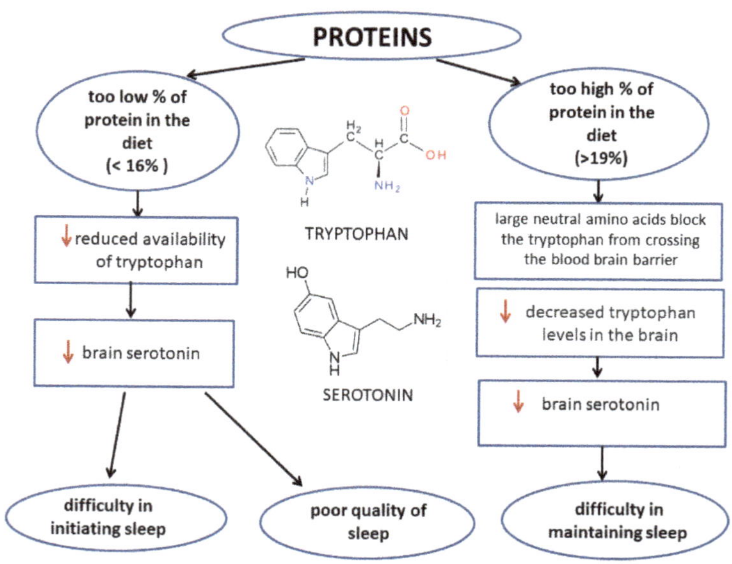

Figure 3. Protein intake and sleep quality.

3.1.4. Dietary Carbohydrates and Sleep Quality

Dietary carbohydrates, and the degree of their processing, significantly affect the quality of sleep. Both low-carbohydrate and high-carbohydrate diets affect the sleep architecture [35]. Carbohydrates have been shown to affect mainly the NREM phase (slow-wave sleep) and the REM phase. Moreover, dietary carbohydrates may also delay the onset of the REM phase and may delay the onset of sleep [34,68].

The carbohydrate quality is even more important for sleep quality than the amount of dietary carbohydrates. A study that was conducted in a group of 12 healthy subjects who were aged 18–35 years, and who consumed a meal that contained carbohydrates with a high glycemic index (GI) four hours before bedtime, showed a significant reduction in the delay in falling asleep, compared to a meal that contained low GI products [69].

Other studies suggest that a diet with a high glycemic index is a factor that increases the risk of insomnia [70]. The "Women's Health Initiative Observational Study", which was conducted with the participation of postmenopausal women, examined the probability of insomnia after consuming carbohydrates with different glycemic indexes, glycemic loads, and fiber contents. The risk of insomnia was increased by products with higher glycemic indexes and higher amounts of added sugars, refined grains, and starches. By contrast, a higher consumption of dietary fiber, whole grains, fruit, and vegetables was associated with a lower risk of insomnia [70]. Higher dietary fiber content in food lowers the glycemic index and slows down the metabolism of carbohydrates [71].

The mechanism of carbohydrate consumption on the occurrence of insomnia is not yet fully understood, but potential mechanisms have been suggested. By consuming foods with high glycemic indexes, the concentration of insulin increases, which may change the ratio of tryptophan to other large neutral amino acids (LNAAs), such as leucine, isoleucine, phenylalanine, valine, methionine, and tyrosine [72]. Insulin affects the higher selective uptake of LNAAs by the muscles, which thus causes a higher ratio of tryptophan compared to these amino acids. LNAAs compete with tryptophan for transport to the brain, and the greater muscle uptake of amino acids may lead to increased levels of tryptophan in the brain [73]. Tryptophan, on the other hand, is a precursor of serotonin, which affects sleep; therefore, the consumption of a large amount of carbohydrates with a high glycemic index may improve sleep wellbeing. However, in a proper diet, carbohydrates with a high glycemic index should be avoided because they contribute to the development of NCD, such as diabetes type 2 [69]. For a meal to have such an effect on the body, it should contain only carbohydrates. Even if only 5% of a meal comprises protein, it may inhibit the increase in the tryptophan concentration in the brain [70].

A diet with a high glycemic index may cause hyperglycemia, and the resulting hyperinsulinemia may induce the release of hormones such as cortisol, growth hormone, glucagon, and insulin, which contribute to sleep disorders [71]. A diet with a high glycemic index may deteriorate the quality of sleep by stimulating the inflammatory immune response, which leads to changes in the intestinal microbiome [34].

Some studies have examined the influence of diet on sleep quality. In a group of 26 adults, who usually slept for 7–9 h per day, the consumption of products that contained little dietary fiber and high amounts of saturated fatty acids resulted in less deep sleep [74]. The consumption of refined carbohydrates and sugar resulted in an increased number of awakenings during sleep [74].

Sleep quality is also influenced by the relationship between the percentage of energy that is consumed from sugar and nonfiber carbohydrates during the day. An increased probability of the reduced regularity of sleep and wakefulness has been shown in people who consume higher amounts of carbohydrates (i.e., $\geq$70.7% of energy comes from carbohydrates), compared to people who consume a moderate amount of carbohydrates (61 to 66% of energy) [75].

A meal that is abundant in carbohydrates and that is eaten in the evening reduces the nocturnal secretion of melatonin and delays the circadian rhythm of the basal body temperature [75]. The consumption of fiber was associated with more regenerative and deeper sleep. It is possible that a diet that is high in crude carbohydrates, contains more fiber, and with a reduced consumption of nonfibrous carbohydrates and sugar may significantly improve the quality of sleep of people with insomnia [34]. A study of 410 young women found that those who slept less consumed more carbohydrates and less dietary fiber [76].

The effects of carbohydrates on sleep quality are shown in Figure 4.

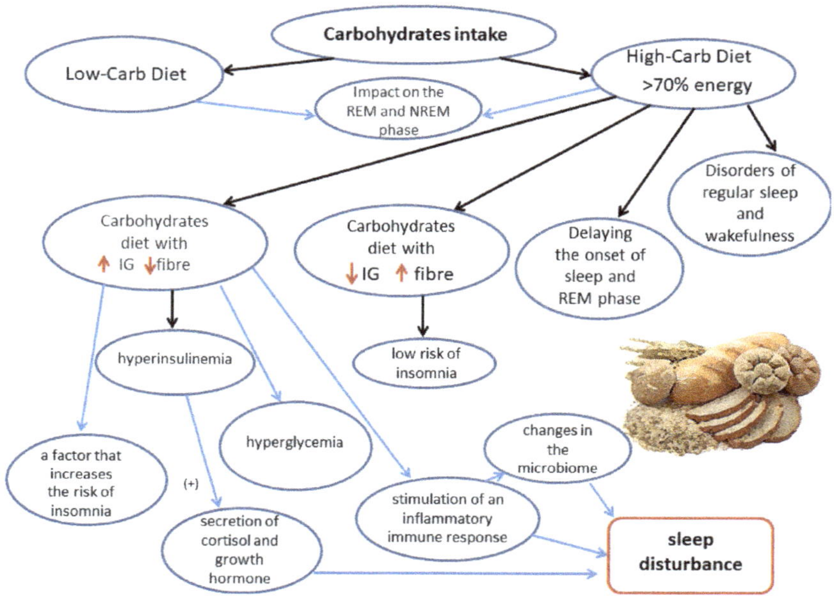

Figure 4. Carbohydrates and sleep quality.

3.1.5. Caffeine

One of the most commonly consumed stimulants is caffeine, which is found in coffee, tea, chocolate, energy drinks, and carbonated drinks. These products are widely used and are also consumed by children and adolescents [34].

Caffeine is, among other things, a substance that is taken to relieve fatigue; however, abuse can have a negative effect on sleep [37,77,78]. Therefore, it is very important to eat an adequate and balanced diet in order to enjoy complete mental and physical health [40,41].

The half-life of caffeine is between 2 and 10 h, depending on exogenous and endogenous factors. Nicotine can change (accelerate) the rate of caffeine metabolism by up to 50%. The residual effect of caffeine may last longer than 10 h, or even up to 20 h [37].

The mechanism of action of caffeine on the central nervous system is adenosine-receptor antagonism. Consequently, the effect of caffeine on sleep quality appears to be mainly due to adenosine receptors [38,79]. Most adults consume daily caffeine (coffee, energy drinks, tea, or other drinks), with an average intake of 200 mg/day. Self-reported studies tend to underestimate the actual caffeine consumption. This is because many people miscalculate the caffeine content of their diets by stating only the amount of caffeine in the coffee that they consume, without indicating the caffeine content in cold medicines, painkillers, tea, chocolate, hot chocolate, and energy drinks. It is difficult to develop a completely decaf diet, considering how widespread and easily accessible caffeine is [37].

Caffeine acts, in particular, on the A1 and A2A receptors, which influence sleep, cognitive functions, and the arousal of the brain. The absorption of caffeine takes place in the small intestine and stomach. This process is fast and effective, with the maximum plasma concentration being reached within the first 30 min after caffeine absorption [37,38].

For most people in the world, sleep is the time when they stop consuming caffeine. The negative effects of caffeine consumption (too much or too late) can only be felt the next day [37]. The intake of caffeine-containing coffee results in a decreased secretion of 6-sulfatoxymelatonin, which is the main metabolite of melatonin [77]. This is one of the mechanisms that causes sleep interruption [37].

The administration of four cups of brewed coffee (equivalent to 400 mg of caffeine) up to 6 h before bedtime leads to a significant deterioration in sleep quality. Caffeine consumption, even in the morning, shifts the REM phase of sleep to the early night [37].

In one study, nine healthy male volunteers were administered 200 mg of caffeine in the morning (at 7:00) [80]. The sleep episodes were monitored by electroencephalography, while the caffeine concentrations were measured in the saliva. A sharp increase in the caffeine levels was observed up to one hour after taking 200 mg of caffeine. A decrease in the caffeine concentration to less than one fifth of the peak level occurred 16 h later. Despite the decrease in the amount of caffeine in the saliva at the time of falling asleep, the overall time and efficiency of the sleep was reduced. This study shows that even a moderate dose of caffeine taken in the morning negatively affects the sleep quality during the subsequent night.

A study of 309 children aged 8 to 12 years investigated the relationship between the sleep quality, caffeine consumption, and daytime behavior [81]. It looked at caffeine intakes of between 0 and 151 mg (with the latter corresponding to an intake of 500 mL) of the energy drink Red Bull, or 750 g of milk chocolate. The intake of caffeine affected the sleep quality, the morning fatigue, and the sleep routine. The most common sources of caffeine among the children were coffee and tea (41%) and carbonated beverages (40%). No effect of caffeine on sleep delay in children has been observed.

Caffeine consumption causes an increased number of naps, a shortened total sleep time, a poor subjective assessment of the sleep quality, and daytime sleepiness [80].

3.1.6. Vitamin D

Vitamin D is a fat-soluble vitamin that plays a crucial role in calcium absorption. The main source of vitamin D is skin synthesis (ultraviolet B). It can also be supplied through food (fatty fish is the main source). On the basis of the serum concentration of the major active metabolite of vitamin D (25-hydroxyvitamin), vitamin D deficiency is widespread [82–84].

Low serum levels of 25-hydroxyvitamin D may impair sleep quality. The relationship between sleep disorders and vitamin D deficiency is not fully understood [85].

Vitamin D receptors are widespread in almost all tissues, including in the central nervous system [86]. Vitamin D receptors are found in the human brain in the prefrontal cortex, hypothalamus, and in black or grey brain matter, all of which play an important role in the regulation of sleep [87].

Vitamin D deficiency can cause nonspecific pain, which can impair sleep and worsen sleep quality. People who complained of nonspecific pain of an unknown cause had an increased risk of shortened sleep duration and worsening sleep quality. A study of 28 US veterans with vitamin D deficiencies showed that vitamin D supplementation can improve sleep quality and duration, relieve pain, and improve quality of life [88].

Vitamin D deficiency is associated with a higher risk of insomnia, including short sleep duration, poor quality of sleep, and daytime sleepiness. Studies suggest a correlation between a deterioration in the sleep quality and a deficiency of 25-hydroxyvitamin D in serum [89].

3.1.7. Tryptophan, Serotonin, and Melatonin

Tryptophan is an essential amino acid that the body cannot synthesize itself and, thus, must be supplied through food. Sources of tryptophan include chicken, turkey, eggs, milk, fish, cheese, beans, and pumpkin seeds [90]. Tryptophan is a precursor of melatonin and serotonin that can cross the blood–brain barrier and compete with other major neutral amino acids [91,92]. The conversion of tryptophan into serotonin occurs under conditions in which tryptophan is sufficiently available in the brain. Increased levels of tryptophan in the brain occur when the ratio of free tryptophan to branched-chain amino acids is increased. Melatonin is formed in the process of tryptophan conversion into serotonin [92,93].

Relatively low doses of tryptophan in the diet increase sleep performance, shorten the waking time during the night, and increase the subjective assessment of the sleep quality [92]. A study was conducted with 35 subjects aged 55 to 75 years who had trouble sleeping. The participants received a meal of 30 g of flakes with 22.5 mg of tryptophan in the first week, 30 g of flakes with 60 mg tryptophan in the second week, and a normal meal in the third week. The study showed a significant improvement in the sleep quality in the second week with a flake diet of 60 mg tryptophan, compared to the first and third weeks. An improvement in the quality of sleep was particularly evident in the performance of sleep, the increase in the actual duration of sleep, and the time during which one was immobile [92].

The soothing hormone that is released by the pineal gland is melatonin. Increased exogenous melatonin levels can improve sleep quality by increasing the body temperature [93]. Melatonin is mainly produced in the dark by the pineal gland from serotonin, and it stimulates the circadian rhythm. With increasing age, the melatonin level decreases, which leads to a disturbance in the circadian sleep rhythm [92]. Melatonin can be supplemented. Melatonin from supplements is characterized by very low toxicity, although no additional benefits have been observed at doses above 3 mg. The absorption of melatonin has a positive effect on the quality of sleep by increasing the propensity to sleep and by increasing the duration of sleep [42].

3.1.8. Gamma Aminobutyric Acid

Gamma-aminobutyric acid (GABA) is a nonprotein amino acid that has a positive effect on many metabolic disorders. The main producers of gamma-aminobutyric acid are lactic acid bacteria [94]. High concentrations of GABA in food can be achieved by the use of *Lactobacillus brevis* or *Lactococcus lactis*, which are present in fermented dairy products. GABA occurs naturally in small amounts in rice, oat, wheat, soya beans, raw spinach, potatoes, and many vegetables [95,96]. Naturally occurring gamma-aminobutyric acid can also increase sleep efficiency and can have hypnotic effects [97].

One study investigated the objective effect of the uptake of naturally occurring gamma-aminobutyric acid by means of serial polysomnography. The study included adults with one or more symptoms of insomnia. The patients received gamma aminobutyric acid tablets or a placebo one hour before bedtime. The patients who received gamma aminobutyric acid tablets had a significantly reduced sleep delay compared to the patients who took placebo tablets. A reduction in the symptoms associated with insomnia and a subjective improvement in the quality of sleep has also been observed in patients taking GABA tablets [97]. Although the study did not investigate whether GABA can cross the blood–brain barrier, it is suspected that gamma aminobutyric acid may inhibit stimulation neurons [97,98].

Gamma aminobutyric acid tablets can affect sleep by enhancing central GABA-ergic neurotransmission. Gamma-aminobutyric acid is involved in the regulation of both the REM and the NREM sleep phases [96,99].

3.2. *Stimulants and Drugs That Affect Sleep Quality*

Sleep hygiene is a set of environmental and behavioral recommendations that are aimed at the promotion of healthy sleep. It should be used in the entire population, and not only in persons with insomnia [100]. Patients are instructed to adhere to the rules of proper sleep hygiene (stopping smoking, avoiding alcohol, regular sleeping hours, regular exercise, and noise avoidance). There are people who do not have access to sleep therapy, even though they meet the criteria for sleep disorders. Such people will more often look for materials to help deal with insomnia themselves and turn to basic care officials. Information about proper sleep hygiene is easily accessible and widespread, as it can be disseminated without the direct involvement of a doctor. As a result, it can also be accessible to people who are not seeking medical help for sleep disorders [101].

Education about proper sleep hygiene is relatively inexpensive and may be the first intervention for people looking to improve their sleep quality. The most common sleep hygiene recommendations refer to smoking, alcohol, caffeine, daytime naps, stress, noise, time outdoors, and exercise [102]. It is important to avoid the factors that influence the deterioration of sleep quality, such as smoking, alcohol consumption, excessive stress, or excessive caffeine intake in the diet [103].

3.2.1. Alcohol

Alcohol is not recommended before going to sleep in order to ensure good sleep hygiene. Alcohol consumption in the late evening prolongs slow-wave sleep in the early part of the night [103] and affects the REM phase and sleep continuity. Alcohol is often used as a trigger for sleep, although the mechanism for reducing sleep delay is more complicated. High doses of alcohol (1 g/kg body weight) within one hour before bedtime inhibits the REM phase, but the reduction in the REM phase disappears with continued alcohol consumption. With high and moderate alcohol consumption, slow-wave sleep is prolonged. Repeated nighttime alcohol consumption leads to a decrease in the NREM phase (slow-wave sleep). Alcohol has a stimulating effect when consumed in low doses (0.16 g/kg body weight) in the first hour after consumption, while it has a sedative effect in large quantities. Alcohol consumption up to 6 h before bedtime impairs the quality of sleep, which indicates a relatively long effect [104].

Increasing alcohol consumption decreases the delay in falling asleep. Two to three hours after drinking, the blood alcohol level drops, which increases arousal. There is a prolongation of the REM phase in the second half of the night, which contributes to the fragmentation of sleep [104]. In long-term users, only a small improvement in sleep can be observed after prolonged abstinence [105–107]. Long-term alcohol consumption may have different effects compared to people who drink little or no alcohol. Alcohol abuse can lead to changes in physiological sleep and waking, which play a role in the regulation of sleep [104,106].

Alcohol is considered a good sleeping remedy, not least because of its easy availability and low cost. Women use alcohol less frequently than men [105]. A total of 67% of patients who complained of insomnia (age range: 18–79 years) and used alcohol reported that it had a positive effect on their sleep quality [108]. However, alcohol quickly loses its sleep-promoting effect and retains its sleep-disrupting properties. Alcohol users are also sleepier during the day than nonalcohol users [105].

Low alcohol consumption can cause snoring and obstructive sleep apnea in healthy people. Alcohol in combination with obstructive sleep apnea increases the risk of strokes, heart attacks, and sudden death. The onset of sleep apnea deteriorates the quality of sleep and causes the feeling of fatigue during the next day [105,109].

Alcohol consumption aggravates movement disorders, which impairs sleep behavior. People who consume two or more drinks a day have a two to threefold increase in periodic leg movements, which leads to the increased fragmentation of sleep [108].

Alcohol consumption can also cause other sleep-disturbing symptoms, such as gastritis, gastroesophageal reflux, and polyuria. Increased thirst and polyuria cause frequent awakening, which also affects sleep quality [105].

3.2.2. Nicotine

Smokers have a higher risk of sleep disorders such as sleep apnea, sleep disturbances, poor sleep quality (increased sleep delay, shorter time, and greater difficulty in maintaining sleep, as well as daytime sleepiness), and insomnia [110].

Nicotine disturbs the balance of neurotransmitters that are involved in the regulation of sleep. In addition, nicotine withdrawal occurs during sleep, which affects the onset of insomnia. In exploratory studies, a significant interaction was observed between evening nicotine intake and the reported occurrence of insomnia. In individuals with symptoms of insomnia, nicotine intake at bedtime has also been associated with a 40 min reduction in sleep duration [110].

Nicotine promotes excitement and alertness by stimulating the cholinergic neurons in the basal region of the forebrain. The intake of nicotine in the form of a patch, a pill, or smoking is associated with sleep disturbances. The administration of nicotine in any form reduces the total sleep time, increases sleep delay, suppresses slow-wave and REM sleep, and increases early morning awakening. Therefore, it is recommended to avoid nicotine in order to maintain good quality sleep [111].

People who use nicotine and are addicted to it should be individually investigated for the direct effects of nicotine withdrawal on sleep quality. The withdrawal of nicotine in the early stages of smoking cessation is often associated with the onset of sleep disorders. A deterioration in sleep quality may occur up to 3–4 weeks after quitting smoking. The most noticeable discomfort during nicotine depletion is the more frequent and prolonged awakening during sleep [110,112].

3.2.3. Cannabis

Cannabis (also known as marijuana) is one of the most commonly used drugs. Cannabis has a positive effect on sleep quality. It is sleep-promoting and hypnotic, but it also reduces the waking time after falling asleep, reduces the delay in falling asleep, shortens the REM phase, and prolongs the duration of slow-wave sleep [113–116].

A 2017 New England study on 1500 patients found a reduction in sleeping-pill use of two-thirds among patients using medicinal marijuana [117]. The hypnotic effects of cannabis are often the reason for its use in people with sleep disorders. However, in cannabis users, the hypnotic effect may be tolerated because of neurological changes in the endocannabinoid system [118].

Cannabis, and especially if used for a short time, can have a soothing effect on sleep disorders in terms of subjective sensations. With the prolonged use of cannabis, however, negative effects on sleep quality have been noted, and most notably during withdrawal. The negative effects of cannabis use can also be observed in individuals who take low doses of cannabis [119,120].

People who use cannabis on a daily basis are more likely to report sleep disturbances compared to people who use cannabis rarely or never [121].

The medical use of cannabis for the treatment of post-traumatic stress disorder, pain, and multiple sclerosis can improve sleep quality. However, long-term cannabis use causes tolerance, long-term sleep disorders, and withdrawal symptoms, which may lead to the worsening of post-traumatic stress disorder [118].

A study was conducted with eight subjects who orally received a fluid that contained naturally occurring substances in marijuana. The volunteers received a polysomnography to assess the sleep and morning functions. Decreased performance on the following day (mood swings and memory deficits) and a deterioration in the sleep quality (decreased sleep performance) were observed [118].

In a sleep study that used polysomnography, it was shown that the administration of 10, 20, or 30 mg of one of the substances contained in marijuana (THC) resulted in a shortening of the latency of falling asleep, and a reduction in the total time of falling asleep. However, not all studies show this effect of marijuana, which may be due to the soporific effects of THC and the stimulant effects of cannabidiol, which is one of the active chemicals that is identified in cannabis [119]. Some studies have also shown that cannabis shortens the rapid-eye sleep phase, reduces the density of the rapid-eye sleep phase, and lengthens the NREM (slow-wave sleep) phase. Long-term marijuana users develop a tolerance to the effects of cannabis, but they also experience a deterioration in sleep performance [119].

Numerous studies on marijuana withdrawal have shown an increase in wakefulness after falling asleep, an increase in sleep latency with rapid eye movement, an increase in the delay of falling asleep, and decreases in slow-wave sleep, sleep performance, and total sleep duration. Such effects are more pronounced in people who use marijuana intensively (marijuana use ≥5 days a week for the last 3 months). Symptoms can persist for more than 45 days [119].

More and more people are complaining about a decrease in the quality of sleep or the occurrence of insomnia. Proper nutrition that is rich in tryptophan, vitamin D, and gamma-aminobutyric acid can improve the quality of sleep. By using foods that are rich in these substances, the effectiveness and the actual sleep time are improved. In addition, there is a noticeable delay in the subjective assessment of sleep.

Substances such as alcohol, nicotine, excess caffeine, and cannabis negatively affect the quality of sleep. They cause, among other things, an increase in the waking time after falling asleep, a shorter sleeping time, and greater difficulty in maintaining sleep. A nonpharmacological method of treating insomnia is to eliminate the consumption of the abovementioned substances [87,104,110,121]. The effects of caffeine, alcohol, nicotine, and marijuana on sleep quality are shown in Figure 5.

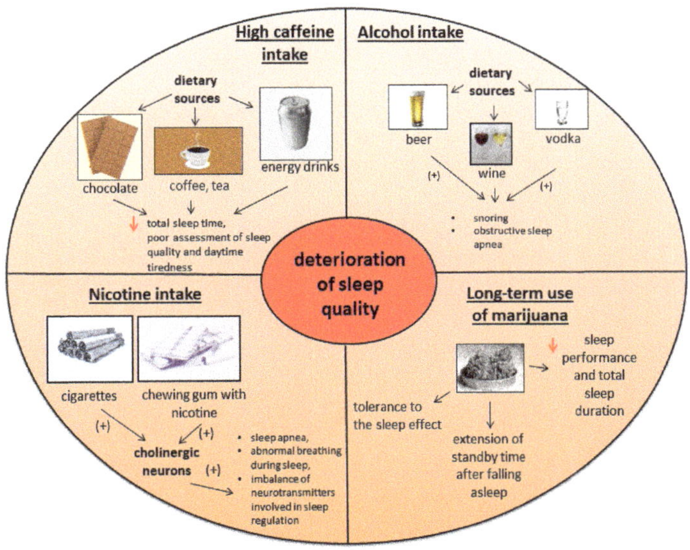

Figure 5. Stimulants and drugs that affect sleep quality.

3.3. The Impact of Physical Activity on the Quality of Sleep

Sleep and physical activity are related to cognitive functions, and especially executive control and memory consolidation. However, it has not been discovered how physical activity is related to the executive-control processes that are responsible for monitoring, initiating, and planning target-oriented behaviors (e.g., working memory) [122,123]. The quality of sleep in the elderly is affected by physical activity. Longer daily activity is associated with better sleep quality. A study was conducted in which three age groups were included: young (21–29 years), middle-aged (36–64 years), and older (65–81 years). The older people who engaged in more physical activity scored lower on the PSQI (Pittsburgh Sleep Quality Index), which indicates better sleep quality. The older people who slept better reported less fatigue. A relationship between sleep quality and training intensity has also been observed. Moderate and intense physical activity has a positive effect on sleep quality, while light physical activity has no effect on sleep quality [124]. People who practice physical activity sleep better and longer than those with a sedentary lifestyle. By introducing the appropriate amount of physical activity and the time spent outdoors, and by engaging in activities such as walking, we can nonpharmacologically improve the quality of sleep [125]. Long-term physical activity has a positive effect on the quality of sleep. An improvement in the sleep quality occurs with an increase in the activity time and the number of steps, and so even moderate physical activity has a positive effect [126]. High-intensity exercise during the nighttime period affects the secretion of melatonin and can quickly change its concentration in the body within a few minutes [126]. The concentration of melatonin depends on the intensity, duration, and type of exercise that is performed. Physical exercise late in the evening, when melatonin is physiologically secreted, can cause a decrease in its concentration. On the other hand, night exercise, both of high and moderate intensities, causes a delay in the secretion of melatonin on the following evening. Physical exercise during the day, regardless of intensity, does not have a quick and constant effect on melatonin secretion [127].

The effects of physical activity on sleep quality are shown in Figure 6.

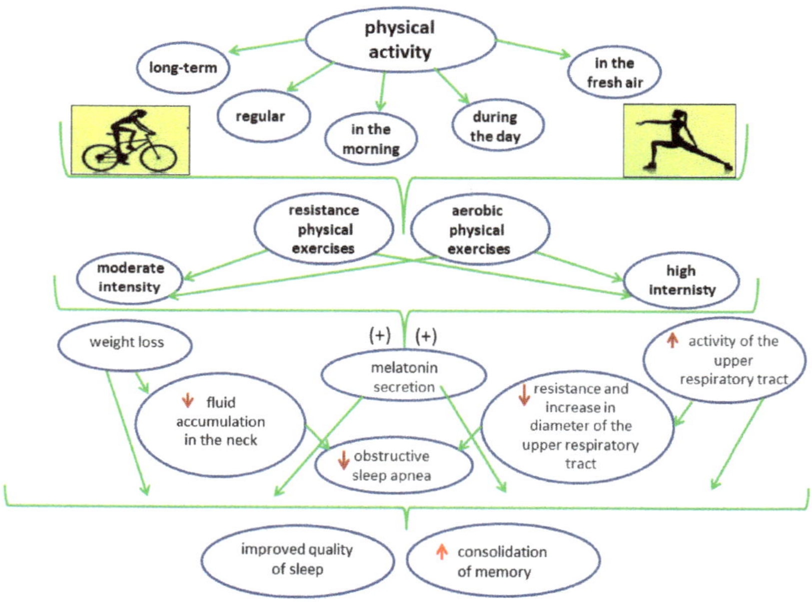

Figure 6. Effects of physical activity on sleep quality.

3.3.1. The Effect of Recreational Physical Activity on Sleep Quality

Physical activity, physical fitness, and exercise are interrelated, but are, at the same time, separate collective terms. Physical activity is any movement that causes energy expenditure, and it includes, among other things, daily duties, such as household activities and commuting. Exercise is a structured and repetitive activity that is aimed at improving health or at maintaining it at a constant level. Physical fitness is the ability to perform physical activities without the excessive fatigue of the body [128].

Physical activity and sleep positively correlate with cognitive functions, and especially with executive control and memory consolidation (i.e., the processes that consolidate the acquired information in the brain). Physical activity has a positive effect on the quality of sleep, and especially on its depth, latency, and performance [122,123].

Physical activity is an important element of public health that is used both in the prevention and treatment of various diseases. Regular exercise reduces the risk of cancer, diabetes, and coronary heart disease, as well as the onset of neurodegenerative disorders. The currently available data also indicate a positive effect of exercise and physical activity on sleep quality. Both relatively high-intensity and low-intensity exercise produce significant benefits that are associated with sleep quality [129,130].

Physical activity, and especially regular exercise, can improve the quality of sleep by affecting the adenosine levels and the body temperature; however, when performed too late in the evening, it can cause sleep disruption by increasing physiological arousal. There are also studies that examine the likelihood of sleep improvement through late physical activity that is due to the induced antidepressant, anxiolytic, and body-warming effects [131]. The effect of exercise on the body temperature can be extremely important late in the evening, as there is a decrease in the body temperature when falling asleep, and exercise causes an initial increase in the deep body temperature and it increases the rate of the decrease in the body temperature [129]. The timing of exercises for sleep quality is therefore unclear, as there are many conflicting arguments.

3.3.2. The Effect of Physical Exercise on the Quality of Sleep in People with Insomnia

A study was conducted on 48 patients suffering from insomnia, who were divided into four groups: a control group; a second group, in which people performed moderate-intensity aerobic exercise; a third group, in which high-intensity aerobic exercise was performed; and a fourth group, in which moderate-intensity resistance exercise was performed [129]. In the group in which the participants performed moderate-intensity aerobic exercise, the data from the polysomnogram showed a reduction in the total wakefulness, a delay in falling asleep, and an increase in the efficiency and overall sleep duration. A reduction in anxiety was also observed in those who engaged in moderate-intensity aerobic exercise [129]. The patients with primary insomnia experienced a decrease in their rate of sleep anxiety. Moderate-intensity resistance exercise can cause a reduction in anxiety for up to five hours [132].

Obstructive sleep apnea (OSA) is a condition that occurs during sleep that is characterized by the obstruction of the upper respiratory tract. Obstructive sleep apnea, despite its frequent occurrence, is a clinical condition that is rarely diagnosed [133]. Obstructive sleep apnea is diagnosed by polysomnography, in which the ratio of the total number of apneas and the respiratory shallowness to the total sleep time is assessed [134]. OSA is a growing burden on health care funds, as it is a major source of cardiovascular morbidity and mortality [134]. Patients with obstructive sleep apnea have significantly reduced work performance and quality of life because they often suffer from morning headaches, urination at night, decreased libido, impaired concentration and attention, irritability, depression, sleep fragmentation, neurocognitive disorders, deterioration of the sleep quality, and the occurrence of excessive daytime sleepiness [133].

An easy and cheap method of treating OSA is physical exercise, which alleviates several consequences that are caused by the disease, such as fatigue and cardiovascular disorders. Physical exercise affects weight loss, which also affects the alleviation or resolution of OSA. The mechanisms that attenuate obstructive sleep apnea through physical activity are not yet fully understood, but there are several plausible hypotheses. During physical exertion, the respiratory muscles are stimulated, which leads to structural and metabolic adaptations that increase the resistance to fatigue. It is likely that endurance exercises cause increased activity in the upper respiratory tract, which results in a decrease in the resistance and an increase in the diameter of the upper respiratory tract. Endurance exercises also counteract pharyngeal collapse during sleep [133].

During sleep, fluid accumulates in the neck, which causes an increase in the pressure on the larynx, which can cause the onset of OSA. After aerobic physical exercise, there is a significant reduction in the amount of fluid in the neck [135].

People with OSA have reduced durations of slow-wave sleep and greater difficulty achieving slow-wave sleep, as well as increased daytime sleepiness [136]. Physical exercise increases the body temperature, which can make it easier to fall asleep. During physical activity, there is also an increase in the energy expenditure, which affects the extension of the NREM sleep phase [133]. Adult patients (a study of 129 participants) with OSA who practiced physical activity experienced a decrease in daytime sleepiness and increases in the peak oxygen consumption and the sleep performance [137].

3.3.3. The Impact of Physical Activity on the Quality of Sleep of Children and Adolescents

Sleep is a key element for the proper health and development of children. A short sleep duration among preschool children is associated with a higher prevalence of obesity with age [138]. In recent years, the number of obese children has increased sharply. Research conducted by the World Health Organization has shown that about 41 million children under 5 years of age are overweight or obese. It is especially important to maintain a healthy weight in childhood, since overweight and obesity at a young age can adversely affect the mental, physical, and social development of a child. Abnormal body weight in children has a serious impact on the development of diseases in adulthood, such as diabetes, cardiovascular diseases, and cancer [139].

International guidelines recommend that infants sleep up to 17 h a day, and that children aged 1–5 years sleep from 10 to 14 h/day [21]. Currently, children sleep less than children did a century ago, and parents report reduced sleep quality observed in their children [138].

A study that analyzed the incidence of insomnia included a group of 700 children aged 5 to 12 years. It was shown that the incidence of insomnia among the children was 19.3%. Insomnia among boys was at a similar level in all the age groups (5–7 years, 8–10 years, and 11–12 years), while, in girls, the 11–12-year age group showed the highest incidence of insomnia symptoms, compared to the 5–7-year and 8–10-year groups. Children with symptoms of insomnia take longer to fall asleep, and they have an increased delay in the REM phase and reduced slow-wave sleep, compared to children without sleep disorders [140].

Higher levels of total physical activity in infants are associated with poorer sleep performance, a shorter total sleep duration, and fewer naps throughout the day. In young children and preschoolers, a higher degree of physical activity has a positive effect on the quality of sleep, as it results in the better quality and the stability of sleep. The intensity of physical activity also affects the quality of sleep. Light physical activity in preschool children is associated with a later bedtime. By contrast, moderate to intense levels of physical activity are associated with a later time and a shorter total duration of sleep. In children aged 1–3 years, higher levels of physical activity are associated with better stability, a shorter overall time to fall asleep, and better sleep quality [138].

Playing outdoors, for children aged 1–3 years, is associated with shorter falling-asleep times, fewer wakeups, a shorter total sleeping time, and an earlier bedtime [138]. The physical activity of preschoolers in the open air in the form of play is associated with fewer night awakenings, a shorter time to fall asleep, an earlier sleeping time, and a longer total sleep time. Preschool children who engage in sports have better sleep quality (better sleep performance and earlier time to fall asleep) [138,141].

A study of 91 adolescents (11–19 years old) found that 73.6% of them had trouble maintaining sleep, and that 60.5% had trouble falling asleep [142]. Adherence to physical activity guidelines (60 or more minutes per day of moderate or intense physical activity) resulted in improved quality of sleep and shortened sleep duration [142].

3.3.4. The Effect of Physical Activity on Sleep Quality in Adults

Physical activity and spending time outdoors can be a nonpharmacological means of maintaining proper sleep quality and fighting insomnia [124]. People who practice physical activity have better sleep quality (latency, depth, and sleep performance). Regular physical activity and sleep positively correlate with cognitive functions (executive control and memory consolidation). The available data indicate that women are more likely to engage in nonpharmacological measures to treat insomnia. Furthermore, physically active women had better sleep quality than sedentary women [124].

One study, which involved 305 participants over the age of 40 years, assessed the effects of physical activity on sleep quality. The participants took part in an exercise program that consisted of high-intensity resistance exercise and moderate-intensity aerobic exercise. Pooled analyses of the results showed that physical activity had a positive effect on the sleep quality, which was indicated by a decrease in the PSQI and in the subjective feelings of the participants. An improvement in the quality and latency of the sleep was noted. Those participating in physical activity did not sleep any longer, but they experienced better sleep quality [143,144].

Sleep deprivation is correlated with the occurrence of hypertension, obesity, and stroke. Hypertension is a significant factor in increasing the risk of coronary artery disease, stroke, heart failure, and end-stage renal failure. With an increase in blood pressure above 115/75 mmHg, the risk of cardiovascular disease increases, and it doubles with each increase by 20/10 mmHg. Blood pressure should decrease during the night by 10–20% of the level of the daily blood pressure. In a study that involved 20 hypertensive patients, the participants first performed an exercise test until exhaustion, and then, in random order, they engaged in 30 min of exercise on the treadmill at 7 a.m., 1 p.m., and 7 p.m. This study showed lower blood pressure at night after moderate aerobic exercise at 7 a.m. Physical activity also resulted in improved sleep wellbeing. The deep-sleep phase was increased as a result of the increased energy expenditure during the day, and especially after physical exercise at 7:00 a.m. Therefore, aerobic exercise may be a nonpharmacological method of improving sleep quality [143].

A study was conducted in which a total of 377 women took part [124]. The physical activity was measured by using accelerometers. It was shown that there was a high probability of improving the sleep quality and circadian rhythms through morning exercise. However, the duration of physical activity should be refined, and the clinical significance of the time of day during which physical activity is practiced should be assessed in order to clarify the recommendations for sleep optimization [124].

An important factor that influences the maintenance of proper health in older people is physical activity. Older people are more likely to have mental and physical disabilities and are more likely to have physical limitations compared to younger people. With age, the quality of sleep deteriorates, which is associated with a greater feeling of fatigue during the day, and with reduced comfort and increased mortality [145].

The World Health Organization recommends that people practice moderate-intensity aerobic exercise for at least 150 min per week, or high-intensity aerobic exercise for at least 75 min per week [146].

It is assumed that half of the population over the age of 65 years has reduced quality of sleep. A study that surveyed a group of 60 people (22 younger people aged 21–29 years, 16 middle-aged people aged 36–64 years, and 22 older people aged 65–81 years) showed a positive relationship between physical activity and sleep wellbeing in older people [123]. The quality of sleep in the elderly was not related to physical fitness, but to the level of physical activity. Moderate to intense physical activity is particularly associated with improved sleep quality [123].

Older people who are physically active are less likely to report symptoms that indicate poor sleep quality. Being more active results in increased sleep efficiency, longer sleep duration, and a reduced delay in falling asleep [147].

4. Discussion

Studies show that sleep duration has decreased significantly in all age groups [46,47], and sleep disorders and insomnia are diagnosed at all ages [4,9,24,44–46]. Currently, many studies suggest that sleep disorders and insomnia increase the risk of cardiovascular disease, obesity, depression, cancer, and infectious diseases [9–12,50,54].

A reduced sleep time influences poor dietary choices, such as skipping breakfast, eating processed foods that contain fewer vitamins, and eating excessively fatty foods [76], which lead to excessive calorie intake [50]. A major problem is the increase in overweight and obesity in children and adolescents. Studies show that this age group often does not meet the requirements for adequate sleep time [45,46]. The prevention of sleep disorders and insomnia is therefore crucial to prevent noncommunicable diseases, which often start in early childhood.

Many studies have shown that proper nutrition, physical activity, and fewer stimulants have a positive effect on sleep quality. On the other hand, poor nutrition can lead, in the long term, to inflammation, which is closely associated with insomnia [15]. The nutritional factors that regulate sleep may have different mechanisms of action [35]. Sleep may be affected by individual ingredients (e.g., caffeine), or by a complex of food metabolites. Foods may also influence the commensal microflora, which may lead to the formation of certain bioactive metabolites [41]. Gamma-aminobutyric acid (GABA), which is one of the metabolites of the bacteria, can increase sleep performance and promote sleep [97].

Studies show that a balanced diet has a positive effect on sleep quality [65,66,75]. Foods and meals that contain sufficient protein, carbohydrates, and fats are essential for maintaining the quality of sleep [57,58]. Not only the quantity, but also the quality of the nutrients is important. A sufficient amount of the amino acid tryptophan, which is the precursor of melatonin, has a positive effect on sleep [92]. The scientific evidence points to the role of omega-3 fatty acids, which may positively influence the regulation of serotonin secretion [58]. To improve sleep quality, it is recommended that individuals eat carbohydrate-containing meals with low glycemic indexes, low glycemic loads, and high fiber contents [69,70]. In order to ensure an adequate quality of sleep, processed foods that are high in saturated fatty acids and refined carbohydrates and low in fiber should be avoided [59].

An important factor for sleep quality is its hygiene [101]. Proper sleep hygiene can prevent insomnia and can be a nonmedicinal treatment method. Studies show that the use of stimulants, such as alcohol, nicotine, caffeine, and cannabis, significantly reduces sleep quality [77,84,105,110,119].

Physical activity plays an important role in the maintenance of good sleep quality. A sufficient amount of moderate- to high-intensity exercise can improve the quality of sleep and prevent insomnia [126,129]. Physical activity in the late evening, when melatonin is released, may, however, lower melatonin levels [126]. Night exercises, both at high and medium intensities, can even delay the release of melatonin the next evening. However, exercise during the day, regardless of its intensity, does not have a rapid and continuous effect on melatonin secretion [127].

Although there is scientific evidence of an association between disease occurrence and sleep problems, there is little research on the sleep quality in people with noncommunicable diseases in the context of nutrition [68].

5. Conclusions

Sleep-related issues are a broad and open topic that require further research, and especially because sleep disorders may contribute to the emergence of many chronic diseases. Studies that combine an assessment of the relationship between sleep and diet, physical activity, and the health of the population should be conducted on a wide group of respondents, and especially among people at risk of noncommunicable diseases.

Author Contributions: Conceptualization, M.S.; investigation, M.S.; writing—original draft preparation, M.S.; writing—review and editing, A.M.W. and I.M.-C.; visualization, I.M.-C.; supervision, A.M.W. and I.M.-C. All authors have read and agreed to the published version of the manuscript.

Funding: This research received no external funding.

Institutional Review Board Statement: Not applicable.

Informed Consent Statement: Not applicable.

Data Availability Statement: Not applicable.

Conflicts of Interest: The authors declare no conflict of interest.

References

1. Krueger, J.M.; Frank, M.G.; Wisor, J.P.; Roy, S. Sleep function: Toward elucidating an enigma. *Sleep Med. Rev.* **2016**, *28*, 46–54. [CrossRef] [PubMed]
2. Irwin, M.R. Why Sleep is important for health: A psychoneuroimmunology perspective. *Annu. Rev. Psychol.* **2016**, *66*, 143–172. [CrossRef] [PubMed]
3. Saidi, O.; Rochette, E.; Doré, É.; Maso, F.; Raoux, J.; Andrieux, F.; Fantini, M.L.; Merlin, E.; Pereira, B.; Walrand, S.; et al. Randomized double-blind controlled trial on the effect of proteins with different tryptophan/large neutral amino acid ratios on sleep in adolescents: The protmorpheus study. *Nutrients* **2020**, *12*, 1885. [CrossRef] [PubMed]
4. Murawski, B.; Plotnikoff, R.C.; Rayward, A.T.; Vandelanotte, C.; Brown, W.J.; Duncan, M.J. Randomised controlled trial using a theory-based m-health intervention to improve physical activity and sleep health in adults: The Synergy Study protocol. *BMJ Open* **2018**, *8*, e018997. [CrossRef]
5. Dashti, H.S.; Jones, S.E.; Wood, A.R.; Lane, J.M.; van Hees, V.T.; Wang, H.; Rhodes, J.A.; Song, Y.; Patel, K.; Anderson, S.G.; et al. Genome-wide association study identifies genetic loci for self-reported habitual sleep duration supported by accelerometer-derived estimates. *Nat. Commun.* **2019**, *10*, 1100. [CrossRef]
6. Morsing, J.A.; Smith, M.G.; Ögren, M.; Thorsson, P.; Pedersen, E.; Forssén, J.; Waye, K.P. Wind turbine noise and sleep: Pilot studies on the influence of noise characteristics. *Int. J. Environ. Res. Public Health* **2018**, *15*, 2573. [CrossRef]
7. Hu, X.; Cheng, L.Y.; Chiu, M.H.; Paller, K.A. Promoting memory consolidation during sleep: A meta-analysis of targeted memory reactivation. *Psychol. Bull.* **2020**, *146*, 218–244. [CrossRef]
8. Rakowska, M.; Abdellahi, M.E.; Bagrowska, P.; Navarrete, M.; Lewis, P.A. Long term effects of cueing procedural memory reactivation during NREM sleep. *NeuroImage* **2021**, *244*, 118573. [CrossRef] [PubMed]
9. Yu, J.; Rawtaer, I.; Fam, J.; Jiang, M.-J.; Feng, L.; Kua, E.H.; Mahendran, R. Sleep correlates of depression and anxiety in an elderly Asian population. *Psychogeriatrics* **2016**, *16*, 191–195. [CrossRef]
10. Kwok, C.S.; Kontopantelis, E.; Kuligowski, G.; Gray, M.; Muhyaldeen, A.; Gale, C.P.; Peat, G.M.; Cleator, J.; Chew-Graham, C.; Loke, Y.K.; et al. Self-reported sleep duration and quality and cardiovascular disease and mortality: A dose-response meta-analysis. *J. Am. Heart Assoc.* **2018**, *7*, e008552. [CrossRef]
11. Zhai, L.; Zhang, H.; Zhang, D. Sleep duration and depression among adults: A meta-analysis of prospective studies. *Depress. Anxiety* **2015**, *32*, 664–670. [CrossRef] [PubMed]
12. Prather, A.A.; Janicki-Deverts, D.; Hall, M.H.; Cohen, S. Behaviorally assessed sleep and susceptibility to the common cold. *Sleep* **2015**, *38*, 1353–1359. [CrossRef] [PubMed]
13. Khan, M.S.; Aouad, R. The effects of insomnia and sleep loss on cardiovascular disease. *Sleep Med. Clin.* **2017**, *12*, 167–177. [CrossRef]
14. Gobbi, G.; Comai, S. Differential function of melatonin MT1 and MT2 receptors in REM and NREM sleep. *Front. Endocrinol.* **2019**, *10*, 87. [CrossRef] [PubMed]
15. Irwin, M.R. Sleep and inflammation: Partners in sickness and in health. *Nat. Rev. Immunol.* **2019**, *19*, 702–715. [CrossRef] [PubMed]
16. Valko, P.O.; Hunziker, S.; Graf, K.; Werth, E.; Baumann, C.R. Sleep-wake misperception. A comprehensive analysis of a large sleep lab cohort. *Sleep Med.* **2021**, *88*, 96–103. [CrossRef]
17. Brown, R.E.; Basheer, R.; McKenna, J.T.; Strecker, R.E.; McCarley, R.W. Control of sleep and wakefulness. *Physiol. Rev.* **2012**, *92*, 1087–1187. [CrossRef]
18. Abbasi, S.F.; Jamil, H.; Chen, W. EEG-based neonatal sleep stage classification using ensemble learning. *Comput. Mater. Contin.* **2022**, *70*, 4619–4633. [CrossRef]
19. Watson, N.F.; Buchwald, D.; Vitiello, M.; Noonan, C.; Goldberg, J. A twin study of sleep duration and body mass Index. *J. Clin. Sleep Med.* **2010**, *6*, 11–17. [CrossRef]
20. Watson, N.F.; Safwan Badr, M.; Belenky, G.; Bliwise, G.L.; Buxton, O.M.; Buysse, D.; Dinges, D.F.; Gangwisch, J.; Grandner, M.A.; Kushida, C.; et al. Recommended amount of sleep for a healthy adult: A joint consensus statement of the american academy of sleep medicine and sleep research society. *Sleep* **2015**, *38*, 843–844. [CrossRef]
21. Hirshkowitz, M.; Whiton, K.; Albert, S.M.; Alessi, C.; Bruni, O.; DonCarlos, L.; Hazen, N.; Herman, J.; Katz, E.S.; Kheirandish-Gozal, L.; et al. National sleep foundation's sleep time duration recommendations: Methodology and results summary. *Sleep Health* **2015**, *1*, 40–43. [CrossRef] [PubMed]
22. Buysse, D.J. Insomnia. *JAMA* **2013**, *309*, 706–716. [CrossRef] [PubMed]
23. Bollu, P.C.; Kaur, H. Sleep medicine: Insomnia and sleep. *Mo. Med.* **2019**, *116*, 68–75. [PubMed]
24. Reed, D.L.; Sacco, W.P. Measuring sleep efficiency: What should the denominator be? *J. Clin. Sleep Med.* **2016**, *12*, 263–266. [CrossRef] [PubMed]
25. Klingelhoefer, L.; Bhattacharya, K.; Reichmann, H. Restless legs syndrome. *Clin. Med.* **2016**, *16*, 379–382. [CrossRef]
26. de Zambotti, M.; Goldstone, A.; Colrain, I.M.; Baker, F. Insomnia disorder in adolescence: Diagnosis, impact, and treatment. *Sleep Med. Rev.* **2018**, *39*, 12–24. [CrossRef]
27. Besedovsky, L.; Lange, T.; Born, J. Sleep and immune function. *Pflügers Arch. Eur. J. Physiol.* **2012**, *463*, 121–137. [CrossRef]
28. Scammell, T.E.; Arrigoni, E.; Lipton, J.O. Neural circuitry of wakefulness and sleep. *Neuron* **2017**, *93*, 747–765. [CrossRef]

29. Jessen, N.A.; Munk, A.S.F.; Lundgaard, I.; Nedergaard, M. The glymphatic system: A beginner's guide. *Neurochem. Res.* **2015**, *40*, 2583–2599. [CrossRef]
30. Adam, K. Sleep as a restorative process and a theory to explain why. *Prog. Brain Res.* **1980**, *53*, 289–305. [CrossRef]
31. Rasmussen, M.K.; Mestre, H.; Nedergaard, M. The glymphatic pathway in neurological disorders. *Lancet Neurol.* **2018**, *17*, 1016–1024. [CrossRef]
32. Chen, Y.; Tan, F.; Wei, L.; Li, X.; Lyu, Z.; Feng, X.; Wen, Y.; Guo, L.; He, J.; Dai, M.; et al. Sleep duration and the risk of cancer: A systematic review and meta-analysis including dose–response relationship. *BMC Cancer* **2018**, *18*, 1149. [CrossRef] [PubMed]
33. Baglioni, C.; Battagliese, G.; Feige, B.; Spiegelhalder, K.; Nissen, C.; Voderholzer, U.; Lombardo, C.; Riemann, D. Insomnia as a predictor of depression: A meta-analytic evaluation of longitudinal epidemiological studies. *J. Affect. Disord.* **2011**, *135*, 10–19. [CrossRef] [PubMed]
34. St-Onge, M.-P.; Mikic, A.; Pietrolungo, C.E. Effects of diet on sleep quality. *Adv. Nutr. Int. Rev. J.* **2016**, *7*, 938–949. [CrossRef]
35. Binks, H.; Vincent, G.E.; Gupta, C.; Irwin, C.; Khalesi, S. Effects of diet on sleep: A narrative review. *Nutrients* **2020**, *12*, 936. [CrossRef]
36. Martínez-Rodríguez, A.; Rubio-Arias, J.; Ramos-Campo, D.J.; Reche-García, C.; Leyva-Vela, B.; Nadal-Nicolás, Y. Psychological and sleep effects of tryptophan and magnesium-enriched mediterranean diet in women with fibromyalgia. *Int. J. Environ. Res. Public Health* **2020**, *17*, 2227. [CrossRef]
37. O'Callaghan, F.; Muurlink, O.; Reid, N. Effects of caffeine on sleep quality and daytime functioning. *Risk Manag. Health Policy* **2018**, *11*, 263–271. [CrossRef]
38. Shilo, L.; Sabbah, H.; Hadari, R.; Kovatz, S.; Weinberg, U.; Dolev, S.; Dagan, Y.; Shenkman, L. The effects of coffee consumption on sleep and melatonin secretion. *Sleep Med.* **2002**, *3*, 271–273. [CrossRef]
39. He, S.; Hasler, B.P.; Chakravorty, S. Alcohol and sleep-related problems. *Curr. Opin. Psychol.* **2019**, *30*, 117–122. [CrossRef]
40. Gérard, C.; Vidal, H. Impact of gut microbiota on host glycemic control. *Front. Endocrinol.* **2019**, *10*, 29. [CrossRef]
41. Zhao, M.; Tuo, H.; Wang, S.; Zhao, L. The effects of dietary nutrition on sleep and sleep disorders. *Mediat. Inflamm.* **2020**, *2020*, 3142874. [CrossRef] [PubMed]
42. Doherty, R.; Madigan, S.; Warrington, G.; Ellis, J. Sleep and nutrition interactions: Implications for athletes. *Nutrients* **2019**, *11*, 822. [CrossRef]
43. Ogilvie, R.P.; Patel, S. The epidemiology of sleep and obesity. *Sleep Health* **2018**, *3*, 383–388. [CrossRef] [PubMed]
44. Basch, C.E.; Basch, C.H.; Ruggles, K.V.; Rajan, S. Prevalence of sleep duration on an average school night among 4 nationally representative successive samples of american high school students, 2007–2013. *Prev. Chronic Dis.* **2014**, *11*, E216. [CrossRef] [PubMed]
45. Matricciani, L.; Olds, T.; Petkov, J. In search of lost sleep: Secular trends in the sleep time of school-aged children and adolescents. *Sleep Med. Rev.* **2012**, *16*, 203–211. [CrossRef] [PubMed]
46. Ford, E.S.; Cunningham, T.J.; Croft, J.B. Trends in self-reported sleep duration among US adults from 1985 to 2012. *Sleep* **2015**, *38*, 829–832. [CrossRef] [PubMed]
47. Gohil, A.; Hannon, T.S. Poor sleep and obesity: Concurrent epidemics in adolescent youth. *Front. Endocrinol.* **2018**, *9*, 364. [CrossRef]
48. Taheri, S.; Lin, L.; Austin, D.; Young, T.; Mignot, E. Short sleep duration is associated with reduced leptin, elevated ghrelin, and increased body mass index. *PLoS Med.* **2004**, *1*, e62. [CrossRef]
49. Liu, X.; Forbes, E.; Ryan, N.D.; Rofey, D.; Hannon, T.S.; Dahl, R.E. Rapid eye movement sleep in relation to overweight in children and adolescents. *Arch. Gen. Psychiatry* **2008**, *65*, 924–932. [CrossRef]
50. Ding, C.; Lim, L.L.; Xu, L.; Kong, A.P.S. Sleep and obesity. *J. Obes. Metab. Syndr.* **2018**, *27*, 4–24. [CrossRef]
51. Grandner, M.A.; Kripke, D.F.; Naidoo, N.; Langer, R.D. Relationships among dietary nutrients and subjective sleep, objective sleep, and napping in women. *Sleep Med.* **2010**, *11*, 180–184. [CrossRef] [PubMed]
52. Boeke, S.C.E.; Storfer-Isser, A.; Redline, S.; Taveras, E.M. Childhood sleep duration and quality in relation to leptin concentration in two cohort studies. *Sleep* **2014**, *37*, 613–620. [CrossRef] [PubMed]
53. Bailey, M.A.; Holscher, H.D. Microbiome-mediated effects of the mediterranean diet on inflammation. *Adv. Nutr.* **2018**, *9*, 193–206. [CrossRef]
54. Al-Rashed, F.; Sindhu, S.; Al Madhoun, A.; Alghaith, A.; Azim, R.; Al-Mulla, F.; Ahmad, R. Short sleep duration and its association with obesity and other metabolic risk factors in kuwaiti urban adults. *Nat. Sci. Sleep* **2021**, *13*, 1225–1241. [CrossRef] [PubMed]
55. Solfrizzi, V.; Colacicco, A.M.; D'Introno, A.; Capurso, C.; Torres, F.; Rizzo, C.; Capurso, A.; Panza, F. Dietary intake of unsaturated fatty acids and age-related cognitive decline: A 8.5-year follow-up of the Italian longitudinal study on aging. *Neurobiol. Aging* **2006**, *27*, 1694–1704. [CrossRef] [PubMed]
56. de Souza, R.; Mente, A.; Maroleanu, A.; Cozma, A.I.; Ha, V.; Kishibe, T.; Uleryk, E.; Budylowski, P.; Schünemann, H.; Beyene, J.; et al. Intake of saturated and trans unsaturated fatty acids and risk of all cause mortality, cardiovascular disease, and type 2 diabetes: Systematic review and meta-analysis of observational studies. *BMJ* **2015**, *351*, h3978. [CrossRef]
57. Hansen, A.L.; Dahl, L.; Olson, G.; Thornton, D.; Graff, I.E.; Frøyland, L.; Thayer, J.F.; Pallesen, S. Fish consumption, sleep, daily functioning, and heart rate variability. *J. Clin. Sleep Med.* **2014**, *10*, 567–575. [CrossRef]

58. Lavialle, M.; Champeil-Potokar, G.; Alessandri, J.M.; Balasse, L.; Guesnet, P.; Papillon, C.; Pévet, P.; Vancassel, S.; Vivien-Roels, B.; Denis, I. An (n-3) polyunsaturated fatty acid–deficient diet disturbs daily locomotor activity, melatonin rhythm, and striatal dopamine in syrian hamsters. *J. Nutr.* **2008**, *138*, 1719–1724. [CrossRef]
59. St-Onge, M.-P.; Roberts, A.; Shechter, A.; Choudhury, A.R. Fiber and saturated fat are associated with sleep arousals and slow wave sleep. *J. Clin. Sleep Med.* **2016**, *12*, 19–24. [CrossRef]
60. Sacks, F.M.; Lichtenstein, A.H.; Wu, J.H.Y.; Appel, L.J.; Creager, M.A.; Kris-Etherton, P.M.; Miller, M.; Rimm, E.B.; Rudel, L.L.; Robinson, J.G.; et al. Dietary fats and cardiovascular disease: A presidential advisory from the american heart association. *Circulation* **2017**, *136*, e1–e23. [CrossRef]
61. Surani, S.; Brito, V.; Surani, A.; Ghamande, S. Effect of diabetes mellitus on sleep quality. *World J. Diabetes.* **2015**, *6*, 868–873. [CrossRef]
62. Urade, Y.; Hayaishi, O. Prostaglandin D2 and sleep/wake regulation. *Sleep Med. Rev.* **2011**, *15*, 411–418. [CrossRef] [PubMed]
63. Shan, Z.; Rehm, C.D.; Rogers, G.; Ruan, M.; Wang, D.D.; Hu, F.B.; Mozaffarian, D.; Fang Zhang, F.; Bhupathiraju, S.N. Trends in dietary carbohydrate, protein, and fat intake and diet quality among US adults, 1999–2016. *JAMA J. Am. Med. Assoc.* **2019**, *322*, 1178–1187. [CrossRef]
64. Brandhorst, S.; Longo, V.S.D. Protein quantity and source, fasting-mimicking diets, and longevity. *Adv. Nutr. Int. Rev. J.* **2019**, *10*, S340–S350. [CrossRef]
65. Tanaka, E.; Yatsuya, H.; Uemura, M.; Murata, C.; Otsuka, R.; Toyoshima, H.; Tamakoshi, K.; Sasaki, S.; Kawaguchi, L.; Aoyama, A. Associations of protein, fat, and carbohydrate intakes with insomnia symptoms among middle-aged Japanese workers. *J. Epidemiol.* **2013**, *23*, 132–138. [CrossRef]
66. Sutanto, C.N.; Loh, W.W.; Toh, D.W.K.; Lee, D.P.S.; Kim, J.E. Association between dietary protein intake and sleep quality in middle-aged and older adults in Singapore. *Front. Nutr.* **2022**, *9*, 832341. [CrossRef] [PubMed]
67. Trommelen, J.; Van Loon, L.J.C. Pre-sleep protein ingestion to improve the skeletal muscle adaptive response to exercise training. *Nutrients* **2016**, *8*, 763. [CrossRef] [PubMed]
68. Daneshzad, E.; Keshavarz, S.A.; Qorbani, M.; Larijani, B.; Azadbakht, L. Association between a low-carbohydrate diet and sleep status, depression, anxiety, and stress score. *J. Sci. Food Agric.* **2020**, *100*, 2946–2952. [CrossRef]
69. Afaghi, A.; O'Connor, H.; Chow, C.M. High-glycemic-index carbohydrate meals shorten sleep onset. *Am. J. Clin. Nutr.* **2007**, *85*, 426–430. [CrossRef]
70. Gangwisch, J.; Hale, L.; St-Onge, M.-P.; Choi, L.; LeBlanc, E.S.; Malaspina, D.; Opler, M.G.; Shadyab, A.H.; Shikany, J.M.; Snetselaar, L.; et al. High glycemic index and glycemic load diets as risk factors for insomnia: Analyses from the women's health initiative. *Am. J. Clin. Nutr.* **2020**, *111*, 429–439. [CrossRef]
71. Jenkins, D.J.A.; Jenkins, A.L. Dietary fiber and the glycemic response. *Proc. Soc. Exp. Biol. Med.* **1985**, *180*, 422–431. [CrossRef] [PubMed]
72. Wurtman, R.J.; Wurtman, J.J.; Regan, M.M.; McDermott, J.M.; Tsay, R.H.; Breu, J.J. Effects of normal meals rich in carbohydrates or proteins on plasma tryptophan and tyrosine ratios. *Am. J. Clin. Nutr.* **2003**, *77*, 128–132. [CrossRef] [PubMed]
73. Kim, Y.; Chen, J.; Wirth, M.D.; Shivappa, N.; Hebert, J.R. Lower dietary inflammatory index scores are associated with lower glycemic index scores among college students. *Nutrients* **2018**, *10*, 182. [CrossRef] [PubMed]
74. Kräuchi, K.; Cajochen, C.; Werth, E.; Wirz-Justice, A. Alteration of internal circadian phase relationships after morning versus evening carbohydrate-rich meals in humans. *J. Biol. Rhythm.* **2002**, *17*, 364–376. [CrossRef]
75. Haghighatdoost, F.; Karimi, G.; Esmaillzadeh, A.; Azadbakht, L. Sleep deprivation is associated with lower diet quality indices and higher rate of general and central obesity among young female students in Iran. *Nutrition* **2012**, *28*, 1146–1150. [CrossRef]
76. Nishiura, C.; Noguchi, J.; Hashimoto, H. Dietary patterns only partially explain the effect of short sleep duration on the incidence of obesity. *Sleep* **2010**, *33*, 753–757. [CrossRef]
77. Watson, E.J.; Coates, A.M.; Kohler, M.; Banks, S. Caffeine consumption and sleep quality in Australian adults. *Nutrients* **2016**, *8*, 479. [CrossRef]
78. Hindmarch, I.; Rigney, U.; Stanley, N.; Quinlan, P.; Rycroft, J.; Lane, J. A naturalistic investigation of the effects of day-long consumption of tea, coffee and water on alertness, sleep onset and sleep quality. *Psychopharmacology* **2000**, *149*, 203–216. [CrossRef]
79. Van Dongen, H.P.; Price, N.J.; Mullington, J.M.; Szuba, M.P.; Kapoor, S.C.; Dinges, D.F. Caffeine eliminates psychomotor vigilance deficits from sleep inertia. *Sleep* **2001**, *24*, 813–819. [CrossRef]
80. Landolt, H.-P.; Werth, E.; Borbély, A.A.; Dijk, D.-J. Caffeine intake (200 mg) in the morning affects human sleep and EEG power spectra at night. *Brain Res.* **1995**, *675*, 67–74. [CrossRef]
81. Ward, A.L.; Jospe, M.; Morrison, S.; Reynolds, A.N.; Kuroko, S.; Fangupo, L.J.; Smith, C.; Galland, B.C.; Taylor, R.W. Bidirectional associations between sleep quality or quantity, and dietary intakes or eating behaviors in children 6–12 years old: A systematic review with evidence mapping. *Nutr. Rev.* **2021**, *79*, 1079–1099. [CrossRef] [PubMed]
82. Garcion, E.; Wion-Barbot, N.; Montero-Menei, C.N.; Berger, F.; Wion, D. New clues about vitamin D functions in the nervous system. *Trends Endocrinol. Metab.* **2002**, *13*, 100–105. [CrossRef]
83. Lagunova, Z.; Porojnicu, A.; Aksnes, L.; Holick, M.; Iani, V.; Bruland, O.S.; Moan, J. Effect of vitamin D supplementation and ultraviolet B exposure on serum 25-hydroxyvitamin D concentrations in healthy volunteers: A randomized, crossover clinical trial. *Br. J. Dermatol.* **2013**, *169*, 434–440. [CrossRef] [PubMed]

84. Vaes, A.; Brouwer-Brolsma, E.; van der Zwaluw, N.; van Wijngaarden, J.; Berendsen, A.; van Schoor, N.; van der Velde, N.; Uitterlinden, A.; Lips, P.; Dhonukshe-Rutten, R.; et al. Food sources of vitamin D and their association with 25-hydroxyvitamin D status in Dutch older adults. *J. Steroid Biochem. Mol. Biol.* **2017**, *173*, 228–234. [CrossRef] [PubMed]
85. Jung, Y.S.; Chae, C.H.; Kim, Y.O.; Son, J.S.; Kim, C.W.; Park, H.O.; Lee, J.H.; Shin, Y.H.; Kwak, H.S. The relationship between serum vitamin D levels and sleep quality in fixed day indoor field workers in the electronics manufacturing industry in Korea. *Ann. Occup. Environ. Med.* **2017**, *29*, 25. [CrossRef]
86. Musiol, I.M.; Stumpf, W.E.; Bidmon, H.-J.; Heiss, C.; Mayerhofer, A.; Bartke, A. Vitamin d nuclear binding to neurons of the septal, substriatal and amygdaloid area in the siberian hamster (phodopus sungorus) brain. *Neuroscience* **1992**, *48*, 841–848. [CrossRef]
87. Eyles, D.W. Vitamin D: Brain and behavior. *JBMR Plus* **2021**, *5*, e10419. [CrossRef]
88. Watson, E.J.; Banks, S.; Coates, A.M.; Kohler, M.J. The relationship between caffeine, sleep, and behavior in children. *J. Clin. Sleep Med.* **2017**, *13*, 533–543. [CrossRef]
89. Majid, M.S.; Ahmad, H.S.; Bizhan, H.; Hosein, H.Z.M.; Mohammad, A. The effect of vitamin D supplement on the score and quality of sleep in 20–50 year-old people with sleep disorders compared with control group. *Nutr. Neurosci.* **2018**, *21*, 511–519. [CrossRef]
90. Watford, M.; Wu, G. Protein. *Adv. Nutr. Int. Rev. J.* **2018**, *9*, 651–653. [CrossRef]
91. He, W.; Wu, G. Metabolism of amino acids in the brain and their roles in regulating food intake. *Adv. Exp. Med. Biol.* **2020**, *1265*, 167–185. [CrossRef] [PubMed]
92. Bravo, R.; Matito, S.; Cubero, J.; Paredes, S.D.; Franco, L.; Rivero, M.; Rodríguez, A.B.; Barriga, C. Tryptophan-enriched cereal intake improves nocturnal sleep, melatonin, serotonin, and total antioxidant capacity levels and mood in elderly humans. *AGE* **2012**, *35*, 1277–1285. [CrossRef] [PubMed]
93. Zhao, D.; Yu, Y.; Shen, Y.; Liu, Q.; Zhao, Z.; Sharma, R.; Reiter, R.J. Melatonin synthesis and function: Evolutionary history in animals and plants. *Front. Endocrinol.* **2019**, *10*, 249. [CrossRef] [PubMed]
94. Howatson, G.; Bell, P.G.; Tallent, J.; Middleton, B.; McHugh, M.P.; Ellis, J. Effect of tart cherry juice (Prunus cerasus) on melatonin levels and enhanced sleep quality. *Eur. J. Nutr.* **2012**, *51*, 909–916. [CrossRef]
95. Diana, M.; Quílez, J.; Rafecas, M. Gamma-aminobutyric acid as a bioactive compound in foods: A review. *J. Funct. Foods* **2014**, *10*, 407–420. [CrossRef]
96. Briguglio, M.; Dell'Osso, B.; Panzica, G.; Malgaroli, A.; Banfi, G.; Dina, C.Z.; Galentino, R.; Porta, M. Dietary neurotransmitters: A narrative review on current knowledge. *Nutrients* **2018**, *10*, 591. [CrossRef] [PubMed]
97. Byun, J.-I.; Shin, Y.Y.; Chung, S.-E.; Shin, W.C. Safety and efficacy of gamma-aminobutyric acid from fermented rice germ in patients with insomnia symptoms: A randomized, double-blind trial. *J. Clin. Neurol.* **2018**, *14*, 291–295. [CrossRef]
98. Yamatsu, A.; Yamashita, Y.; Maru, I.; Yang, J.; Tatsuzaki, J.; Kim, M. The improvement of sleep by oral intake of GABA and apocynum venetum leaf extract. *J. Nutr. Sci. Vitaminol.* **2015**, *61*, 182–187. [CrossRef]
99. Luppi, P.-H.; Peyron, C.; Fort, P. Not a single but multiple populations of GABAergic neurons control sleep. *Sleep Med. Rev.* **2017**, *32*, 85–94. [CrossRef]
100. Irish, L.A.; Kline, C.E.; Gunn, H.E.; Buysse, D.J.; Hall, M.H. The role of sleep hygiene in promoting public health: A review of empirical evidence. *Sleep Med. Rev.* **2015**, *22*, 23–36. [CrossRef]
101. Chen, P.H.; Kuo, H.Y.; Chueh, K.H. Sleep hygiene education efficacy on sleep quality in working women. *J. Nurs. Res.* **2010**, *18*, 283–289. [CrossRef] [PubMed]
102. Hershner, S.; O'Brien, L.M. The impact of a randomized sleep education intervention for college students. *J. Clin. Sleep Med.* **2018**, *14*, 337–347. [CrossRef] [PubMed]
103. Zengin, L.; Aylaz, R. The effects of sleep hygiene education and reflexology on sleep quality and fatigue in patients receiving chemotherapy. *Eur. J. Cancer Care* **2019**, *28*, e13020. [CrossRef]
104. Stein, M.D.; Friedmann, P.D. Disturbed sleep and its relationship to alcohol use. *Subst. Abus.* **2006**, *26*, 1–13. [CrossRef]
105. Drummond, S.P.; Gillin, J.C.; Smith, T.L.; DeModena, A. The sleep of abstinent pure primary alcoholic patients: Natural course and relationship to relapse. *Alcohol. Clin. Exp. Res.* **1998**, *22*, 1796–1802. [CrossRef]
106. Costa e Silva, J.A.; Chase, M.; Sartorius, N.; Roth, T. Special report from a symposium held by the world health organization and the world federation of sleep research societies: An overview of insomnias and related disorders-recognition, epidemiology, and rational management. *Sleep* **1996**, *19*, 412–416. [CrossRef]
107. Zhang, R.; Tomasi, D.; Manza, P.; Shokri-Kojori, E.; Demiral, S.B.; Feldman, D.E.; Kroll, D.S.; Biesecker, C.L.; McPherson, K.L.; Wang, G.-J.; et al. Sleep disturbances are associated with cortical and subcortical atrophy in alcohol use disorder. *Transl. Psychiatry* **2021**, *11*, 428. [CrossRef] [PubMed]
108. Aldrich, M.S.; Shipley, J.E. Alcohol use and periodic limb movements of sleep. *Alcohol. Clin. Exp. Res.* **1993**, *17*, 192–196. [CrossRef]
109. Nigri, D.H.; Addor, G.; De Barros Franco, C.A.; Martins, R.C.; Gomes, M.P.V.F.; Strollo, P.J. Alcohol induced apnea. *J. Clin. Sleep. Med.* **2005**, *1*, 424–426. [CrossRef]
110. Purani, H.; Friedrichsen, S.; Allen, A.M. Sleep quality in cigarette smokers: Associations with smoking-related outcomes and exercise. *Addict. Behav.* **2018**, *90*, 71–76. [CrossRef]
111. AlRyalat, S.A.; Kussad, S.; El Khatib, O.; Hamad, I.; Al-Tanjy, A.; Alshnneikat, M.; AbuMahfouz, B. Assessing the effect of nicotine dose in cigarette smoking on sleep quality. *Sleep Breath.* **2021**, *25*, 1319–1324. [CrossRef] [PubMed]

112. Wetter, D.W.; Fiore, M.C.; Baker, T.B.; Young, T.B. Tobacco withdrawal and nicotine replacement influence objective measures of sleep. *J. Consult. Clin. Psychol.* **1995**, *63*, 658–667. [CrossRef]
113. Kesner, A.J.; Lovinger, D.M. Cannabinoids, endocannabinoids and sleep. *Front. Mol. Neurosci.* **2020**, *13*, 125. [CrossRef] [PubMed]
114. Short, N.A.; Zvolensky, M.J.; Schmidt, N.B. A pilot randomized clinical trial of brief behavioral treatment for insomnia to reduce problematic cannabis use among trauma-exposed young adults. *J. Subst. Abus. Treat.* **2021**, *131*, 108537. [CrossRef] [PubMed]
115. Chakravorty, S.; Vandrey, R.G.; He, S.; Stein, M.D. Sleep management among patients with substance use disorders. *Med. Clin. N. Am.* **2018**, *102*, 733–743. [CrossRef]
116. Kaul, M.; Zee, P.C.; Sahni, A.S. Effects of cannabinoids on sleep and their therapeutic potential for sleep disorders. *Neurotherapeutics* **2021**, *18*, 217–227. [CrossRef]
117. Piper, B.J.; DeKeuster, R.M.; Beals, M.L.; Cobb, C.M.; Burchman, C.A.; Perkinson, L.; Lynn, S.T.; Nichols, S.D.; Abess, A.T. Substitution of medical cannabis for pharmaceutical agents for pain, anxiety, and sleep. *J. Psychopharmacol.* **2017**, *31*, 569–575. [CrossRef]
118. Angarita, G.A.; Emadi, N.; Hodges, S.; Morgan, P.T. Sleep abnormalities associated with alcohol, cannabis, cocaine, and opiate use: A comprehensive review. *Addict. Sci. Clin. Pract.* **2016**, *11*, 9. [CrossRef]
119. Conroy, D.A.; Kurth, M.E.; Strong, D.R.; Brower, K.J.; Stein, M.D. Marijuana use patterns and sleep among community-based young adults. *J. Addict. Dis.* **2016**, *35*, 135–143. [CrossRef]
120. Haney, M.; Ward, A.S.; Comer, S.D.; Foltin, R.W.; Fischman, M.W. Abstinence symptoms following oral THC administration to humans. *Psychopharmacology* **1999**, *141*, 385–394. [CrossRef]
121. Diaz, K.M.; Shimbo, D. Physical activity and the prevention of hypertension. *Curr. Hypertens. Rep.* **2013**, *15*, 659–668. [CrossRef] [PubMed]
122. Wilckens, K.A.; Erickson, K.I.; Wheeler, M.E. Physical Activity and Cognition: A Mediating Role of Efficient Sleep. *Behav. Sleep Med.* **2018**, *16*, 569–586. [CrossRef] [PubMed]
123. Christie, A.D.; Seery, E.; Kent, J.A. Physical activity, sleep quality, and self-reported fatigue across the adult lifespan. *Exp. Gerontol.* **2016**, *77*, 7–11. [CrossRef]
124. Murray, K.; Godbole, S.; Natarajan, L.; Full, K.; Hipp, J.A.; Glanz, K.; Mitchell, J.; Laden, F.; James, P.; Quante, M.; et al. The relations between sleep, time of physical activity, and time outdoors among adult women. *PLoS ONE* **2017**, *12*, e0182013. [CrossRef]
125. Tsunoda, K.; Kitano, N.; Kai, Y.; Uchida, K.; Kuchiki, T.; Okura, T.; Nagamatsu, T. Prospective study of physical activity and sleep in middle-aged and older adults. *Am. J. Prev. Med.* **2015**, *48*, 662–673. [CrossRef] [PubMed]
126. Bisson, A.N.S.; Robinson, S.; Lachman, M.E. Walk to a better night of sleep: Testing the relationship between physical activity and sleep. *Sleep Health* **2019**, *5*, 487–494. [CrossRef]
127. Buxton, O.M.; L'Hemite-Balériaux, M.; Hirschfeld, U.; Cauter, E. Acute and delayed effects of exercise on human melatonin secretion. *J. Biol. Rhythm.* **1997**, *12*, 568–574. [CrossRef]
128. Caspersen, C.J.; Powell, K.E.; Christenson, G.M. Physical activity, exercise, and physical fitness: Definitions and distinctions for health-related research. *Public Health Rep.* **1985**, *100*, 126–131.
129. Passos, G.S.; Poyares, D.; Santana, M.G.; Garbuio, S.A.; Tufik, S.; de Mello, M.T. Effect of acute physical exercise on patients with chronic primary insomnia. *J. Clin. Sleep Med.* **2010**, *6*, 270–275. [CrossRef]
130. Chen, L.-J.; Fox, K.R.; Ku, P.-W.; Chang, Y.-W. Effects of aquatic exercise on sleep in older adults with mild sleep impairment: A randomized controlled trial. *Int. J. Behav. Med.* **2015**, *23*, 501–506. [CrossRef] [PubMed]
131. Vitale, J.A.; Torre, A.L.; Banfi, G.; Bonato, M. Acute sleep hygiene strategy improves objective sleep latency following a late-evening soccer-specific training session: A randomized controlled trial. *J. Sports Sci.* **2019**, *37*, 2711–2719. [CrossRef] [PubMed]
132. Focht, B.C.; Koltyn, K.F. Influence of resistance exercise of different intensities on state anxiety and blood pressure. *Med. Sci. Sports Exerc.* **1999**, *31*, 456–463. [CrossRef] [PubMed]
133. Dias de Andrade, F.M.; Pinto Pedrosa, R. The role of physical exercise in obstructive sleep apnea. *J. Bras. Pneumol.* **2016**, *42*, 457–464. [CrossRef] [PubMed]
134. Pham, L.V.; Schwartz, A.R. The pathogenesis of obstructive sleep apnea. *J. Thorac. Dis.* **2015**, *7*, 1358–1372.
135. Mendelson, M.; Lyons, O.; Yadollahi, A.; Inami, T.; Oh, P.; Bradley, T.D. Effects of exercise training on sleep apnoea in patients with coronary artery disease: A randomised trial. *Eur. Respir. J.* **2016**, *48*, 142–150. [CrossRef]
136. McSharry, D.G.; Saboisky, J.P.; DeYoung, P.; Matteis, P.; Jordan, A.S.; Trinder, J.; Smales, E.; Hess, L.; Guo, M.; Malhotra, A. A Mechanism for upper airway stability during slow wave sleep. *Sleep* **2013**, *36*, 555–563. [CrossRef]
137. Iftikhar, I.H.; Kline, C.E.; Youngstedt, S.D. Effects of exercise training on sleep apnea: A meta-analysis. *Lung* **2014**, *192*, 175–184. [CrossRef]
138. Janssen, X.; Martin, A.; Hughes, A.R.; Hill, C.M.; Kotronoulas, G.; Hesketh, K.R. Associations of screen time, sedentary time and physical activity with sleep in under 5 s: A systematic review and meta-analysis. *Sleep Med. Rev.* **2020**, *49*, 101226. [CrossRef]
139. Ji, M.; Tang, A.; Zhang, Y.; Zou, J.; Zhou, G.; Deng, J.; Yang, L.; Li, M.; Chen, J.; Qin, H.; et al. The relationship between obesity, sleep and physical activity in Chinese preschool children. *Int. J. Environ. Res. Public Health* **2018**, *15*, 527. [CrossRef]
140. Calhoun, S.L.; Fernandez-Mendoza, J.; Vgontzas, A.N.; Liao, D.; Bixler, E.O. Prevalence of insomnia symptoms in a general population sample of young children and preadolescents: Gender effects. *Sleep Med.* **2014**, *15*, 91–95. [CrossRef]

141. Yoong, S.L.; Grady, A.; Stacey, F.; Polimeni, M.; Clayton, O.; Jones, J.; Nathan, N.; Wyse, R.; Wolfenden, L. A pilot randomized controlled trial examining the impact of a sleep intervention targeting home routines on young children's (3–6 years) physical activity. *Pediatr. Obes.* **2019**, *14*, e12481. [CrossRef] [PubMed]
142. Collier, S.; Fairbrother, K.; Cartner, B.; Alley, J.R.; Curry, C.D.; Dickinson, D.L.; Morris, D.M. Effects of exercise timing on sleep architecture and nocturnal blood pressure in prehypertensives. *Vasc. Health Risk Manag.* **2014**, *10*, 691–698. [CrossRef] [PubMed]
143. Štefan, L.; Vrgoč, G.; Rupčić, T.; Sporiš, G.; Sekulić, D. Sleep duration and sleep quality are associated with physical activity in elderly people living in nursing homes. *Int. J. Environ. Res. Public Health* **2018**, *15*, 2512. [CrossRef] [PubMed]
144. Hartescu, I.; Morgan, K.; Stevinson, C.D. Increased physical activity improves sleep and mood outcomes in inactive people with insomnia: A randomized controlled trial. *J. Sleep Res.* **2015**, *24*, 526–534. [CrossRef] [PubMed]
145. Yang, P.-Y.; Ho, K.-H.; Chen, H.-C.; Chien, M.-Y. Exercise training improves sleep quality in middle-aged and older adults with sleep problems: A systematic review. *J. Physiother.* **2012**, *58*, 157–163. [CrossRef]
146. WHO. A Technical Package for Increasing Physical Activity. Available online: https://apps.who.int/iris/bitstream/handle/10665/275415/9789241514804-eng.pdf?sequence=1&isAllowed=y. (accessed on 1 March 2022).
147. Reid, K.J.; Baron, K.G.; Lu, B.; Naylor, E.; Wolfe, L.; Zee, P.C. Aerobic exercise improves self-reported sleep and quality of life in older adults with insomnia. *Sleep Med.* **2010**, *11*, 934–940. [CrossRef] [PubMed]

nutrients MDPI

Article

Changes in Sleep Patterns during Pregnancy and Predictive Factors: A Longitudinal Study in Saudi Women

Sara Al-Musharaf

Department of Community Health Sciences, College of Applied Medical Sciences, King Saud University, Riyadh 11451, Saudi Arabia; salmosharruf@ksu.edu.sa; Tel.: +966-55-424-3033

Abstract: This study aimed to assess sleep patterns during the three trimesters of pregnancy and whether vitamin D concentrations, along with other risk factors, are associated with these alterations. In a longitudinal study, 140 pregnant women (age 18 to 39 years) were followed throughout their first, second, and third trimesters. Sleep was measured using the Pittsburgh Sleep Quality Index (PSQI) at each trimester, along with an assessment of biochemical parameters, including serum vitamin D levels. The information that was collected included anthropometric data, socio-economic status, dietary intake, and physical activity. The PSQI was higher in mid and late pregnancy than in early pregnancy (both p = 0.001), and the sleep duration was also higher in late versus early pregnancy. Linear regression analyses revealed independent predictors of deteriorating sleep quality from early to late pregnancy, including low income (B ± SE −0.60 ± 0.26, p = 0.03) and low serum vitamin D levels in the second trimester (B ± SE −0.20 ± 0.01, p = 0.04). Energy intake and sitting in the second half of pregnancy were positively associated with changes in the PSQI score from the second to third trimesters (B ± SE 0.15 ± 0.07, p = 0.048) and (B ± SE 0.01 ± 0.00, p = 0.044), respectively. Low socio-economic status, low serum vitamin D levels, greater energy intake, and sitting time were associated with worsening patterns of sleep quality from early to late pregnancy.

Keywords: sleep; PSQI; poor sleep; short sleep duration; vitamin D; pregnancy trimesters; women

Citation: Al-Musharaf, S. Changes in Sleep Patterns during Pregnancy and Predictive Factors: A Longitudinal Study in Saudi Women. *Nutrients* **2022**, *14*, 2633. https://doi.org/10.3390/nu14132633

Academic Editors: Egeria Scoditti and Sergio Garbarino

Received: 1 June 2022
Accepted: 22 June 2022
Published: 25 June 2022

1. Introduction

Sleep needs vary with age, gender, lifestyle, and other factors, such as work schedules and stress [1,2]. Poor sleep is linked to neuroendocrine, metabolic, and inflammatory changes [3]. Studies have also demonstrated that poor sleep quality and inadequate sleep duration may lead to various health complications [4,5]. Poor sleep is common during pregnancy due to physiological and psychological changes [6,7]. However, the prevalence of altered sleep patterns and disturbances has noticeably increased in recent times, i.e., such patterns and disturbances are now considered an epidemic among the general population [8], especially among pregnant women [9]. According to the National Sleep Foundation in 2017, 78% of women reported more disturbed sleep during pregnancy than at any other time in their lives [10]. A meta-analysis showed that the prevalence of poor sleep quality, defined by a Pittsburgh Sleep Quality Index (PSQI) global score of 5 or above, was approximately 44.5% during pregnancy [11,12]. Furthermore, numerous studies have shown a connection between abnormal sleep patterns and a broad spectrum of adverse pregnancy outcomes, including low birthweight, preterm birth, intrauterine growth retardation, caesarean delivery, gestational hypertension, and gestational diabetes [13–16], along with a decreased quality of life and higher levels of depressive symptoms [17].

Despite the high prevalence of abnormal and disturbed sleep during pregnancy and the adverse health ramifications, studies evaluating changes in sleep patterns and the associated predictive factors across pregnancy trimesters are scarce. Few prospective cohort studies have analyzed changes in sleep patterns during pregnancy in the same cohort of women [16,18–22], with the majority assessing only two time points [12].

Frequent urination, body aches, and sleep positions are among the most well-known reported causes of poor sleep during pregnancy and often vary depending on the trimester [6,23]. However, a broad range of other factors, such as gestational age [12,23], maternal age [11,16], income [24], parity, and race [18], have also been associated with changes in sleep patterns during pregnancy. Obesity and gestational weight gain (GWG) have also been shown to correlate with worsening sleep quality [19,25,26]. Although biochemical markers, such as abnormal glucose and lipid levels, have been linked to poor sleep quality in the general population [25,27,28], to our knowledge, few studies have investigated the relationship between hyperglycemia and gestational diabetes during pregnancy and changes in patterns of sleep [29,30].

Few studies have examined the role that excess macronutrients [31] or deficiencies in micronutrients [32], such as vitamin D deficiency [33], may play in contributing to poor sleep. Vitamin D deficiency is common during pregnancy both globally [34] and in Saudi Arabia [35] and it has been independently linked to pregnancy-related complications that adversely affect the long-term health of mothers and their infants [13,36].

A recent meta-analysis, in which two of the nine included studies examined pregnant women, found that vitamin D deficiency was associated with a 1.5-times increased chance of women experiencing poor sleep quality, short sleep duration, and sleepiness [37]. To our knowledge, limited studies have evaluated the association between vitamin D levels and sleep patterns in a pregnant population [33,38–40]. Such studies were conducted in the second or third trimester and did not provide insights into the presence or type of sleep disturbances in early pregnancy, nor did they evaluate changes in sleep patterns throughout pregnancy. Furthermore, most longitudinal studies did not explore biochemical factors, such as glucose, lipid, and vitamin D levels, nor did they consider levels of physical activity or dietary intake in relation to sleep patterns.

Consequently, this study aims to expand our understanding of the types of poor sleep that are experienced by women over the entire pregnancy and identify the predictive factors that increase the risk of poor sleep. Changes in sleep patterns among Saudi women during the first, second, and third trimester of pregnancy were assessed using the PSQI, and the relationship between sleep alterations and biochemical markers such as vitamin D, lipids, glycated hemoglobin (HbA1c), and glucose levels, as well as socio-demographic factors, such as age, parity, body mass index (BMI), GWG, dietary intake, and physical activity, were investigated. Routinely screening pregnant women for sleep-related problems and the presence of potentially modifiable risk factors may contribute substantially to improvements in maternal and fetal/neonatal health outcomes.

2. Materials and Methods

2.1. Study Design

Pregnant women who previously participated in an earlier prospective cohort study (E-13-1013) made up 24% (141/578) of the participants. The parent study (n = 578) assessed serum vitamin D levels with pregnancy-related complications. Ethical approval was obtained from the Ethics Committees of King Saud University Hospital (KSUH) and the Ministry of Health, Riyadh, Kingdom of Saudi Arabia (KSA) (IRB Approval, 14/4067/IRB). Prior to the commencement of the research, written informed consent was obtained from all participating women. The study was conducted in three hospitals in Riyadh city (King Fahad Medical City Hospital, King Khalid University Hospital, and King Salman Hospital) (latitude: 24°42′ N, 46°43′ E) between March 2014 and December 2017, as detailed previously [35]. The current analysis was a secondary analysis of the sleep data that were collected as part of this protocol.

A follow-up was carried out at three different points: early pregnancy (also referred to as the first trimester; 12 ± 3 weeks); mid-pregnancy (also referred to as the second trimester; 26 ± 4.8 weeks); and late pregnancy (also referred to as the third trimester; 34 ± 3 weeks).

2.2. Participants

A total of 141 women who attended the first visit (8–12 weeks) and second visit (24–28 weeks) and successfully participated in a third visit (29–40 weeks) were selected to be included in this secondary analysis. Details of the inclusion and exclusion criteria in the original study are presented in Al-Musharaf et al. [35]. Women were excluded from this analysis if they did not attend the third visit or were not interviewed to complete the sleep questionnaire. Additional exclusion criteria included the presence of systemic or psychiatric disorders; a previous diagnosis by a physician of a sleep disorder, such as sleep apnea syndrome, restless leg syndrome, insomnia, or parasomnia; use of any sleep medication; and involvement in nightshift work. Inclusion criteria included Saudi nationality between 18 and 39 years of age. Based on 80% power and a 95% CI, a sample size of 137 patients was calculated to detect a difference of 1.0 in the PSQI score from the first to third trimester [18].

2.3. Data Collection

Vital signs, a medical examination, and anthropometric measurements, including pre-pregnancy BMI, GWG, waist–hip ratio (WHR), and skin-fold thickness, were recorded for all three trimesters [35]. Additionally, data regarding socio-demographics, such as maternal age, employment, monthly income, parity, and past medical and treatment history were collected by a trained interviewer, as well as information from completed International Physical Activity Questionnaires (IPAQs) and a food frequency questionnaire, as detailed previously [35]. Blood samples were obtained from all pregnant women at the three visits for the parent study. The flow chart below describes the data that were collected at all three visits (Figure 1).

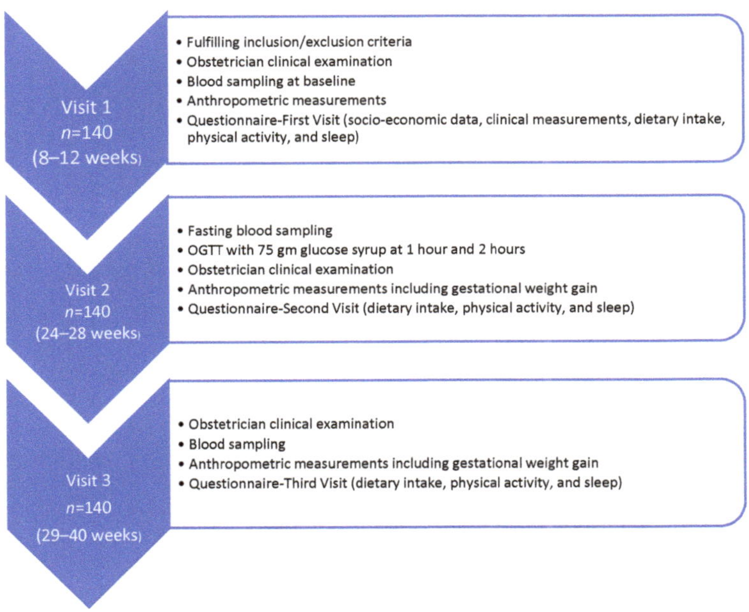

Figure 1. Flow chart of data collection across time.

2.3.1. Sleep Assessment

The PSQI was used to collect information about sleep quality over the previous month. The validated Arabic version of the questionnaire was used [41]. All the women were interviewed by a trained interviewer to complete the PSQI at all three visits. The PSQI is a subjective sleep-quality questionnaire that contains 19 multiple choice questions to address sleep quality, duration, latency (the time taken to fall asleep), efficiency, disturbances, medications, and daytime sleepiness that interferes with daily activities. These seven component scores are summed to obtain a global PSQI score that reflects the overall sleep quality (range 0–21). Each item receives a score from 0 to 3, and a mean global score of 5 or greater indicates poor sleep quality [42]. Thus, higher scores reflect a poorer quality of sleep [42]. The PSQI has good internal consistency and convergent and divergent reliability in pregnant populations [42]. This study defined short sleep duration with a cut-off of <7 h [9].

2.3.2. Biochemical Assessment

Blood samples were obtained at all three visits. At the first and third visits, the blood samples were collected in a non-fasting state. At the second visit, the women were fasting. The blood samples were immediately transported to the Chair for Biomarkers in Chronic Diseases (CBCD) at King Saud University, where they were processed, aliquoted, and stored at the recommended temperature for further analysis. The lipid profile, including total serum cholesterol (TC); high-density lipoprotein-cholesterol (HDL-C); triglycerides (TGs); HbA1c; and glucose levels were measured at all three visits by a colorimetric method using an automated chemistry analyzer (Konelab, ThermoFisher, Vantaa, Finland), as previously detailed [43]. The intra- and inter-assay coefficients of variation (CV) were TC: 0.7% and 1.5%; HDL-C: 0.6% and 1.2%; TG: 0.9% and 1.8%; glucose: 0.8% and 2.6%, respectively. Low-density lipoprotein cholesterol (LDL-C) was calculated using the Friedewald formula [44]. Dyslipidemia was defined as one or more of the following: TC level > 5.172 mmol/L; low LDL-C level $\geq$ 3.36 mmol/L; HDL-C level < 1.29 mmol/L; and TG level $\geq$ 1.7 mmol/L [45]. The HbA1c in whole blood was measured using the DCA Vantage Analyzer (Siemens Healthcare, Erlangen, Germany) point-of-care device.

Participants' 25-hydroxy vitamin D (25(OH)D) serum levels were measured at all three visits using the ECLIA (Roche Diagnostics GmbA, Mannheim, Germany) and commercially available IDS kits (IDS Ltd., Boldon Colliery, Tyne & Wear, UK). The inter- and intra-assay coefficients of variation (CV) for 25 (OH)D ELISA were 5.3% and 4.6%, respectively, with 100% cross-reactivity to 25(OH) D3 and 75% cross-reactivity to 25(OH) D_2. The Biomarker Research Program (BRP) laboratory is a contributing body in the vitamin D External Quality Assessment Scheme (DEQAS), and Quality Assurance (QA) standards are retained by ISO 9000 and 17025. The QA department checks the BRP laboratory at consistent times. Participants' 25(OH)D levels were categorized into deficient (less than 50 nmol/L) or non-deficient (50 nmol/L and greater) groups for a meaningful statistical analysis [46,47].

2.4. Statistical Analysis

The data were analyzed using SPSS version 22.0 (IBM Corp, Armonk, NY, USA). Continuous variables are presented as the mean $\pm$ standard deviation, while categorical variables are presented as frequencies and percentages. Statistical differences between visits involving continuous variables were analyzed using repeated measures with an ANOVA and a Friedman's test. Statistical differences between independent groups were analyzed using the independent sample *t*-test and the Mann–Whitney U-test. Statistical differences between visits involving categorical variables were analyzed using the McNemar test. The relationship between sleep scores and other parameters was analyzed using correlation coefficients, as well as linear and logistic regression analyses. A *p*-value of less than 0.05 was considered to be statistically significant.

3. Results

3.1. General Characteristics of Participants

A total of 141 pregnant women in their first trimester participated in this study. All the participants were married with an average age of 28.0 $\pm$ 5.2 (range 18–39) years and a mean BMI of 26.8 $\pm$ 6.2 kg/m^2 at the beginning of pregnancy. Eighty-six of the participants

(63.7%) were university graduates or post-graduates, 40 participants (28%) were employed, and 43 participants (30%) were earning more than USD 1300/month. Of the 141, ninety-one (65%) were multipara. The participants' general characteristics are summarized in Table 1.

Table 1. Demographic and clinical characteristics.

Characteristics	Values
Age, years	28.0 ± 5.2
University graduate or higher	86 (63.7)
Employment	40 (28.0)
Income	
>1300 USD	43 (30.0)
Parity	
Multipara	91 (65.0)
Pre-pregnancy BMI	26.8 ± 6.2
Overweight	37 (26.0)
Obesity	29 (21.0)

Values are presented as mean ± SD for continuous variables and *n* (%) for categorical values. United state dollars is represented by USD and Body mass index is represented by BMI. the main parameters subtitles.

3.2. Changes in Biochemical and Physical Parameters during Pregnancy

Changes in anthropometric, biochemical, dietary parameters, and physical activity during the course of pregnancy are presented in Table 2. Lipid profiles were significantly elevated from early to late pregnancy (all $p < 0.001$). Although serum vitamin D levels improved significantly in mid and late pregnancy, as compared to the first trimester, most of the patients remained vitamin D-deficient throughout pregnancy (Table 2). The energy intake (percentage of total kcal/day) decreased as pregnancy progressed and was significantly lower in late pregnancy than in early pregnancy ($p = 0.01$).

3.3. Percentage and Characterization of Abnormal Sleep Patterns during Pregnancy

The percentage of poor sleep ranged from 38% to 55% and was higher in late pregnancy than in early and mid-pregnancy ($p = 0.001$) (Figure 2). There was an overall poor sleep pattern in the first trimester, which worsened somewhat in the second trimester and then worsened more in the third trimester (Figure 2). The percentage of women reporting short sleep duration ranged from 46% to approximately 65%, with a significant increase from early to late pregnancy. Short sleep duration scores were higher in late pregnancy than in early and mid-pregnancy ($p < 0.001$) (Figure 2).

The mean scores of the individual components of the PSQI at each trimester time point are presented in Table 3. The scores increased progressively as pregnancy advanced for the global PSQI (5.1 ± 2.6, 5.3 ± 2.6, 6.1 ± 2.4; $p = 0.001$); sleep duration (0.8 ± 1.0, 0.9 ± 1.1, 1.3 ± 1.2; $p = 0.001$); and sleep quality (0.7 ± 0.6, 0.9 ± 0.7, 1.2 ± 0.9; $p < 0.001$). Specifically, the total PSQI score was higher in mid and late pregnancy than in early pregnancy (both $p = 0.001$), and the sleep quality score was higher in late pregnancy than in early and mid-pregnancy (both $p < 0.001$).

3.4. Factors Predictive of Poor Sleep

3.4.1. Predictors for Sleep in Each Trimester

For each trimester, independent risk factors were associated with poor sleep when using logistic regression (poor sleep, PSQI ≥ 5) as dependent variables against all variables, while adjusting for confounders (not shown in Tables). Multiparity was a significant risk factor for poor sleep in the first trimester (OR, 1.62; CI, 1.11–2.36; $p = 0.012$). Alternatively, higher education was protective against poor sleep in the first trimester (OR, 0.71; CI, 0.51–0.99; $p = 0.043$). All other variables in this study did not show significance, including the biochemical profile and other variables in all three visits.

Table 2. Change in general parameters at the three visits.

	First Trimester	Second Trimester	Third Trimester	*p*-Value
n	141	141	141	
Gestational age (weeks)	12.3 ± 3.1	26.0 ± 4.8 A	34.4 ± 3.0 AB	<0.001
Anthropometric parameters				
BMI (Kg/m²)	27.6 ± 6.0	29.8 ± 6.0 A	30.5 ± 5.3 AB	<0.001
Waist–hip ratio	0.84 ± 0.08	0.93 ± 0.09 A	0.97 ± 0.08 AB	<0.001
Body fat %	35.3 ± 5.2	37.8 ± 2.8 A	38.2 ± 2.5 AB	<0.001
Gestational weight gain (kg)		0.35 ± 0.21	0.35 ± 0.16	0.867
Systolic blood pressure (mmHg)	113.2 ± 12.3	110.6 ± 11.2	110.7 ± 10.9	0.154
Diastolic blood pressure (mmHg)	67.0 ± 8.8	66.7 ± 10.5	68.3 ± 9.1	0.544
Dietary parameters				
Energy intake (%)	60.1 ± 12.3	58.5 ± 14.8	51.3 ± 12.1 A	0.010
Carbohydrates intake (%)	90.6 ± 20.7	94.9 ± 30.0	87.5 ± 34.4	0.353
Protein intake (%)	95.7 ± 32.1	89.9 ± 30.9	76.0 ± 31.6	0.093
Fat intake (%)	93.9 ± 28.5	90.3 ± 30.7	77.9 ± 21.9	0.130
Vitamin D intake (IU/day)	149.5 ± 146.4	174.1 ± 167.8	152.3 ± 201.7	0.740
Calcium intake (mg/day)	301.9 ± 380.8	353.4 ± 416.0	262.4 ± 396.2	0.710
Water (mL/day)	1407.8 ± 743.5	1422.9 ± 691.8	1562.5 ± 786.9	0.735
Tea (mL/day)	533.3 ± 696.4	187.3 ± 110.1 A	177.0 ± 90.4 A	0.005
Coffee (mL/day)	78.9 ± 65.7	78.9 ± 75.5	121.3 ± 80.0	0.161
Physical activity				
Sitting (min/wk)	1137.0 ± 728.0	1296.0 ± 721.0	1311.0 ± 601.0	0.440
Low physical activity (min/wk)	351.0 ± 474.0	532.0 ± 660.0	109.0 ± 62.0	0.140
Biochemical parameters				
Calcium (mmol/L)	2.1 ± 0.2	2.1 ± 0.2	2.1 ± 0.3	0.174
Total cholesterol (mmol/L)	5.2 ± 1.1	6.6 ± 1.5 A	6.6 ± 1.3 A	< 0.001
HDL-cholesterol (mmol/L)	1.3 ± 0.4	1.6 ± 0.5 A	1.4 ± 0.5 A	< 0.001
LDL-cholesterol (mmol/L)	3.2 ± 0.8	4.1 ± 1.2 A	4.1 ± 1.1 A	< 0.001
Glucose (mmol/L)	4.9 ± 1.1	4.8 ± 1.0	5.1 ± 1.5 B	0.020
Triglycerides (mmol/L)	1.4 ± 0.5	2.1 ± 0.8 A	2.4 ± 1.0 AB	< 0.001
Vitamin D (nmol/L)	32.9 ± 20.2	40.2 ± 25.6 A	38.3 ± 22.9 A	< 0.001
HbA1c	5.1 ± 0.5	4.8 ± 0.5 A	5.1 ± 0.6 B	< 0.001

Data presented as mean ± SD for continuous variables and *n* (%) for categorical variables; *p*-values obtained from repeated measures with ANOVA and Friedman's tests. Superscripts A, B indicate significance from the 1st, 2nd trimester, respectively. Body mass index is represented by BMI, glycated hemoglobin by HbA1c. Bold font represents the main parameters subtitles.

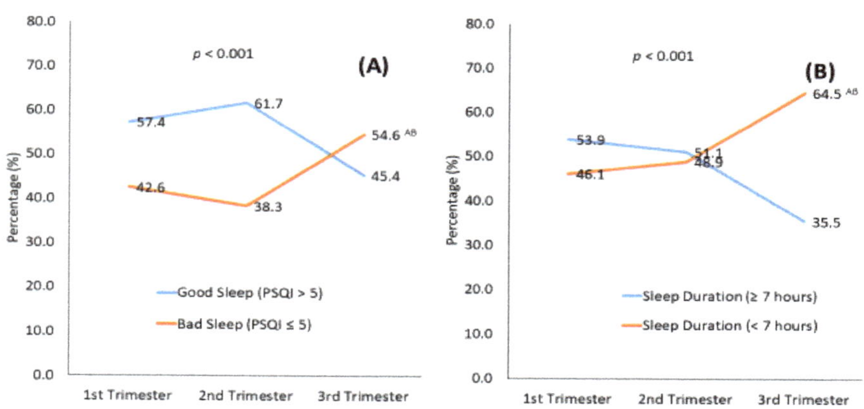

Figure 2. Pittsburgh Sleep Quality Index (PSQI) and score (**A**) and sleep duration (**B**) over the course of pregnancy. A PSQI score of 5 or less indicated poor sleep, and sleep duration less than 7 h indicated a short sleep duration. Superscripts A, B indicate significance from the 1st, 2nd trimester, respectively and AB from both.

Table 3. Mean PSQI scores at each trimester visit.

	First Trimester	Second Trimester	Third Trimester	*p*-Value
n	141	141	141	
Week of gestation	12.3 ± 3.1	26.0 ± 4.8 ^{A}	34.4 ± 3.0 ^{AB}	<0.001
Sleep components				
Habitual sleep efficiency	1.3 ± 1.4	1.2 ± 1.3	1.5 ± 1.3	0.104
Sleep duration	0.8 ± 1.0	0.9 ± 1.1	1.3 ± 1.2 ^{A}	0.001
Sleep latency	1.0 ± 1.1	1.1 ± 1.1	1.1 ± 1.1	0.514
Sleep disturbance	0.8 ± 0.7	0.9 ± 0.7	0.7 ± 0.7	0.064
Sleep quality	0.7 ± 0.6	0.9 ± 0.7	1.2 ± 0.9 ^{AB}	<0.001
Sleep medication	0.0 ± 0.2	0.0 ± 0.0	0.1 ± 0.2	0.532
Day dysfunction	0.4 ± 0.7	0.3 ± 0.6	0.3 ± 0.6	0.465
Fall asleep (in minutes)	51.9 ± 41.6	30.7 ± 27.0 ^{A}	38.0 ± 31.0 ^{A}	<0.001
Total sleep hours (hours/day)	7.9 ± 2.8	7.1 ± 2.5 ^{A}	8.8 ± 4.9 ^{B}	<0.001
Total PSQI score	5.1 ± 2.6	5.3 ± 2.6 ^{B}	6.1 ± 2.4 ^{A}	0.001

Data presented as mean ± SD for continuous variables; *p*-values for continuous variables are obtained from repeated measures with ANOVA and Friedman's tests. Pittsburgh Sleep Quality Index (PSQI). Superscripts A, B indicate significance from the 1st, 2nd trimester, respectively and AB from both. Bold font represents the subtitles.

3.4.2. Predictors for Sleep Changes across Time

- *Vitamin D levels*

The above results show that the PSQI index increased significantly throughout pregnancy, indicating that subjects experienced worsening sleep ($p = 0.001$) (Table 3). Specifically, subjects reported a worsening PSQI index in the third trimester, which differed significantly from the PSQI index in both the first and second trimesters ($p < 0.05$ each). These worsening sleeping index scores—Δ_{31}PSQI and Δ_{32}PSQI—were negatively correlated with the second and first trimester vitamin D concentrations (Figure 3). This finding suggests that patients with a higher second-trimester vitamin D concentration experienced fewer sleep problems than those with a lower vitamin D concentration. Further analysis revealed that the PSQI score decreased significantly in the second trimester for subjects with sufficient first-trimester levels of vitamin D (Δ_{21}PSQI score of −0.9 ± 3.5), compared with subjects who were vitamin D deficient (Δ_{21}PSQI score of 0.5 ± 2.7) (Table 4). Similarly, the PSQI score decreased significantly in the third trimester for subjects with sufficient second-trimester vitamin D levels (Δ_{31}PSQI score of −0.9 ± 3.0), compared with subjects who were vitamin D deficient (Δ_{31}PSQI score of 1.6 ± 3.3).

Table 4. Vitamin D and changes in sleep index.

Change in PSQI	Vitamin D Status in 1st Trimester			Vitamin D Status in 2nd Trimester		
	Sufficient	Deficient	*p*-Value	Sufficient	Deficient	*p*-Value
Δ_{31}PSQI	−0.3 ± 3.8	1.2 ± 3.3	0.060	−0.9 ± 3.0	1.6 ± 3.3	<0.001
Δ_{21}PSQI	−0.9 ± 3.5	0.5 ± 2.7	0.034	−0.8 ± 3.2	0.7 ± 2.6	0.008
Δ_{32}PSQI	0.6 ± 3.2	0.7 ± 3.1	0.926	−0.1 ± 3.3	0.9 ± 3.0	0.113

Δ_{31}, Δ_{21}, and Δ_{32} indicate changes in the PSQI score from the 3rd to the 1st, the 2nd to the 1st, and the 3rd to the 2nd trimester.

After adjusting for confounders, a linear regression analysis revealed that a 1 nmol increase in vitamin D concentration in the second trimester was associated with a reduction of 0.2 units in the PSQI score in the third, compared with the first trimester (Table 5).

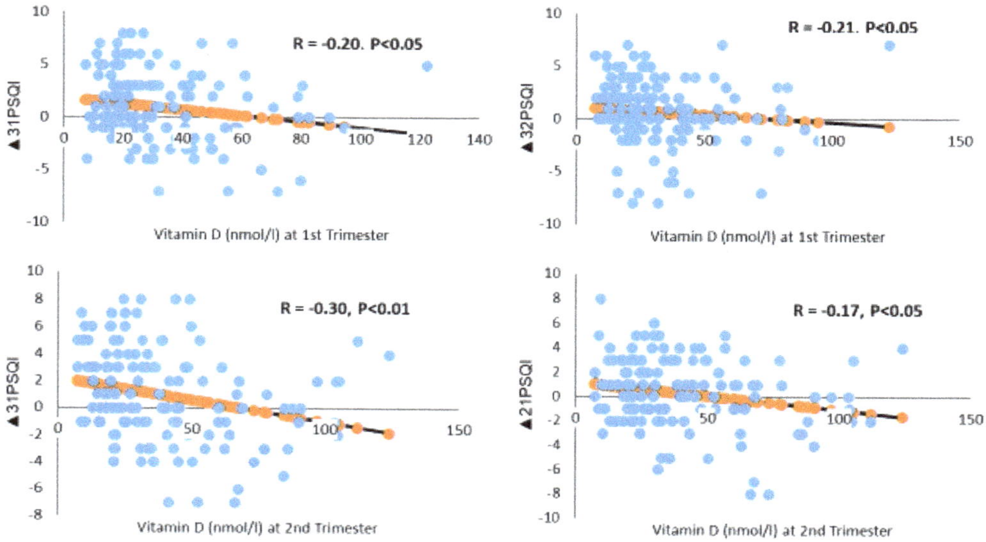

Figure 3. Vitamin D status in relation to changes in Pittsburgh Sleep Quality Index (PSQI) scores across pregnancy. Δ_{31}, Δ_{21}, and Δ_{32} indicate changes in the PSQI score from the 3rd to the 1st, 2nd to 1st, and 3rd to 2nd trimester.

Table 5. Predictors of poor sleep (total PSQI) changes during pregnancy.

	Δ_{31}PSQI		Δ_{21}PSQI		Δ_{32}PSQI	
	B ± SE	*p*-Value	B ± SE	*p*-Value	B ± SE	*p*-Value
High income	−0.60 ± 0.26	**0.025**	−0.11 ± 0.22	0.611	−0.49 ± 0.25	0.054
First trimester						
NS						
Second trimester						
Vitamin D (nmol/L) #	−0.20 ± 0.01	**0.039**	−0.01 ± 0.01	0.621	−0.03 ± 0.02	0.107
Sitting (in min/wk) #	0.00 ± 0.00	0.216	0.00 ± 0.00	0.373	0.01 ± 0.00	**0.044**
Third trimester						
Energy intake (%) Kcal/day #	0.01 ± 0.04	0.771	−0.03 ± 0.05	0.589	0.15 ± 0.07	**0.048**

Data are Beta ± standard error obtained from linear regression with # indicating adjustment for age, BMI, parity, income, lipid levels, diet, and physical activity at the respective visits. PSQI Δ_{31}, Δ_{21}, and Δ_{32} indicate changes in the PSQI score from the 3rd to the 1st, 2nd to 1st, and 3rd to 2nd trimester. Not significant indicated by NS; *p*-value < 0.05 considered significant.

- *Income*

Income was inversely correlated with changes in the global PSQI and sleep duration scores from the first to the third trimester (r = −0.2, $p < 0.05$) and (r = −0.3, $p < 0.001$), respectively. After adjusting for confounders, a linear regression analysis revealed that low income was predictive of the change in the global PSQI score from the first to the third trimester (Table 5). The significant coefficient of high income indicated that increased income reduced the PSQI index by 0.6 units in the third trimester.

- *Energy Intake*

After adjusting for covariates, a lower energy intake (kcal/day) in the third trimester was associated with the change in global PSQI score from the second to the third trimester (Table 5). A one-unit increase in energy intake in the third trimester was associated with an increase in the PSQI score of 0.15 units.

- *Sitting*

In addition, lack of physical activity (sitting) in the second trimester was found to be positively correlated with worsening sleep changes (Figure 4). Both the Δ_{31}PSQI and Δ_{32}PSQI scores were positively correlated with sitting time in the second trimester, with a correlation coefficient of 0.18 and 0.23, respectively ($p < 0.05$). After adjusting for confounders, a linear regression analysis revealed that an increased sitting time of 1 unit during the second trimester was associated with an increase in the third trimester PSQI score of 0.01 unit (Table 5).

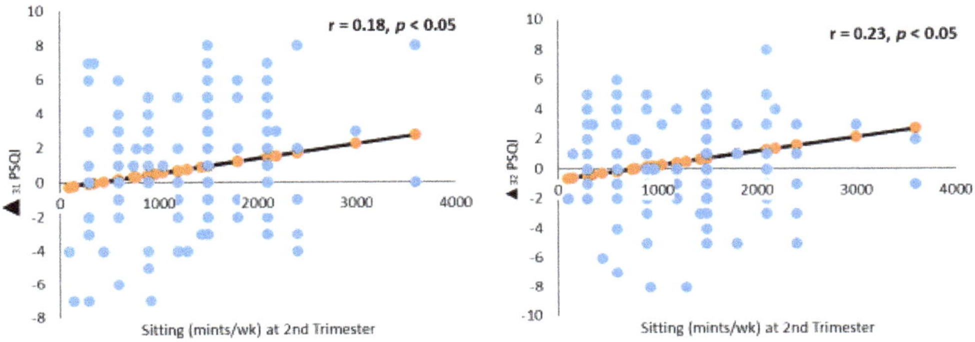

Figure 4. Sitting (minutes/week) in relation to changes in Pittsburgh Sleep Quality Index (PSQI) score across pregnancy. PSQI Δ_{31}, Δ_{21}, and Δ_{32} indicate changes in the PSQI score from the 3rd to the 1st, 2nd to 1st, and 3rd to 2nd trimester.

There were no associations between changes in the total PSQI and changes in BMI, WHR, GWG, lipid profile, glucose, and Hba1c. All the variables were incorporated in the regression but did not show significance.

4. Discussion

This prospective study reveals a high percentage of poor sleep and short sleep duration throughout pregnancy among pregnant Saudi women. Poor sleep (indicated by a PSQI of 5 or greater) and short sleep duration (less than 7 h per night) ranged from 38% to 55% and 46% to 65%, respectively, with a trend toward a worsening sleep index from early to late pregnancy. The independent predictors of worsening changes in sleep patterns as pregnancy progressed were low income, lower serum vitamin D levels, greater energy intake, and more sitting. Additionally, independent risk factors for poor sleep in the first trimester included multiparity, while possessing university graduate or post-graduate degrees prevented poor sleep.

4.1. Alterations in Sleep Patterns

To our knowledge, this is the first study to investigate sleep patterns among a cohort of Saudi women throughout all three trimesters of pregnancy. Our finding that up to 55% of women reported poor sleep, with the highest incidence occurring in the third trimester, is consistent with some but not all previous studies. A recent meta-analysis that included pregnant women (25–30 gestational weeks) from different countries, including the United States, China, and Turkey, reported that 44.5% experienced poor sleep during the second half of pregnancy [11]. Several studies also support our finding that sleep patterns worsen as pregnancy progresses, as reflected by the increasing total PSQI scores over time [9,18,48,49]. A study involving a cohort of 283 Iranian pregnant women found that poor sleep, as assessed by the PSQI, increased steadily from 48% in the first trimester to 63% in the second trimester and 75% in the third trimester [49]. This progressive decrease in sleep quality was further quantified in a recent systematic review of longitudinal studies in which sleep quality decreased by 1.68 points from the second to the third trimester of pregnancy [12]. In contrast to our findings, other studies have reported no significant difference in sleep quality throughout pregnancy [17]. Indeed, Liu et al. reported more disrupted sleep in early and late pregnancy compared with mid-pregnancy, suggesting that the second trimester represented a kind of honeymoon phase [16]. Different conclusions in these studies may, in part, be explained by differences in the population enrolled and the assessment tools. Furthermore, some studies assessed sleep at only two time points during pregnancy, either early, mid, or late pregnancy.

In our cohort of women, short sleep duration (less than 7 h per night) was alarmingly high, reaching 65% in the last trimester of pregnancy. This is almost twice the prevalence that is reported in non-pregnant Saudi women (37%) [50]. Short sleep duration also increased throughout pregnancy which is in line with findings from several other studies [9,16,17]. The incremental increases in poor sleep (both poor quality and short duration) as pregnancy progresses are problematic. Previous studies have demonstrated a relationship between poor sleep and specific poor maternal/fetal/neonatal health outcomes [13–16,51,52]. Thus, screening for poor sleep in early pregnancy may provide an opportunity to introduce interventions, as the problem only seems to worsen as pregnancy progresses.

4.2. Factors Associated with Worsening Sleep Patterns

4.2.1. Demographic Factors and Parity

Few studies have investigated the relationship between demographic factors and poor sleep at different time points over the course of pregnancy, and the findings were inconsistent. Mindell et al. showed that unemployment, low income, and low education were significant predictors of poor sleep during pregnancy [9]. We also found that high income decreased sleep deterioration from early to late pregnancy by 0.6 units in late pregnancy. As birth becomes more imminent, suboptimal sleep environments combined with greater perceived stress and financial strain associated with low income may contribute to worsening sleep in this group of women [53].

Our study also found a higher educational level to be protective against poor sleep during early pregnancy. Having a university or post-graduate degree initially appeared to be protective against poor sleep during the first trimester (OR, 0.71; CI, 0.51–0.99; $p = 0.043$). However, as pregnancy progressed, women with university and post-graduate degrees reported better sleep quality than women without degrees. This finding is similar to Mindell et al. [9] and may, in part, be due to greater health awareness among this group. Although the above-referenced studies support our findings, Colon et al. and Hedman et al. found no association between sleep changes and education or income [20,54]. This may be due to different populations, different sleep, and socio-demographic assessment tools.

Poor sleep quality has been shown to occur with greater frequency in multiparous than in nulliparous women during each trimester of pregnancy [18], and multiparity has also been associated with worsening alterations in sleep patterns throughout pregnancy in some [55], but not all [9,18,20,54], studies. Sleep disturbances in multiparous women may partially be explained by the external demands related to child-bearing and discomfort resulting from possible health complications [56]. In our study, multiparous women reported a modest increase in poor sleep that occurred only in the first trimester, by 1.62 times. This finding differs from other studies that reported poor sleep among multiparous women in the first, second [18], and third trimesters [21] of pregnancy. This discrepancy may be because, as pregnancy progresses, parity-related differences may be obscured by other factors, such as fetal growth and frequent urination, and difficulty finding a comfortable position may considerably disturb sleep, irrespective of parity [9].

4.2.2. Low Vitamin D Levels

Consistent with results from previous studies [57,58], we found that vitamin D levels were deficient at all three time points assessed during pregnancy, though serum levels were higher in mid and late rather than early pregnancy. To the best of our knowledge, this is the first study to assess vitamin D deficiency prospectively and its association with poor sleep patterns in each trimester of pregnancy. We found that pregnant women with lower second-trimester serum vitamin D levels had higher global PSQI scores in late versus early pregnancy. Furthermore, increased vitamin D in the second trimester was associated with a significant reduction in the PSQI score in the third trimester. This suggests that early vitamin D screening could help to identify women that are at risk of developing poor sleep and potentially offer therapeutic interventions to lower the risk.

There are few published studies on the association between poor sleep quality and serum vitamin D levels in the general population [37,59–61]. To date, there have been only four studies investigating the correlation between serum vitamin D levels and sleep quality during pregnancy, but the measurements were restricted to only one time point, providing no insight into changes throughout pregnancy [33,38,40]. In a cross-sectional study by Cheng et al. involving 890 Singaporean pregnant women in their second trimester (26–28 weeks), plasma 25(OH)D deficiency was found to be associated with a three-fold increase in the chance of poor sleep quality (PSQI greater than 5), (OR, 3.49; 95% CI, 1.84–6.63) [33]. More recently, Woo et al. assessed 115 African American and Latina pregnant women in their third trimester (29–32 weeks) and found that serum 25(OH)D concentration levels accounted for 17% of sleep quality variance using the PSQI after controlling for race, pre-pregnancy BMI, gestational age, and maternal age [40]. The most recent Turkish study of 153 pregnant women (27–28 weeks) found that poor sleep reached 85.6% (PSQI > 5) among pregnant women with vitamin D deficiency [38]. In contrast, a Turkish study of 91 pregnant women (36 weeks of gestation, third trimester) found no association between plasma 25(OH)D and sleep quality [39]. This discrepancy in the results may be due to a small sample size, limited seasonal variation through the study period, and a high-latitude geographical location.

The mechanism underlying the relationship between poor sleep and serum vitamin D levels is still not clear. One potential mechanism may involve vitamin D hormonal functions and the presence of vitamin D receptors in specific areas of the brain and spinal cord, some of which are thought to play a role in sleep [62,63]. It has been proposed that vitamin D may have direct effects on the initiation and maintenance of sleep by targeting neurons in the diencephalon and other brainstem nuclei, which are linked with circadian clocks [64]. Alternatively, the link between vitamin D deficiency and sleep disturbances may also result from bone disorders that cause non-specific pain [65] and conditions such as non-inflammatory skeletal myopathy [66]. Moreover, studies have proposed that vitamin D deficiency increases inflammation and infection, including those types that interfere with sleep regulation [67].

Some studies have reported that vitamin D supplementation among the general population improves sleep. A randomized clinical trial found that supplementation of 50,000 IU of vitamin D over two months improved sleep duration and quality in people with sleep disorders [68]. Furthermore, vitamin D deficiency among pregnant women results in maternal and fetal complications [13,36]. Consequently, combined with our findings, these results suggest that enhanced vitamin D levels may improve sleep. Additional studies are needed to investigate the role of vitamin D and the potential therapeutic benefit of vitamin D supplementation in improving sleep and potentially reducing poor maternal/fetal/neonatal outcomes.

4.2.3. Energy Intake and Sitting

This study showed that increments in energy intake during the third trimester worsened sleep in the second half of pregnancy by 0.15 units. Short sleep among the general population has been linked to cravings and increased appetite and hunger [69]. Similar to our result, a recent study showed that pregnant obese women with shortened sleep had a greater energy intake in early and late pregnancy [30]. We and others [54,70] found no correlation between BMI or GWG and sleep patterns throughout pregnancy, although recent studies found that excess weight gain from early to late pregnancy was associated with worsening sleep quality [19,30].

Regular physical activity during pregnancy has been shown to improve sleep duration and quality [70,71]. This is consistent with our findings that an increase in sedentary activity, measured by sitting time, was associated with worsening sleep quality in the second half of pregnancy. A recent meta-analysis on different factors concerning sleep showed that poor sleep was associated with less physical activity, with an effect size of 0.13 [70]. However, the authors did not identify any trimester-related associations between sleep disturbance and activity.

The main strength of this study is the prospective design and longitudinal assessment of women across all trimesters of pregnancy. This is also the only study to assess biochemical parameters, in addition to factors that are related to sleep, diet, and physical activity during all three visits. Hence, we adjusted for all potential confounders that may have impacted sleep along with the other variables assessed. Lastly, using trained personnel to carry out the interviews and ensure the completion of the PSQI questionnaire ensured consistency in the follow-up and completeness of the data collection.

There were several limitations to our study. First, we did not assess sleep quality objectively. However, the PSQI has been shown to have good internal consistency and construct validity when used in a pregnant population [9]. Second, the study did not assess other variables that may affect sleep quality, such as frequent urination, nausea, vomiting, difficulty breathing due to increasing abdominal girth, and general physical discomfort, though these factors may be addressed to some extent in the sleep efficiency component of the PSQI. Finally, the questionnaires that were used in this study, including the IPAQ and food frequency questionnaire, may be subject to recall bias.

5. Conclusions

In conclusion, poor sleep and short sleep duration are common and poorly recognized sleep disturbances that frequently occur among pregnant women at different time points during their pregnancy. Since poor sleep has been associated with poor maternal, fetal, and neonatal health outcomes, understanding these patterns in different populations, including Middle Eastern women, is crucial. Our results strengthen previously reported correlations between low income, lower education, sitting, and greater energy intake during pregnancy and an increased prevalence of poor sleep. Additionally, we identified a pattern of progressive deterioration in sleep quality as pregnancy advanced. The association between low serum vitamin D levels and worsening sleep quality from early to late pregnancy is an important and unique finding of this study and offers a potential avenue for intervention, both pre-conception and during pregnancy. Additional studies are needed to confirm this association and better elucidate the role that vitamin D deficiency plays in contributing to poor sleep quality during pregnancy and the mechanisms of action underlying the effect. Screening for poor sleep quality and the predictive factors identified in this study could lead to therapeutic interventions to improve maternal and neonatal health outcomes related to abnormal sleep patterns.

Funding: The authors extend their appreciation to the Deputyship for Research & Innovation, "Ministry of Education" in Saudi Arabia for funding this research work through the Directed Research Projects number IFKSUDR_H137.

Institutional Review Board Statement: The study was conducted in accordance with the Declaration of Helsinki and approved by the Institutional Review Board of King Khalid university hospital and the Ministry of Health, Riyadh, Kingdom of Saudi Arabia (KSA) (IRB Approval, 14/4067/IRB).

Informed Consent Statement: Informed consent was obtained from all the subjects involved in the study.

Data Availability Statement: Not applicable.

Acknowledgments: The authors extend their appreciation to Syed Danish Hussain for helping with the biostatistics. The author extends their appreciation for all participants in the study.

Conflicts of Interest: The authors declare no conflict of interest.

References

1. Lee, K.A. Alterations in Sleep during Pregnancy and Postpartum: A Review of 30 Years of Research. *Sleep Med. Rev.* **1998**, *2*, 231–242. [CrossRef]
2. Fatima, Y.; Doi, S.A.; Najman, J.M.; Mamun, A.A. Exploring Gender Difference in Sleep Quality of Young Adults: Findings from a Large Population Study. *Clin. Med. Res.* **2016**, *14*, 138–144. [CrossRef]
3. Nodine, P.M.; Matthews, E.E. Common Sleep Disorders: Management Strategies and Pregnancy Outcomes. *J. Midwifery Womens Health* **2013**, *58*, 368–377. [CrossRef]
4. Yannakoulia, M.; Anastasiou, C.A.; Karfopoulou, E.; Pehlivanidis, A.; Panagiotakos, D.B.; Vgontzas, A. Sleep Quality Is Associated with Weight Loss Maintenance Status: The MedWeight Study. *Sleep Med.* **2017**, *34*, 242–245. [CrossRef]
5. Stamatakis, K.A.; Punjabi, N.M. Effects of Sleep Fragmentation on Glucose Metabolism in Normal Subjects. *Chest* **2010**, *137*, 95–101. [CrossRef] [PubMed]
6. Hertz, G.; Fast, A.; Feinsilver, S.H.; Albertario, C.L.; Schulman, H.; Fein, A.M. Sleep in Normal Late Pregnancy. *Sleep* **1992**, *15*, 246–251. [CrossRef] [PubMed]
7. Hall, W.A.; Hauck, Y.L.; Carty, E.M.; Hutton, E.K.; Fenwick, J.; Stoll, K. Childbirth Fear, Anxiety, Fatigue, and Sleep Deprivation in Pregnant Women. *J. Obstet. Gynecol. Neonatal. Nurs.* **2009**, *38*, 567–576. [CrossRef]
8. Tan, X.; Cook, J.D.; Cedernaes, J.; Benedict, C. Consumer Sleep Trackers: A New Tool to Fight the Hidden Epidemic of Obstructive Sleep Apnoea? *Lancet Respir. Med.* **2019**, *7*, 1012. [CrossRef]
9. Mindell, J.A.; Cook, R.A.; Nikolovski, J. Sleep Patterns and Sleep Disturbances across Pregnancy. *Sleep Med.* **2015**, *16*, 483–488. [CrossRef] [PubMed]
10. National Sleep Foundation. Pregnancy and Sleep. Available online: http://sleepfoundation.org/sleep-topics/pregnancy-and-sleep (accessed on 25 May 2022).
11. Yang, Y.; Li, W.; Ma, T.J.; Zhang, L.; Hall, B.J.; Ungvari, G.S.; Xiang, Y.T. Prevalence of Poor Sleep Quality in Perinatal and Postnatal Women: A Comprehensive Meta-analysis of Observational Studies. *Front. Psychiatry* **2020**, *11*, 161. [CrossRef]
12. Sedov, I.D.; Cameron, E.E.; Madigan, S.; Tomfohr-Madsen, L.M. Sleep Quality during Pregnancy: A Meta-analysis. *Sleep Med. Rev.* **2018**, *38*, 168–176. [CrossRef]
13. August, E.M.; Salihu, H.M.; Biroscak, B.J.; Rahman, S.; Bruder, K.; Whiteman, V.E. Systematic Review on Sleep Disorders and Obstetric Outcomes: Scope of Current Knowledge. *Am. J. Perinatol.* **2013**, *30*, 323–334. [CrossRef] [PubMed]
14. Facco, F.L.; Grobman, W.A.; Kramer, J.; Ho, K.H.; Zee, P.C. Self-Reported Short Sleep Duration and Frequent Snoring in Pregnancy: Impact on Glucose Metabolism. *Am. J. Obstet. Gynecol.* **2010**, *203*, 142.e1–142.e5. [CrossRef] [PubMed]
15. Lu, Q.; Zhang, X.; Wang, Y.; Li, J.; Xu, Y.; Song, X.; Su, S.; Zhu, X.; Vitiello, M.V.; Shi, J.; et al. Sleep Disturbances during Pregnancy and Adverse Maternal and Fetal Outcomes: A Systematic Review and Meta-analysis. *Sleep Med. Rev.* **2021**, *58*, 101436. [CrossRef] [PubMed]
16. Liu, H.; Li, H.; Li, C.; Chen, L.; Zhang, C.; Liu, Z.; Wu, Y.; Huang, H. Associations between Maternal Sleep Quality throughout Pregnancy and Newborn Birth Weight. *Behav. Sleep Med.* **2021**, *19*, 57–69. [CrossRef]
17. Tsai, S.Y.; Lee, P.L.; Lin, J.W.; Lee, C.N. Cross-Sectional and Longitudinal Associations between Sleep and Health-Related Quality of Life in Pregnant Women: A Prospective Observational Study. *Int. J. Nurs. Stud.* **2016**, *56*, 45–53. [CrossRef]
18. Christian, L.M.; Carroll, J.E.; Porter, K.; Hall, M.H. Sleep Quality across Pregnancy and Postpartum: Effects of Parity and Race. *Sleep Health* **2019**, *5*, 327–334. [CrossRef]
19. Balieiro, L.; Gontijo, C.A.; Fahmy, W.M.; Maia, Y.; Crispim, C.A. Does Sleep Influence Weight Gain during Pregnancy? A Prospective Study. *Sleep Sci.* **2019**, *12*, 156. [CrossRef] [PubMed]
20. Hedman, C.; Pohjasvaara, T.; Tolonen, U.; Suhonen-Malm, A.S.; Myllylä, V.V. Effects of Pregnancy on Mothers' Sleep. *Sleep Med.* **2002**, *3*, 37–42. [CrossRef]
21. Román-Gálvez, R.M.; Amezcua-Prieto, C.; Salcedo-Bellido, I.; Martínez-Galiano, J.M.; Khan, K.S.; Bueno-Cavanillas, A. Factors Associated with Insomnia in Pregnancy: A Prospective Cohort Study. *Eur. J. Obstet. Gynecol. Reprod. Biol.* **2018**, *221*, 70–75. [CrossRef]
22. Zhao, P.; Bedrick, B.S.; Brown, K.E.; McCarthy, R.; Chubiz, J.E.; Ju, Y.S.; Raghuraman, N.; Fay, J.C.; Jungheim, E.S.; Herzog, E.D.; et al. Sleep Behavior and Chronotype before and throughout Pregnancy. *Sleep Med.* **2022**, *94*, 54–62. [CrossRef] [PubMed]
23. Mindell, J.A.; Jacobson, B.J. Sleep Disturbances during Pregnancy. *J. Obstet. Gynecol. Neonatal. Nurs.* **2000**, *29*, 590–597. [CrossRef] [PubMed]
24. Silva-Perez, L.J.; Gonzalez-Cardenas, N.; Surani, S.; Etindele Sosso, F.A.; Surani, S.R. Socioeconomic Status in Pregnant Women and Sleep Quality during Pregnancy. *Cureus* **2019**, *11*, e6183. [CrossRef] [PubMed]
25. Rutters, F.; Besson, H.; Walker, M.; Mari, A.; Konrad, T.; Nilsson, P.M.; Balkau, B.; Dekker, J.M. The Association between Sleep Duration, Insulin Sensitivity, and B-cell Function: The Egir-Risc Study. *J. Clin. Endocrinol. Metab.* **2016**, *101*, 3272–3280. [CrossRef]

26. Gay, C.L.; Richoux, S.E.; Beebe, K.R.; Lee, K.A. Sleep Disruption and Duration in Late Pregnancy Is Associated with Excess Gestational Weight Gain among Overweight and Obese Women. *Birth* **2017**, *44*, 173–180. [CrossRef]
27. Chua, E.C.; Shui, G.; Cazenave-Gassiot, A.; Wenk, M.R.; Gooley, J.J. Changes in Plasma Lipids during Exposure to Total Sleep Deprivation. *Sleep* **2015**, *38*, 1683–1691. [CrossRef]
28. Kaneita, Y.; Uchiyama, M.; Yoshiike, N.; Ohida, T. Associations of Usual Sleep Duration with Serum Lipid and Lipoprotein Levels. *Sleep* **2008**, *31*, 645–652. [CrossRef]
29. Xu, Y.H.; Shi, L.; Bao, Y.P.; Chen, S.J.; Shi, J.; Zhang, R.L.; Lu, L. Association between Sleep Duration during Pregnancy and Gestational Diabetes Mellitus: A Meta-analysis. *Sleep Med.* **2018**, *52*, 67–74. [CrossRef]
30. Flanagan, E.W.; Most, J.; Broskey, N.T.; Altazan, A.D.; Beyl, R.A.; Keadle, S.K.; Drews, K.L.; Singh, P.; Redman, L.M. Identification of Changes in Sleep across Pregnancy and the Impact on Cardiometabolic Health and Energy Intake in Women with Obesity. *Sleep Med.* **2021**, *77*, 120–127. [CrossRef]
31. Bennett, C.J.; Cain, S.W.; Blumfield, M.L. Monounsaturated Fat Intake Is Associated with Improved Sleep Quality in Pregnancy. *Midwifery* **2019**, *78*, 64–70. [CrossRef]
32. Lee, K.A.; Zaffke, M.E.; Baratte-Beebe, K. Restless Legs Syndrome and Sleep Disturbance during Pregnancy: The Role of Folate and Iron. *J. Womens Health Gend. Med.* **2001**, *10*, 335–341. [CrossRef] [PubMed]
33. Cheng, T.S.; Loy, S.L.; Cheung, Y.B.; Cai, S.; Colega, M.T.; Godfrey, K.M.; Chong, Y.S.; Tan, K.H.; Shek, L.P.; Lee, Y.S.; et al. Plasma Vitamin D Deficiency Is Associated with Poor Sleep Quality and Night-Time Eating at Mid-pregnancy in Singapore. *Nutrients* **2017**, *9*, 340. [CrossRef] [PubMed]
34. Tous, M.; Villalobos, M.; Iglesias, L.; Fernández-Barrés, S.; Arija, V. Vitamin D Status during Pregnancy and Offspring Outcomes: A Systematic Review and Meta-analysis of Observational Studies. *Eur. J. Clin. Nutr.* **2020**, *74*, 36–53. [CrossRef] [PubMed]
35. Al-Musharaf, S.; Fouda, M.A.; Turkestani, I.Z.; Al-Ajlan, A.; Sabico, S.; Alnaami, A.M.; Wani, K.; Hussain, S.D.; Alraqebah, B.; Al-Serehi, A.; et al. Vitamin D Deficiency Prevalence and Predictors in Early Pregnancy among Arab Women. *Nutrients* **2018**, *10*, 489. [CrossRef] [PubMed]
36. Barrett, H.; McElduff, A. Vitamin D and Pregnancy: An Old Problem Revisited. *Best Pract. Res. Clin. Endocrinol. Metab.* **2010**, *24*, 527–539. [CrossRef]
37. Gao, Q.; Kou, T.; Zhuang, B.; Ren, Y.; Dong, X.; Wang, Q. The Association between Vitamin D Deficiency and Sleep Disorders: A Systematic Review and Meta-analysis. *Nutrients* **2018**, *10*, 1395. [CrossRef]
38. Sağlam, G.; Pektaş, G.; Karakullukçu, S.; Pektaş, A.; Aykut, S.D. Vitamin D Deficiency Is Associated with Depression, Anxiety and Sleep Disturbance in Pregnant Women. *J. Acad. Res. Med.* **2021**, *11*, 51. [CrossRef]
39. Gunduz, S.; Kosger, H.; Aldemir, S.; Akcal, B.; Tevrizci, H.; Hizli, D.; Celik, H.T. Sleep Deprivation in the Last Trimester of Pregnancy and Inadequate Vitamin D: Is There a Relationship? *J. Chin. Med. Assoc.* **2015**, *79*, 34–38. [CrossRef]
40. Woo, J.; Penckofer, S.; Giurgescu, C.; Yeatts, P.E. Vitamin D Deficiency and Sleep Quality in Minority Pregnant Women. *MCN Am. J. Matern. Child Nurs.* **2020**, *45*, 155–160. [CrossRef]
41. Suleiman, K.H.; Yates, B.C.; Berger, A.M.; Pozehl, B.; Meza, J. Translating the Pittsburgh Sleep Quality Index into Arabic. *West. J. Nurs. Res.* **2010**, *32*, 250–268. [CrossRef]
42. Jomeen, J.; Martin, C.R. Assessment and Relationship of Sleep Quality to Depression in Early Pregnancy. *J. Reprod. Infant. Psychol.* **2007**, *25*, 87–99. [CrossRef]
43. Wani, K.; Sabico, S.; Alnaami, A.M.; Al-Musharaf, S.; Fouda, M.A.; Turkestani, I.Z.; Al-Ajlan, A.; Alshingetti, N.M.; Alokail, M.S.; Al-Daghri, N.M. Early-Pregnancy Metabolic Syndrome and Subsequent Incidence in Gestational Diabetes Mellitus in Arab Women. *Front. Endocrinol.* **2020**, *11*, 98. [CrossRef] [PubMed]
44. Friedewald, W.T.; Levy, R.I.; Fredrickson, D.S. Estimation of the Concentration of Low-Density Lipoprotein Cholesterol in Plasma, without Use of the Preparative Ultracentrifuge. *Clin. Chem.* **1972**, *18*, 499–502. [CrossRef] [PubMed]
45. Stone, N.J.; Robinson, J.G.; Lichtenstein, A.H.; Bairey Merz, C.N.; Blum, C.B.; Eckel, R.H.; Goldberg, A.C.; Gordon, D.; Levy, D.; Lloyd-Jones, D.M.; et al. 2013 ACC/AHA Guideline on the Treatment of Blood Cholesterol to Reduce Atherosclerotic Cardiovascular Risk in Adults: A Report of the American College of Cardiology/American Heart Association Task Force on Practice Guidelines. *J. Am. Coll. Cardiol.* **2014**, *63*, 2889–2934. [CrossRef]
46. Al-Daghri, N.M.; Al-Saleh, Y.; Aljohani, N.; Sulimani, R.; Al-Othman, A.M.; Alfawaz, H.; Fouda, M.; Al-Amri, F.; Shahrani, A.; Alharbi, M.; et al. Vitamin D Status Correction in Saudi Arabia: An Experts' Consensus under the Auspices of the European Society for Clinical and Economic Aspects of Osteoporosis, Osteoarthritis, and Musculoskeletal Diseases (ESCEO). *Arch. Osteoporos.* **2017**, *12*, 1. [CrossRef]
47. De-Regil, L.M.; Palacios, C.; Ansary, A.; Kulier, R.; Peña-Rosas, J.P. Vitamin D Supplementation for Women during Pregnancy. *Cochrane Database Syst. Rev.* **2012**, *2*, CD008873.
48. Skouteris, H.; Germano, C.; Wertheim, E.H.; Paxton, S.J.; Milgrom, J. Sleep Quality and Depression during Pregnancy: A Prospective Study. *J. Sleep Res.* **2008**, *17*, 217–220. [CrossRef]
49. Shariat, M.; Abedinia, N.; Noorbala, A.A.; Raznahan, M. The Relationship between Sleep Quality, Depression, and Anxiety in Pregnant Women: A Cohort Study. *J. Sleep Sci.* **2017**, *2*, 20–27.
50. Ahmed, A.E.; Al-Jahdali, F.; AlALwan, A.; Abuabat, F.; Bin Salih, S.A.; Al-Harbi, A.; Baharoon, S.; Khan, M.; Ali, Y.Z.; Al-Jahdali, H. Prevalence of Sleep Duration among Saudi Adults. *Saudi Med. J.* **2017**, *38*, 276. [CrossRef]
51. Okun, M.L.; Schetter, C.D.; Glynn, L.M. Poor Sleep Quality Is Associated with Preterm Birth. *Sleep* **2011**, *34*, 1493–1498. [CrossRef]

52. Naghi, I.; Keypour, F.; Ahari, S.B.; Tavalai, S.A.; Khak, M. Sleep Disturbance in Late Pregnancy and Type and Duration of Labour. *J. Obstet. Gynaecol.* **2011**, *31*, 489–491. [CrossRef] [PubMed]
53. Hall, M.H.; Matthews, K.A.; Kravitz, H.M.; Gold, E.B.; Buysse, D.J.; Bromberger, J.T.; Owens, J.F.; Sowers, M. Race and Financial Strain Are Independent Correlates of Sleep in Midlife Women: The Swan Sleep Study. *Sleep* **2009**, *32*, 73–82. [PubMed]
54. Conlon, R.; Wang, B.; Germeroth, L.J.; Cheng, Y.; Buysse, D.J.; Levine, M.D. Demographic, Pregnancy-Related, and Health-Related Factors in Association with Changes in Sleep among Pregnant Women with Overweight or Obesity. *Int. J. Behav. Med.* **2020**, *28*, 200–206. [CrossRef] [PubMed]
55. Aukia, L.; Paavonen, E.J.; Jänkälä, T.; Tolvanen, M.; Korja, R.; Karlsson, L.; Karlsson, H.; Polo-Kantola, P. Insomnia Symptoms Increase during Pregnancy, but No Increase in Sleepiness-Associations with Symptoms of Depression and Anxiety. *Sleep Med.* **2020**, *72*, 150–156. [CrossRef] [PubMed]
56. Ates, S.; Batmaz, G.; Sevket, O.; Molla, T.; Dane, C.; Dane, B. Pregnancy Outcome of Multiparous Women Aged over 40 Years. *Int. J. Reprod. Med.* **2013**, *2013*, 287519. [CrossRef]
57. Bärebring, L.; Schoenmakers, I.; Glantz, A.; Hulthén, L.; Jagner, Å.; Ellis, J.; Bärebring, M.; Bullarbo, M.; Augustin, H. Vitamin D Status during Pregnancy in a Multi-Ethnic Population-Representative Swedish Cohort. *Nutrients* **2016**, *8*, 655. [CrossRef]
58. Lundqvist, A.; Sandström, H.; Stenlund, H.; Johansson, I.; Hultdin, J. Vitamin D Status during Pregnancy: A Longitudinal Study in Swedish Women from Early Pregnancy to Seven Months Postpartum. *PLoS ONE* **2016**, *11*, e0150385.
59. McCarty, D.E.; Reddy, A.; Keigley, Q.; Kim, P.Y.; Marino, A.A. Vitamin D, Race, and Excessive Daytime Sleepiness. *J. Clin. Sleep Med.* **2012**, *8*, 693–697. [CrossRef]
60. Shiue, I. Low Vitamin D Levels in Adults with Longer Time to Fall Asleep: US NHANES, 2005–2006. *Age* **2013**, *168*, 5074–5075. [CrossRef]
61. Massa, J.; Stone, K.L.; Wei, E.K.; Harrison, S.L.; Barrett-Connor, E.; Lane, N.E.; Paudel, M.; Redline, S.; Ancoli-Israel, S.; Orwoll, E.; et al. Vitamin D and Actigraphic Sleep Outcomes in Older Community-Dwelling Men: The MrOS Sleep Study. *Sleep* **2015**, *38*, 251–257. [CrossRef]
62. Musiol, I.M.; Stumpf, W.E.; Bidmon, H.J.; Heiss, C.; Mayerhofer, A.; Bartke, A. Vitamin D Nuclear Binding to Neurons of the Septal, Substriatal and Amygdaloid Area in the Siberian Hamster (Phodopus Sungorus) Brain. *Neuroscience* **1992**, *48*, 841–848. [CrossRef]
63. Stumpf, W.E.; Bidmon, H.J.; Li, L.; Pilgrim, C.; Bartke, A.; Mayerhofer, A.; Heiss, C. Nuclear Receptor Sites for Vitamin D-Soltriol in Midbrain and Hindbrain of Siberian Hamster (Phodopus sungorus) Assessed by Autoradiography. *Histochemistry* **1992**, *98*, 155–164. [CrossRef] [PubMed]
64. Mizoguchi, A.; Eguchi, N.; Kimura, K.; Kiyohara, Y.; Qu, W.M.; Huang, Z.L.; Mochizuki, T.; Lazarus, M.; Kobayashi, T.; Kaneko, T.; et al. Dominant Localization of Prostaglandin D Receptors on Arachnoid Trabecular Cells in Mouse Basal Forebrain and Their Involvement in the Regulation of Non-rapid Eye Movement Sleep. *Proc. Natl. Acad. Sci. USA* **2001**, *98*, 11674–11679. [CrossRef] [PubMed]
65. Heidari, B.; Shirvani, J.S.; Firouzjahi, A.; Heidari, P.; Hajian-Tilaki, K.O. Association between Nonspecific Skeletal Pain and Vitamin D Deficiency. *Int. J. Rheum. Dis.* **2010**, *13*, 340–346. [CrossRef] [PubMed]
66. Okura, K.; Lavigne, G.J.; Huynh, N.; Manzini, C.; Fillipini, D.; Montplaisir, J.Y. Comparison of Sleep Variables between Chronic Widespread Musculoskeletal Pain, Insomnia, Periodic Leg Movements Syndrome and Control Subjects in a Clinical Sleep Medicine Practice. *Sleep Med.* **2008**, *9*, 352–361. [CrossRef]
67. Khoo, A.L.; Chai, L.Y.; Koenen, H.J.; Sweep, F.C.; Joosten, I.; Netea, M.G.; van der Ven, A.J. Regulation of Cytokine Responses by Seasonality of Vitamin D Status in Healthy Individuals. *J. Clin. Exp. Immunol.* **2011**, *164*, 72–79. [CrossRef] [PubMed]
68. Mohammad Shahi, M.; Hosseini, S.; Helli, B.; Haghighyzade, M.; Abolfathi, M. The Effect of Vitamin D Supplement on Quality of Sleep in Adult People with Sleep Disorders. *Tehran Univ. Med. J.* **2017**, *75*, 443–448.
69. Patterson, R.E.; Emond, J.A.; Natarajan, L.; Wesseling-Perry, K.; Kolonel, L.N.; Jardack, P.; Ancoli-Israel, S.; Arab, L. Short Sleep Duration Is Associated with Higher Energy Intake and Expenditure among African-American and Non-Hispanic White Adults. *J. Nutr.* **2014**, *144*, 461–466. [CrossRef]
70. Pauley, A.M.; Moore, G.A.; Mama, S.K.; Molenaar, P.; Downs, D.S. Systematic Review of the Associations between Prenatal Sleep Behaviours and Components of Energy Balance for Regulating Weight Gain. *J. Sleep Res.* **2022**, e13619. [CrossRef]
71. Sivertsen, B.; Hysing, M.; Dørheim, S.K.; Eberhard-Gran, M. Trajectories of Maternal Sleep Problems before and after Childbirth: A Longitudinal Population-Based Study. *BMC Pregnancy Childbirth* **2015**, *15*, 129. [CrossRef]

nutrients

MDPI

Review

Mediterranean Diet on Sleep: A Health Alliance

Egeria Scoditti [1,*], Maria Rosaria Tumolo [2] and Sergio Garbarino [3]

1 National Research Council (CNR)—Institute of Clinical Physiology (IFC), 73100 Lecce, Italy
2 Department of Biological and Environmental Sciences and Technology, University of Salento, 73100 Lecce, Italy; mariarosaria.tumolo@unisalento.it
3 Department of Neuroscience, Rehabilitation, Ophthalmology, Genetics and Maternal/Child Sciences (DINOGMI), University of Genoa, 16132 Genoa, Italy; sgarbarino.neuro@gmail.com
* Correspondence: egeria.scoditti@ifc.cnr.it; Tel.: +39-0831-298860

Abstract: The Mediterranean diet is a plant-based, antioxidant-rich, unsaturated fat dietary pattern that has been consistently associated with lower rates of noncommunicable diseases and total mortality, so that it is considered one of the healthiest dietary patterns. Clinical trials and mechanistic studies have demonstrated that the Mediterranean diet and its peculiar foods and nutrients exert beneficial effects against inflammation, oxidative stress, dysmetabolism, vascular dysfunction, adiposity, senescence, cognitive decline, neurodegeneration, and tumorigenesis, thus preventing age-associated chronic diseases and improving wellbeing and health. Nocturnal sleep is an essential physiological function, whose alteration is associated with health outcomes and chronic diseases. Scientific evidence suggests that diet and sleep are related in a bidirectional relationship, and the understanding of this association is important given their role in disease prevention. In this review, we surveyed the literature concerning the current state of evidence from epidemiological studies on the impact of the Mediterranean diet on nighttime sleep quantity and quality. The available studies indicate that greater adherence to the Mediterranean diet is associated with adequate sleep duration and with several indicators of better sleep quality. Potential mechanisms mediating the effect of the Mediterranean diet and its foods and nutrients on sleep are described, and gap-in-knowledge and new research agenda to corroborate findings are discussed.

Keywords: Mediterranean diet; sleep quality; sleep quantity; mental health; vasculoprotection; metabolism; inflammation; microbiota; melatonin

Citation: Scoditti, E.; Tumolo, M.R.; Garbarino, S. Mediterranean Diet on Sleep: A Health Alliance. *Nutrients* **2022**, *14*, 2998. https://doi.org/10.3390/nu14142998

Academic Editor: Ramón Estruch

Received: 9 July 2022
Accepted: 20 July 2022
Published: 21 July 2022

1. Introduction

Diet has become a cornerstone in the prevention and treatment of chronic noncommunicable diseases, with clinical areas of influence increasing over time as scientific evidence accumulates [1]. Epidemiological and clinical studies, along with mechanistic findings from cell and animal models, have indeed provided support for causal relationships between specific dietary patterns or foods/nutrients and health outcomes as well as disease development and progression, showing the ability of diet to significantly modify and often determine the lifelong health trajectories and chronic disease risk [2]. Accordingly, a comparative assessment of the disease burden attributable to diet in adult populations among 195 countries showed that suboptimal diets, i.e., diets high in sodium, low in whole grains, low in fruit, low in nuts and seeds, low in vegetables, and low in n-3 fatty acids, are the leading dietary risk factors for deaths and disability-adjusted life-years worldwide [3]. In contrast, diets characterized by increased consumption of vegetables, fruits, legumes, nuts, whole grains, unsaturated vegetable oils, fish, and lean meat or poultry (when meat was included) were associated with decreased risk of all-cause mortality among adults and older adults [4]. This has led to the inclusion of specific dietary recommendations into guidelines by public health entities and their prioritization in the health authority and research agenda to promote health and wellbeing, along with other modifiable lifestyle factors including smoking cessation and physical activity [5].

Among the healthiest diets, the Mediterranean diet is a plant-based dietary pattern that is increasingly become popular worldwide. The traditional Mediterranean diet is the dietary pattern consumed by the populations of the olive tree-growing areas of the Mediterranean basin before the mid-1960s. The traditional Mediterranean diet has entered the medical literature following publications of results from the Seven Countries Study, initiated in the late 1950s and showing that the Mediterranean diet is not simply, or mainly, a cholesterol-lowering diet, but exerts a range of beneficial health effects conferring longevity, better quality of life and preventing major chronic disease such as cardiovascular disease [6,7]. Other observational cohort studies, including the population-based European Prospective Investigation into Cancer and Nutrition (EPIC), confirmed the protection by the Mediterranean diet against chronic diseases, also including cancer. Randomized clinical trials, such as the PREDIMED (PREvencion con DIeta MEDiterranea) study and the Lyon Diet Heart Study, found impressive benefit of the Mediterranean diet in primary and secondary prevention of cardiovascular disease, respectively [6,7]. Later on, exponentially accumulating scientific evidence has consistently corroborated these findings and extended the health benefits of the Mediterranean dietary pattern against metabolic and neurodegenerative diseases, cognitive impairment as well as overall mortality, which is apparent among younger and older generations across Mediterranean and non-Mediterranean populations [7]. Concomitantly, many investigations have tried to identify the health promoting components(s) within the Mediterranean diet and the mechanisms mediating the beneficial biological effects. Evidence has shown the potential contributory role of some foods and nutrients herein present, such as olive oil, fish, fruits and vegetables, red wine, nuts, grains, and legumes [8]. Although the exact mechanism is not known, many interrelated and overlapping pathways have been hypothesized to play a role including: regulation of lipid and glucose metabolism; improvement of insulin sensitivity; protection against oxidative stress, inflammation, and platelet aggregation; enhancement of endothelial function; inhibition of tumorigenesis; and modulation of the gut microbiota [8].

Peculiar features of the traditional Mediterranean diet are: the abundance of olive oil (25–50 mL/day) as the main culinary fat for cooking and seasoning; the high consumption of vegetables (more than two servings per meal), fruits (one or two servings per meal), nuts (either as part of the recipes or as healthy snacks), cereals (one or two servings per meal), and legumes (more than two servings weekly); the moderate to high consumption of fish and shellfish (two or more servings weekly); the moderate consumption of poultry (two servings weekly), eggs (two to four servings weekly), dairy products (e.g., yogurt, low fat cheese, small portions daily), and alcohol, mainly wine, consumed preferably with meals (for women: ≤one drink/day; for men: one to two drinks/day); the low consumption of red meat and processed meats (less than one serving weekly); occasional consumption of foods rich in sugars and saturated fat (typical of Westernized dietary patterns); and use of herbs and spices as a key ingredient in the unique flavor of many Mediterranean dishes [9,10]. The richness in bioactive components such as vitamins, minerals, and phytochemicals (mostly polyphenols) from fresh fruits, vegetables, nuts, and legumes, contributes with synergistic actions to the health benefits of the Mediterranean diet. Though the intake of total fat is relatively high (30–40% of total energy intake), it mainly comes from virgin olive oil, tree nuts, and fatty fish, and therefore is predominantly unsaturated, mostly monounsaturated fatty acid (MUFA, more than 20% of total energy intake) and polyunsaturated fatty acids (PUFA), mainly n-3 fatty acids: consequently, the ratio of unsaturated to saturated fat is high. Furthermore, carbohydrates in the Mediterranean diet come mostly from unrefined, fiber-rich sources such as whole wheat and beans, while high quality proteins are provided by fish, sea foods, poultry, and legumes [9].

Taking into account adaptations to each country's and region's specific realities, the Mediterranean diet, recognized as Intangible Cultural Heritage of Humanity by UNESCO, is thought not only as a way of eating of the countries surrounding the Mediterranean Sea, but also as an integral part of a preserved social, cultural, and lifestyle sustainable model featured by moderation, biodiversity, local production, conviviality, culinary activities, regular physical activity, and adequate rest including nocturnal sleep and after-meal nap [9].

Besides diet, another essential health-promoting factor is sleep, an active physiological process necessary for life and normally occupying one third of our lives, playing a fundamental role for physical, mental, and emotional health [11]. Normal sleep architecture is comprised of nonrapid eye movement (NREM) sleep and REM sleep. NREM sleep is divided into three substages: stage N1, stage N2, and stage N3 (slow wave sleep). Older classification had four stages of NREM sleep. In the current rules, NREM stage 3 and NREM stage 4 are combined as stage N3. Sleep stages occur in cycles lasting 90 to 120 min each, with four to five cycles occurring during a typical night of sleep. Shifting of stages occurs over the course of the night, typically with an increased percentage of NREM sleep in the first half of the night and an increased percentage of REM sleep in the second half of the night [12]. Well recognized indicators of sleep quality are sleep latency, i.e., the length of time, in minutes, it takes to transition from wake to sleep (normal range for good sleep quality: 10–20 min), and sleep efficiency, i.e., the percentage of time in bed that is spent asleep (normal value for good sleep quality: above 85%).

Sleep patterns and sleep need are influenced by a complex interplay between genetic, behavioral, environmental, and social factors [13]. Expert consensus recommendations suggest that adults should obtain good sleep quality and duration (a minimum of 7 h per night) to promote optimal health and wellbeing [14,15]. Poor sleep quality and quantity can result from comorbid clinical conditions or sleep disorders, such as insomnia and sleep apnoea, but may also derive from the modern lifestyle with constant social, work, and family commitments, night- and shift-work, and the late-night use of technology (i.e., smartphones, computer, and television), which lead to circadian sleep–wake cycle disruption, chronic insufficient or poor quality sleep among adults as well as children and adolescents [16–19]. Serious consequences may arise from sleep curtailment ranging from fatigue, excessive daytime sleepiness, depressed mood, poor daytime functioning, and impaired cognitive and safety-related performance, to an increased risk of adverse health outcomes, including weight gain, obesity, type 2 diabetes, hypertension, cardiovascular, and neurodegenerative diseases, cancer as well as all-cause mortality [20–23]. Plausible biological mechanisms linking poor sleep and chronic disease risk involve endocrine, metabolic, and immune-inflammatory pathways [24], which are physiologically influenced by nocturnal sleep and whose dysfunctions play a determinant role in the development and progression of chronic diseases [25,26].

Sleep is therefore a modifiable risk factor for the development of chronic diseases, making it a target for intervention strategies [27,28]. Interestingly, in a prospective cohort study adding adequate sleep ($\geq$7 h/night) to traditional healthy lifestyle factors, such as physical activity, a healthy diet (Mediterranean diet), moderate alcohol consumption, and nonsmoking, conferred further benefit against cardiovascular disease risk compared with no addition [29].

Besides being both key targetable lifestyle determinants of overall health and chronic disease risk, diet and sleep are linked by a bidirectional relationship. Indeed, qualitatively and/or quantitatively insufficient sleep or mistimed sleep may lead to overfeeding, metabolic impairments with weight gain and obesity, and poor diet quality, and, conversely, diet may influence sleep quality and duration so that nutritionally unbalanced diets or mistimed eating patterns negatively impact sleep parameters [30–32].

Accumulating scientific evidence from observational and clinical studies, as synthetized by recent literature reviews [33–37], suggests that dietary factors may impact on—and predict—sleep outcomes in otherwise healthy or clinical populations. In general, it has been shown that foods rich in melatonin (a sleep promoting hormone), or its precursors tryptophan and serotonin, micronutrients (vitamin D and B, magnesium, zinc), carbohydrate-containing foods, food items including cherries and fish, can improve sleep parameters (e.g., sleep latency, time, efficiency). Contrarily, caffeinated and sugar-rich beverages as well as high fat (mainly saturated fat) and processed foods may negatively affect sleep quality and duration [38,39].

However, nutrients and foods are not consumed in isolation but in combination within dietary patterns (as defined a priori through a validated score or defined a posteriori through data-driven methods) that are expected to more extensively and differentially impact on biological and behavioral processes and hence be more predictive of overall health status and disease risk than individual foods or nutrients [40]. Furthermore, a more sustainable lifestyle approach is thought to be via the modification of the dietary pattern as a whole, instead of simply supplementing or depleting some specific foods or nutrients. However, the influence of dietary patterns on sleep has been less explored [41–44], and most of the available evidence in this context has been focused on the effects of the Mediterranean diet on sleep and to see whether sleep improvements might be an additional health benefit of this diet.

The aim of the present scoping review is to synthetize and discuss scientific evidence on the effects of the Mediterranean diet on sleep outcomes (quality and duration), with a focus on potential plausible mechanisms underlying this association.

2. Literature Search Methods

A literature search in PubMed, Scopus, and Web of Science was conducted from inception until April 2022, using the following key search terms, including MeSH terms, which were determined in accordance with the PICO (Population, Intervention, Comparison, and Outcome) method: "adolescents" and "adults" for population; "Mediterranean diet" and "Mediterranean dietary pattern" for intervention; and "sleep", "sleep quality", "sleep quantity", and "sleep duration" for outcomes. Studies were included if conducted in otherwise healthy individuals, assessed sleep using questionnaire or polysomnography/actigraphy, and had cross-sectional, prospective/retrospective, or clinical intervention design. Studies were excluded if they: (1) were conducted in patients with preexisting chronic diseases; (2) did not have full text articles available in the English language; (3) investigated individual components of the Mediterranean diet rather than the whole diet as the exposure; or (4) examined multiple interventions/exposures, e.g., Mediterranean diet plus physical exercise or plus pharmaceuticals, or there were no arms to control for the effect of the combined treatment (Figure 1). The search strategy was supplemented by manually reviewing the reference list of all retrieved articles.

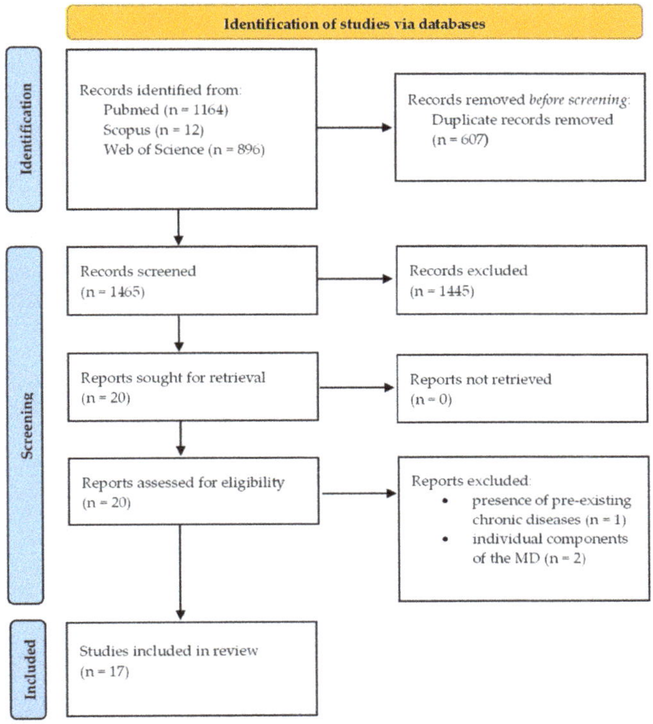

Figure 1. Flow diagram of the study selection process.

3. Main Findings

Following the literature search and articles selection, 17 studies were included in the present review [45–61]. A summary of the included studies and findings is reported in Table S1. Adherence to the Mediterranean diet was assessed using a priori derived scores, which are constructed through evidence from the scientific literature and encapsulate this dietary pattern into a numeric score for assessment of health outcomes [62]. For all scoring systems, a higher score reflects higher compliance to the traditional Mediterranean diet or the recommendations of the Mediterranean diet pyramid. Starting from the first Mediterranean diet score developed by Trichopoulou in 1995 and updated thereafter [63], other scoring systems (either modified or new) including the Mediterranean diet adherence score (MEDAS) [64], the Mediterranean diet adherence score based on the literature (MEDI-LITE) [65], and the alternate Mediterranean diet scale [47], have been developed and adapted to also take into account the cultural and geographical context of the specific populations studied, also in non-Mediterranean countries, thus often limiting interpretation and across-study comparisons. Sleep outcomes were generally self-reported using validated questionnaires, such as the Pittsburgh Sleep Quality Index (PSQI), but also using one or two questions not validated, asking about usual sleep duration, subjective sleep quality, or ease of getting to or maintaining sleep. Only in two studies were objective measures, such as actigraphy, used to assess sleep [48,57], while information on objective sleep architecture that can be derived using polysomnography is lacking.

All but one of the studies were observational, mainly with a cross-sectional design and only two studies with a longitudinal design, one in older adults [47] and another in adult women [61]. One of the main limitations is the paucity of intervention studies with the Mediterranean diet, thus not allowing to infer directionality or causality of the association.

A randomized controlled feeding trial was conducted in overweight and obese adults to assess the short-term effect of two Mediterranean diet patterns with different amounts of lean unprocessed red meat (500 g/wk, which is a typical US intake, or 200 g/wk as recommended in heart-healthy eating patterns) on several indicators of personal wellbeing, including sleep quality and sleep pattern [57]. The study did not find either any significant and robust effects of each Mediterranean diet pattern or a difference between the two patterns on sleep parameters, but the intervention duration was short (5 weeks) and the study not adequately powered [57].

Most studies were conducted in Mediterranean countries, including Spain, Greece, Italy, France, Iran, Jordan, but evidence also came from the USA and Northern Europe. Due to the small sample size or the inclusion criteria of the studies, the potential role of race/ethnicity, as well as postmenopausal women, which are at increased risk of poor sleep [13,66] in association with the Mediterranean diet with sleep, received little attention [48,61]. One study [50] found a positive association between the Mediterranean diet and sleep quality in pregnant women during the pregnancy course, where sleep disturbances are common and may be linked to metabolic anomalies as well as negative birth outcomes [67,68].

Although the adult population was the age group evaluated in the majority of the studies, in a few studies the effects on the Mediterranean diet were also assessed in adolescents [45,46,49,58], who are relevant to study being particularly vulnerable, mostly in recent years, to unhealthy behaviors such as sleep deprivation [17], lower physical activity, poor diet, and irregular eating patterns [69].

Despite the limited number of studies available, overall study results point to a beneficial impact of adherence to the Mediterranean diet pattern on sleep quality and sleep duration in both adolescents and adult individuals at cross-sectional and prospective evaluations. Specific features of sleep quality, such as sleep efficiency, sleep latency, daytime dysfunction due to sleepiness, and sleep disturbances resulted in improved greater adherence to the Mediterranean diet [51,54,61]. Furthermore, the presence of insomnia symptoms, a prevalent disordered pattern of sleep, was evaluated as an outcome in some studies [48,53,60]: here, in accordance with other findings in clinical populations [70], the Mediterranean diet was associated with a reduced risk of insomnia symptoms and of the most severe phenotype of insomnia in conjunction with short sleep duration, which is associated with worse cardiometabolic outcomes compared with insomnia without short sleep duration [71].

Interestingly, the only retrieved study reporting no association of the Mediterranean diet with sleep measures was conducted in Sweden in elderly men [59], where the Mediterranean diet adherence score was adapted to the dietary habits and food products of the Swedish population, and was assessed for association with only two sleep parameters, i.e., self-reported sleep initiation or maintenance problems.

In some studies, subgroups analyses showed that factors including age, gender, and body weight influenced the association between the Mediterranean diet and sleep. Indeed, in the MEAL study [51] the benefit of the Mediterranean diet on sleep quality (in particular sleep latency) was observed only in normal/overweight individuals (highest vs. lowest quartile of adherence score, OR = 2.30, 95% CI: 1.49, 3.54), but not in obese participants (highest vs. lowest quartile of adherence score, OR = 1.12, 95% CI: 0.33, 3.79). This result may be related to the beneficial modulation by the Mediterranean diet and its main components of metabolic profile, adiposity, and body weight status [72–74]. Therefore, the positive association between Mediterranean diet adherence and sleep quality may be mediated by improvement of weight status and obesity, which have a negative impact on sleep quality [55].

Furthermore, in the HELIAD study [54], when the analysis was stratified according to age, sleep quality was positively associated with Mediterranean diet adherence only in individuals aged $\leq$75 years ($p < 0.001$) and not in older (>75 years) individuals ($p = 0.675$), in which the presence of multicomorbidity and related polypharmacy and/or the increased occurrence of sleep problems could have masked or diluted the efficacy of the Mediterranean diet. In the French Three City Study [53], the inverse association between Mediterranean diet adherence and insomnia symptoms was significant only in women (OR = 0.80, 95% CI: 0.64−1.00) and not in men (OR = 0.93, 95% CI: 0.69−1.26), possibly due to hormonal or behavioral factors predisposing women to a higher prevalence of insomnia symptoms compared with men, and to a greater susceptibility to dietary influence. A similar gender difference in the effect of Mediterranean diet adherence on sleep has been observed in another study [52], where lower adherence to the Mediterranean diet was associated with shorter sleep duration in women, mainly linked to lower consumption of legumes, vegetables, and fruits.

4. Potential Mechanisms Underlying the Mediterranean Diet Effects on Sleep

Exploring the mechanisms potentially mediating the beneficial effects of the Mediterranean diet on sleep parameters, several hypotheses have been raised. This dietary pattern has a healthy profile of fat, proteins, and carbohydrates and a peculiar richness in polyphenols and vitamins, mainly provided by the moderate to high intake of fruits, vegetables, nuts, olive oil, cereals, and fish. Mechanisms associated with these foods and nutrients and their possible combinations might explain the benefit of the Mediterranean diet on sleep. On the contrary, a high consumption of red meat, saturated fat, and sugar-rich foods and beverages that are eaten occasionally in the Mediterranean-style diet and characterize unhealthy diets, was associated with negative effects on sleep quality and quantity, and with insomnia symptoms [31,49,50,55,75,76].

Plant-based diets [41,75,77], such as the Mediterranean diet, have been shown to be associated with better sleep quality and/or duration. Indeed, though not consistent, evidence supports the association of intakes of fruits, vegetables, legumes, and their fiber and protein content, with improved sleep parameters [76,78]. These food items as key components of the Mediterranean diet have been found to contribute to the favorable epidemiological association between Mediterranean diet adherence and sleep in both longitudinal [61] and cross-sectional analyses [49,50,55,61]. Interestingly, in some studies [50,51,55] olive oil consumption, which is typical of the Mediterranean diet and a rich source of MUFA (55–83% olive oil fat) and polyphenols (50–800 mg/kg olive oil, about 2% of oil weight), have also emerged as potentially protective for sleep.

Regarding dietary fats, in accordance with cross-sectional findings [31], among major nutrients included in the Mediterranean diet, a higher intake of MUFA to PUFA ratio and of unsaturated fat, predicted better sleep quality at follow-up [61]. Although with mixed results [79], fatty fish (>150 g three times a week), a major food source of unsaturated fat, as well as plasma concentrations of n-3 fatty acids, including docosahexaenoic acid (DHA), and DHA supplementation were found to be positively associated with sleep quality in adults [55,75,80,81] and pediatric populations [82]. Confirmatory results have been reported in a very recent study in which higher plasma DHA concentrations were related to earlier sleep timing and longer sleep duration on the weekends in Mexican adolescents [83].

One of the most characteristic features and bioactive components of the Mediterranean diet is the richness in polyphenols, which are antioxidant compounds naturally present in plant foods and beverages (Table 1). They have gained increasing scientific attention due to their biological activities, their great abundance in human diet, and their role in the prevention of various chronic degenerative diseases, such as cancer, cardiovascular, and neurodegenerative diseases, through mechanisms also independent of their conventional antioxidant activities [84]. Over 500 different molecules having a polyphenol structure (i.e., several hydroxyl groups on aromatic rings) with different properties and bioavailabilities have been identified in edible plants, where they are synthetized as secondary metabolites for defense against biotic and abiotic stresses. They encompass five main groups according to structure: phenolic acids, flavonoids, stilbenes, lignans, and others. The flavonoid class is further divided into six subclasses including flavonols, flavones, isoflavones, flavanones, anthocyanidins, and flavanols [85]. In the European PREDIMED study, the mean total polyphenol intake was 820 ± 323 mg daily, mainly provided by fruits (44%), vegetables (12%), alcoholic (red wine) (6%) and nonalcoholic (coffee) beverages (55%), cereals (5%), olives and olive oil (11%), cocoa products, nuts, and legumes (each food group around 1–2%) [86].

Table 1. Main Mediterranean diet polyphenols, food sources and dietary intake.

Class	Subclass	Main Representatives	Main Food Source	Intake (mg/day)
Flavonoids	Flavanols	Catechin Epicatechin Epigallocatechin	Apples, red wine, tea, peaches, cocoa products, beans	26.7 ± 19.6
	Flavonols	Quercetin Kaempferol Myricetin	Spinach, beans, onions, lettuce	80.4 ± 32.7
	Flavanones	Hesperidin and its aglycone hesperetin Naringenin Didymin	Oranges, orange juice, red wine, tomatoes	132 ± 125
	Flavones	Apigenin Luteolin	Oranges, whole-grain wheat-flour bread, refined-grain wheat-flour bread	41.6 ± 26.1
	Isoflavones	Genistein Daidzen Glycitein	Beans, beer	0.003 ± 0.003
	Anthocyanins	Malvidin Cyanidin Delphinidin	Cherries, red wine, olives, strawberries	38.5 ± 37.4
	Phenolic alcohol and secoiridoids	Hydroxytyrosol, tyrosol Oleuropein Oleacein Oleocanthal	Olive oil	39.46 ± 29.37
Non-flavonoids	Stilbenes	Resveratrol	Red wine, white wine, grapes, strawberries	1.84 ± 3.39
	Phenolic acids	Hydroxycinnamic acids (cinnamic, p-coumaric, ferulic, caffeic, chlorogenic, and rosmarinic acids, verbascoside)	Coffee, potatoes, apples, olives	276 ± 146
		Hydroxybenzoic acids (p-hydroxybenzoic, gallic, syringic, protocatechuic, and vanilic acids)	Olives, red wine, walnuts, beer	19.1 ± 16.8
	Lignans	Secoisolariciresinol Pinoresinol 1-Acetoxypinoresinol	Olive oil, whole-grain wheat-flour bread	0.85 ± 0.36
	Tannins	Condensed tannins or proanthocyanidins (oligomers or polymers of flavanols) Hydrolyzable tannins or gallotannins, ellagitannin	Red wine, apples, peaches, plums, orange, green beans, lentils	117 ± 81

Data on polyphenol content were adapted from [86]. Food sources are reported in decreasing order of specific polyphenols content.

After ingestion, phenolic compounds undergo several transformations in the gastrointestinal tract and also after absorption into the blood [87]. Additionally, a great proportion of nonabsorbed compounds reaches the colon where polyphenols are metabolized by the resident microflora, generating a different array of bioactive metabolites that are absorbed, further transformed, and released into the circulation [87]. Therefore, the bioactivities of ingested polyphenols depend mostly on their metabolites than on the native compounds.

Though experimental animal and in vitro studies have provided some potential mechanisms of sleep regulation by polyphenols (as described below), few human observational studies reported inconclusive results regarding the association of the polyphenol content of the diet and sleep parameters. A prospective study in UK women found that the total polyphenol content (but not polyphenol classes) of fruits and vegetables was inversely associated with sleep duration, in agreement with another study in Chinese adults reporting that soy isoflavone intake was associated with a low risk of long sleep duration [88]. On the contrary, adequate sleep duration and better sleep quality have been documented in association with high intakes of soy isoflavones in a study conducted in Japanese adults [89]. However, soy is not a typical Mediterranean food, and the effect on sleep of isoflavones from specific Mediterranean food items such as legumes has not been assessed. Recently, data from the Italian MEAL cohort study regarding the energy-adjusted total (poly)phenol intake estimated using the Phenol-Explorer database showed that a higher intake of some flavonoid subclasses (flavanones and flavones, mostly contained in fruits, vegetables, cereals, legumes, olive oil, and tea), phenolic acids, such as hydroxycinnamic acids (contained in fruits, vegetables, coffee, nuts, and cereals), and lignans (present in olive oil, cereals, nuts, legumes, fruits, and vegetables) were associated with a significantly lower likelihood of having inadequate sleep quality, only in normal weight individuals [90].

Human interventional studies also tested the effects of some polyphenol supplements on sleep but, again, the results were inconsistent, and many methodological limitations have been recognized including the small sample size, the health status of the participants, the use of supplements instead of food sources having different bioavailability, and the short duration of supplementation (as reviewed by [78]).

However, the combination of foods and nutrients in the frame of the Mediterranean diet pattern, rather than individual components, seems to better predict the sleep improvements associated with the Mediterranean diet in epidemiologic studies [51,55]. Further research, mostly clinical intervention and mechanistic studies are warranted to uncover efficacy by specific Mediterranean diet key components on sleep features.

Potential pathways involved in sleep regulation by the Mediterranean diet foods and nutrients are described below.

4.1. Metabolic and Vascular Improvements

Epidemiological and interventional studies have consistently provided strong evidence in support of Mediterranean diet benefits against metabolic and cardiovascular risk factors, thus preventing cardiometabolic diseases including obesity, type 2 diabetes, heart failure, coronary artery disease, cerebrovascular diseases, and peripheral artery diseases [91]. Improvements in adiposity and body weight, blood pressure, blood lipids, glucose metabolism and insulin sensitivity have been reported in association with adherence to the Mediterranean diet [72,92–94]. These effects may beneficially affect brain function, cognition, and mood [95–97], which are also important to sleep.

The control of body weight and body fat composition by the Mediterranean dietary pattern [72] and its key components including polyphenols [98] is hypothesized as an important mechanism mediating favorable effects on sleep quality and duration exerted by this diet. Obesity, an established independent risk factor for metabolic and cardiovascular diseases, is pathogenically associated with a proinflammatory state featured by elevated circulating levels and altered circadian pattern of inflammatory cytokines, which are implicated in the adverse obesity-associated cardiometabolic consequence [99] and are also sleep-regulating factors [24]. Excess body fat may cause obstructive sleep apnea (OSA), a major sleep disturbance associated with heightened risk of cardiovascular and metabolic diseases [100]. A weight loss intervention based on the Mediterranean diet, compared with a prudent diet, in combination with physical exercise in obese people with sleep apnea showed greater reduction of the apnoea-hypopnoea index (AHI) during REM sleep as well as waist circumference, an indicator of visceral adiposity, although it cannot be determined whether the improvement in sleep parameter was mediated by decreased adiposity [101]. The Mediterranean diet intervention also reduced the levels of inflammatory markers in individuals with obstructive sleep apnea [102]. A lower body mass index and waist circumference, and a higher adherence to the Mediterranean diet were found to be associated with better sleep quality in middle-aged adults [55]. Similarly, an increase in body mass index and fat mass as well as an unhealthy eating behavior (lower adherence to the Mediterranean diet) were associated with short sleep duration and poor sleep in an adolescent population [49]. In the study by Godos et al. [51], the association between higher adherence to the Mediterranean diet and better sleep quality was evident only in normal weight/overweight subjects and not in obese ones, suggesting that the beneficial metabolic effects of the Mediterranean diet may contribute to mediate its favorable influence on sleep. However, this hypothesis needs further investigation, and the control for these metabolic factors should be taken into consideration as potential mediating factors of the epidemiologic associations found.

Protective effects have been documented for the Mediterranean diet [103] and its main components [104,105] on the vascular function and in particular the vascular endothelium, which plays an important role in the modulation of vessel tone, tissue perfusion, dynamic permeability, hemostasis, immune response, and angiogenesis. As such, endothelial dysfunction as a result of several insults, e.g., hypertension, diabetes, dyslipidemia, smoking, as well as ageing, is recognized to contribute to pathophysiology of many disease states, such as cardiovascular, neurological, and neurodegenerative diseases [106–109]. The link between endothelial function and sleep is bidirectional, with sleep disturbances causing endothelial dysfunction as an antecedent to atherosclerosis and cerebro- and cardiovascular disease [110], and endothelial function being crucial for brain health and function, not only by contributing to the systemic vascular tone, immune function and hemostasis, but also by controlling, at the neurovascular unit, blood flow and blood–brain barrier function and interacting with surrounding brain tissue [111]. Furthermore, the recently discovered ability of the cerebral endothelium to contribute to the synthesis and secretion of the neurotrophin brain-derived neurotrophic factor (BDNF), a crucial mediator of synaptic plasticity and synaptic communication, which is implicated in neuronal survival, learning, memory, appetite, and sleep [112–114], makes any improvements of the endothelial function, as observed with the Mediterranean diet, an important contributor to preserving brain function and regulating sleep. Notably, plasma BDNF concentrations, which are lower in individuals with sleep disturbances [113], were improved by a Mediterranean diet intervention, mostly in depressed participants [115].

4.2. Blunting of Inflammation and Oxidative Stress

Chronic low-grade inflammation and an imbalance in the oxidant/antioxidant system are crucial pathogenic processes in the development and progression of chronic diseases such as diabetes, cardiovascular diseases, cancer [116], as well as mental and neurodegenerative diseases [117,118].

While low levels of cytokines regulate physiologic sleep, neuroinflammation, which is characterized by activation of microglia (the resident immune cells of the central nervous system), excessive release of proinflammatory cytokines, oxidative stress and neuronal damage, has been hypothesized to contribute to altered circadian rhythms and to poor sleep quality and quantity [24,119]. Several studies examined the associations between inflammatory markers and sleep health [120,121]. A better profile of circulating biomarkers of inflammation (C reactive protein [CRP]), oxidative stress (γ-glutamyl transferase [GGT]), and antioxidant capacities (bilirubin, carotenoids, uric acid, vitamins A, C, D, and E) has been associated with adequate sleep quality and duration [120,121]. Furthermore, these biomarkers may mediate the relationship between sleep and cardiometabolic health [122], with a particular role of oxidative stress and antioxidants for mediating the sleep duration–waist circumference and sleep duration–blood pressure relationships [122]. In support of the bidirectionality of the sleep-immunity linkage, sleep deprivation and/or inadequate sleep quality as observed in clinical populations and individuals with voluntary sleep curtailment or in experimental human and animal studies may lead to increased tissue and systemic levels of inflammatory markers and oxidative stress, that could mediate the associated adverse health outcomes [24]. The sleep–immunity/inflammation relationship raises relevant clinical implications of promoting sleep health by targeting inflammation and of improving or therapeutically controlling inflammatory response by targeting sleep.

The importance of the inflammatory and antioxidant effects of the diet on sleep has recently emerged from human cross-sectional studies. In one such study, the dietary inflammatory index (DII), a research tool based on literature evidence regarding the effects of diet on six inflammatory biomarkers (CRP, IL-1β, IL-4, IL-6, IL-10, and TNF-α) [123], was calculated based on the dietary intake data, and analyzed for correlation to sleep quality in a cohort of Italian adults [124]. Here, the highest category of DII (i.e., most proinflammatory) was associated with lower likelihood of having adequate sleep quality and in particular impaired sleep latency, and even more strongly after adjusting for the Mediterranean diet adherence underscoring that both DII and the Mediterranean diet were acting through a common denominator that may be inflammation [124]. Similar associations between higher DII scores and poor sleep quality [125] or daytime dysfunction [126] were observed in small cohorts of college students. In another study conducted in OSA patients the DII was positively associated with apnea severity and daytime sleepiness, and predicted rapid eye movement latency [127]. Data on about 30,000 individuals from the National Health and Nutrition Examination Survey (NHANES) found that adults consuming proinflammatory diets, as assessed by the DII, were more likely to present short sleep duration, long sleep duration, and/or self-reported sleep disturbances [128]. In a prospective cohort, changes over time in the DII toward an anti-inflammatory diet were associated with decreased wakening after sleep onset and improved sleep efficiency [129]. In a further in-depth analysis, a blood metabolomic study in participants from the Dietary Approaches to Stop Hypertension (DASH) trial showed that metabolites and pathways known to be implicated in inflammation and oxidative stress were associated with sleep variables (i.e., sleep midpoint, wake time) [130].

Regarding the effect of the antioxidant potential of diet on sleep, one study in post-menopausal women in Iran found that a higher dietary total antioxidant capacity (TAC), a tool for assessment of healthy effects of dietary antioxidants and estimated by measuring the oxygen radical absorbance capacity (ORAC) of selected foods, was associated with a reduction in sleep problems and other menopausal symptoms [131]. Another Iranian study in diabetic women confirmed that a higher dietary TAC, based on the ferric reducing ability of plasma (FRAP) and ORAC databases, was related with a lower risk of poor sleep and psychological disorders [132].

Though these data are only correlative in nature and cannot prove causality, they suggest that an improvement (or worsening) of inflammatory status and oxidative stress by diet could be a plausible mechanism for sleep modulation. A significant part of the health benefit and disease prevention ascribed to the Mediterranean diet has been explained by its antioxidant and anti-inflammatory properties. Indeed, observational and intervention studies [133–135] have shown that the Mediterranean diet is associated with increased serum TAC levels and antioxidant enzymes (superoxide dismutase and catalase) activities. Similarly, human studies have evaluated the impact of the Mediterranean diet on biomarkers of inflammation and immune activation showing that, compared with a Western diet or a prudent low-fat diet, the Mediterranean diet exerts an anti-inflammatory and immunomodulating effect through the downregulation of the levels of leukocyte and endothelial adhesion molecules, proinflammatory cytokines and their receptors, acute phase proteins (e.g., CRP), platelet and leukocyte counts, leukocyte trafficking, and chemoattractant molecules [103,136–139].

The Mediterranean dietary pattern is rich in foods, including olive oil, fruits, vegetables, nuts, fish, red wine, and of nutrients and nutrient combinations including vitamins, polyphenols, and unsaturated fatty acids, mainly MUFA and n-3 PUFA, which have shown to exert antioxidant [140,141] and anti-inflammatory effects [142–146] both at the systemic and cellular levels [147–149], in different cells and tissues including the brain, as discussed in the following paragraph. Interestingly, some of these foods (e.g., olive oil, fish, fruits) and nutrients (e.g., polyphenols, n-3 PUFA) have shown to be positively correlated to sleep quality in observational studies [51,61,90]. Therefore, it seems plausible that there is a contribution of the antioxidant and anti-inflammatory properties of the Mediterranean diet to the observed improvement of sleep, although causation cannot be proven by such findings and should be verified by interventional studies.

4.3. Neuroprotection

Observational data and limited evidence from clinical trials suggest that the Mediterranean dietary patterns improve cognitive performance and reduce the risk for cognitive decline and dementia, including Alzheimer's disease [150–152]. Moreover, high adherence to the Mediterranean diet has been associate with a decreased risk of depressive disorders [153,154].

Besides indirect actions through peripheral effects (e.g., enhancement of endothelial function and cerebrovascular blood flow), actions inside the brain have been suggested as potential mechanisms. Indeed, the antioxidant, anti-inflammatory and vasculoprotective properties of the Mediterranean diet and its components have been hypothesized to contribute to prevent or dampen vascular dysfunction and neurodegenerative processes, and hence to improve brain function, cognition, and mood [152], which are all related to sleep [155–157]. Furthermore, direct neuroprotective effects have been documented for some Mediterranean diet nutrients, such as n-3 PUFA [158] and polyphenols [159].

Longer-chain n-3 fatty acids (eicosapentaenoic acid [EPA], DHA) are synthetized by shorter-chain n-3 fatty acids, such as alpha-linolenic acid (ALA) [160]. However, biological conversion is inefficient, especially during aging, and shorter-chain fatty acids cannot be synthesized by humans. Therefore, diet is the most important source of these fatty acids, mainly in the form of plant-derived ALA (abundant in green leafy vegetable, nuts, legumes) and fish- and marine-derived EPA and DHA, and their supplements [160,161]. The brain is particularly rich in n-3 PUFA, which incorporate into cell membranes and promote a favorable composition of phospholipids, influencing membrane fluidity and related function; as such, the maintenance of a balanced (low) n-6/n-3 PUFA ratio, which is part of the Mediterranean diet due to high dietary intake of n-3 PUFA and low dietary intake of n-6 PUFA, is crucial for brain development, normal neurological function, and in general for reducing the risk of chronic diseases [161–163]. Studies have shown that n-3 PUFA and their endogenous lipid metabolites (i.e., eicosanoids, lipoxins, resolvins, protectins, and maresins) have the ability to decrease [164] and/or resolve [165] neuroinflammation.

Besides this, by affecting gene and protein expression through the modulation of transcription factor pathways, and by positively regulating membrane-bound enzymes, signal transduction, ion channels, receptor activity and neurotransmitter binding, n-3 PUFA affect neurotransmission processes including dopaminergic, serotonergic, cholinergic, and glutamaergic systems [166–171]. Furthermore, animal studies showed that n-3 PUFA exert neuroprotective effects against cerebral ischemia, glial degeneration, neuronal apoptosis, and synaptic loss, and increased neurogenesis and synaptogenesis, as well as executive functions and learning abilities, thus preventing cognitive decline [172]. Concordantly, human observational studies [170,173] and some albeit not consistent intervention studies [174–176] have demonstrated neuroprotective action by high dietary intake of n-3 PUFA, in consonance with the protective effects against cognitive impairment and neurodegeneration exerted by the Mediterranean diet.

Antioxidant nutrients, including vitamins and polyphenols, represent other potential players in the neuroprotective effects of the Mediterranean diet [159]. Recent studies have demonstrated that polyphenols and their metabolites, mostly the less polar (lipophilic) ones, can cross the blood-brain barrier (BBB) and enter the brain at physiologically relevant concentrations, supporting their direct action in a neurological context [177–180]. Furthermore, the BBB endothelial cells have shown to further transform these metabolites into novel bioactive components [177].

In vitro and in vivo neuroprotective actions have been documented for several polyphenols and their metabolites, such as flavonoids, resveratrol, hydroxytyrosol and its derivatives, and hydroxycinnamic acids [181–183]. These include: cytoprotection of neurons, brain capillary endothelial cells, and astrocytes against insults such as oxidative stress, protein aggregates, and inflammatory stimuli [184–187]; reduction in glutamate excitotoxicity [181], and synaptic dysfunction [188,189]; blunting of microglia proinflammatory activation and cytokine production, interfering with mitogen-activated protein kinase (MAPK) and NF-κB pathways and promoting sirtuin 1 (SIRT1) pathway [177,188,189]; inhibition of reactive oxygen overproduction and oxidative stress via the upregulation of nuclear factor (erythroid-derived 2)-like 2 (Nrf-2) pathway and downregulation of pro-oxidant NADPH oxidase [188,190,191]; reduction in brain β amyloid pathology and tau protein aggregation, hallmarks of neurodegeneration [179,191,192]; increase in BDNF [193,194], and stimulation of adult neurogenesis [195]. Evidence also suggests that polyphenols can ameliorate synaptic plasticity [187,196]. These effects were accompanied in some animal and human studies by improvements in motor and/or cognitive parameters such as learning and memory [183,188,197].

With specific reference to sleep, animal studies have found that the hydroxycinnamic acid ferulic acid exerted dose-dependent sedative effects on locomotion activity and promoted sleep in mice by prolonging sleeping time and shortening sleep latency via a serotoninergic system-dependent mechanism [198]. The flavonoid hesperidin was shown to exert a sedative and sleep enhancing effect in mice, with a synergistic action when co-administered with gamma-aminobutyric acid receptor type A (GABA[A]) agonists including medications, such as benzodiazepines, widely used to treat insomnia and anxiety, as well as other flavonoids including apigenin [199]. This result opens up the possibility that flavonoids, besides being potentially valuable single drugs, may also be used with advantage in combination with benzodiazepines, thus achieving the same therapeutic effects with a substantial decrease in the benzodiazepine dose when used in synergistic combination with flavonoids. Apigenin [200] and (-)-epigallocatechin-3-O-gallate [201] have also shown to potentiate the pentobarbital-induced sleep in mice, possibly via chloride channel activation. Moreover, animal studies have demonstrated that dietary polyphenols may counteract the sleep deprivation-induced cognitive impairment, possibly through the inhibition of inflammation [202], or the activation of cAMP-response element-binding protein (CREB) and of mammalian target of rapamycin (mTOR) signaling pathways promoting synaptic plasticity [203,204]. These preclinical findings provide further research avenues for therapeutic exploitation of brain-targeting polyphenols to improve sleep and sleep disorders, and may contribute to the observed benefit of the Mediterranean diet on sleep features.

4.4. Melatonin Biosynthesis

A potential mechanism linking diet and sleep improvement involve the modulation of the tryptophan-serotonin-melatonin system. Serotonin and melatonin are two neurotransmitters involved in sleep regulation: in the central nervous system and in the gut the essential amino acid tryptophan is initially hydroxylated to 5-hydroxytryptophan, which is then decarboxylated with the formation of serotonin (5-hydroxytryptamine). With evening darkness, in the pineal gland of the brain serotonin is converted into melatonin (*N*-acetyl-5-methoxytryptamine), a neurohormone playing a central role in the sleep–wake cycle, sleep induction, and maintenance until dawn [205]. Melatonin supplements are widely used to treat insomnia and sleep apnea, helping in improving sleep quality and duration.

Diet may affect this biosynthetic pathway at different levels. First, some foods and beverages, most of which are typical of the Mediterranean diet, are natural sources of bioavailable melatonin precursors, i.e., tryptophan and serotonin, such as roots, leaves, fruits, seeds, dairy products, and fish, and of melatonin itself, including grapes, red wine, tomato, olive oil, purslane [206]. Accordingly, consumption of a Mediterranean dietary pattern was associated with increased tryptophan plasma concentrations, and these changes in tryptophan levels seemed to confer protection against cardiovascular disease incidence in high-risk individuals [207]. Interestingly, a metabolomic study has shown that the Mediterranean diet interventions, particularly when supplemented with extra virgin olive oil, modified the direct association between plasma metabolites deriving from tryptophan catabolism, i.e., kynurenines, and the risk of heart failure and atrial fibrillation, thus potentially counteracting the detrimental health effects of these metabolites [208]. This effect of the Mediterranean diet on the tryptophan–kynurenine pathway might have implication for sleep regulation, because kynurenines have been shown not only to play a role in peripheral and brain inflammation [209], but also to adversely impact sleep quality and cognition, at least in animal models [210].

Consuming tryptophan-enriched cereals for one-week increased plasma levels of serotonin and melatonin, increased sleep efficiency, and total sleep time, and decreased sleep latency and sleep fragmentation in elderly subjects with sleep difficulties, together with improvements in oxidative stress parameters and mood, in comparison with consumption of no cereals or control cereals [211]. A recent study, however, did not find any improvements in sleep quality in women with fibromyalgia consuming a tryptophan and magnesium-enriched Mediterranean diet, though a beneficial effect against anxiety symptoms, mood disturbance, eating disorders, and dissatisfaction with body image was observed [212]. Although more studies are needed to link the dietary intake of melatonin and its precursors in the frame of the Mediterranean diet to sleep parameters, the available evidence suggests a role for the diet content of tryptophan and their derivatives in the favorable modification of sleep by the Mediterranean diet.

A second pathway through which the Mediterranean diet can influence melatonin biosynthesis is by affecting the transport of tryptophan across the BBB and hence its bioavailability for melatonin synthesis. Of course, protein sources particularly rich in tryptophan may increase the plasma tryptophan and its transport into the brain. Of note, tryptophan enters the brain in a competitive manner with other large chain-neutral amino acids (LCNAA), so that dietary proteins rich in LCNAA may reduce tryptophan transport across BBB. Contrarily, by facilitating the peripheral uptake of LCNAA via the insulin response, a high carbohydrate diet favors tryptophan entry into the brain, thus promoting sleep through increased production of serotonin [213].

Finally, some nutrients characteristic of the Mediterranean dietary pattern, such as n-3 fatty acids (found in fish and nuts, seeds and dried fruit) and B-group vitamins (found in fruit and vegetables), affect melatonin biosynthesis in the brain by stimulating enzymatic reactions, including the conversion of 5-hydroxytryptophan into serotonin, which needs B6 vitamin, and the conversion of serotonin into melatonin which requires n-3 fatty acids [205]. Animal studies have shown that lower intake of DHA is directly related to lower concentrations of DHA in the pineal gland of the brain as well as with irregular melatonin release and the sleep–wake cycle [214]. Lower B vitamins intake has been related to later sleep timing in some studies [215]. Moreover, chronic treatment of aged rats with the stilbene resveratrol prevented the aging-associated reduction in serotonin levels in the pineal gland and other brain regions, indicating increased activities of limiting enzymes tryptophan hydroxylase, in concomitance with a restoration of impaired cognitive functions [216]. Similar results were observed with an antioxidant-rich diet [217]. Though sleep parameters were not measured in these studies, these findings propose some pathways that can mediate the influence of Mediterranean diet nutrients on brain health and sleep [218].

4.5. Microbiota Modulation

Another route for ingested nutrients to affect sleep physiology is the gut microbiota, a complex ecosystem located in the human gastrointestinal tract and able to exert a number of metabolic, immunologic, and neurobehavioral functions by interplaying with different tissues and organs [219]. There are many, important regulatory functions of the gut microbiota for the host so that gut dysbiosis, i.e., alteration of microbiota composition and/or functions, is linked to the pathogenesis of gastrointestinal, metabolic, vascular, neurological, and psychiatric disorders [219].

A bidirectional interaction exists between the gut microbiota and the brain, i.e., the microbiota–gut–brain axis, through various pathways including the vagus nerve, the immune/inflammatory system, the neuroendocrine system and bacterial metabolites, such as tryptophan, melatonin, serotonin, GABA, glutamate, norepinephrine, and short chain fatty acids (SCFA), thus influencing neurotransmission and behavior, including sleep [220].

Cell wall components of bacteria, such as lipopolysaccharide and fragments of peptidoglycans, are known to induce sleep via the stimulation of sleep-regulating cytokines [220]. However, products of live intestinal bacteria may also regulate sleep. Members of the *Firmicutes*, *Bacteroidetes*, and *Actinobacteria* phyla have been shown to produce sleep-regulating metabolites, such as GABA, glutamate, and serotonin [221]. SCFA (acetate, propionate, and butyrate, etc.), which are produced by the intestinal bacteria fermentation of nondigestible polysaccharides present in fibers, may also play a mediating role in sleep modulation by the intestinal microbiota, exerting their influence on brain function possibly through the vagus nerve, downregulation of hypothalamic–pituitary–adrenal (HPA) axis reactivity, anti-inflammatory effects and neurotransmitter regulation [222]. A recent study in rats found that intraportal injection and oral administration of butyrate increased NREM sleep [223].

Studies using germ-free mice have demonstrated the importance of commensal bacteria in regulating brain development, behaviors, and disease states of the central nervous system [224,225]. Furthermore, the gut microbiota composition, diversity, and function are crucial for the maintenance of normal sleep physiology, and sleep disturbance influences the gut microbiota [221,222,226]. Recent findings provide insights into the close, albeit still unknown, relation between gut microbiota and sleep. Studies in antibiotic-treated mice have found that disruption of gut microbiota leads to alteration of gut metabolites related to neurotransmission, with a depletion of serotonin and vitamin B6, a cofactor for dopamine and serotonin synthesis as well as for catecholamine metabolism [227]. Moreover, changes in sleep/wake patterns were observed in gut microbiota-depleted mice compared with normal control, with a shorter duration of NREMS episodes during the sleep phase and more frequent transitions between NREMS and REMS, suggesting alteration of circadian rhythmicity and fragmentation of NREMS episodes in the sleep phase [227].

Human observational studies also found that sleep quality positively correlated with the *Firmicutes/Bacteroidetes* ratio and microbial diversity [228]; in another study, using two metrics of a healthy gut ecosystem, i.e., the richness and diversity, Smith et al. [221] observed that microbiome diversity was positively associated with sleep efficiency and total sleep time, and was negatively associated with sleep fragmentation: in parallel, richness within the *Firmicutes* and *Bacteroidetes* phyla was associated with cognitive outcomes. Furthermore, gut microbiota composition in terms of the *Bacteroidetes* and *Firmicutes* phyla, *Lachnospiraceae*, *Rikenellaceae*, *Sutterellaceae*, *Prevotellaceae*, and *Pseudomonadaceae* families was also found to differ in habitual short sleepers compared with normal sleepers [226]. Interestingly, animal experiments with fecal microbiota transplantation, which may contribute to prove causality in the gut microbiota–sleep relationship, demonstrated that fecal microbiota from intermittent hypoxia-treated mice (a model of OSA) transferred to naive mice was capable of causing the appearance of somnolence in the recipient mice in the active phase [229]. In humans, fecal microbiota transplantation from healthy donors to patients with irritable bowel syndrome led to an increase in gut microbiota diversity along with improvements in subjective sleep quality and psychiatric symptoms (anxiety, depression) [230]. Overall, these results suggest the impact of gut microbiota on sleep health, and the potential role of gut microbiota dysbiosis and its modulation in influencing sleep disturbance.

Regarding the sleep-gut microbiota directionally, studies could not find a common gut bacterial signature associated with different sleep disorders, and results across studies are often inconsistent even within a specific sleep disorder. However, some evidence converges to the observation that disturbed sleep (e.g., sleep deprivation, sleep fragmentation, and OSA) is associated with changes in gut microbial composition and to dysbiosis, with a frequent observation of an increase in *Firmicutes* to *Bacteroidetes* phyla ratio or a reduction in gut microbiota diversity and richness, at least in some types of sleep disturbance. These microbiota alterations have been recognized to adversely affect the intestinal epithelial barrier and to lead to local and systemic inflammation and negative health outcomes [222].

Diet plays a major role in shaping the gut microbiota and, in particular, the Mediterranean diet has been shown to beneficially affect the abundance, composition, and metabolic activity of the gut microflora [231,232]. A seminal observational study showed that high adherence to the Mediterranean diet was associated with increased fecal SCFA levels, likely due to the higher proportion of *Prevotella* among *Bacteroidetes* and *Lachnospira* among *Firmicutes*, which are able to degrade carbohydrates not digestible by the host [233]. Contrarily, lower adherence to the Mediterranean diet, with increased consumption of animal protein, saturated fats, and simple sugars, was associated with higher urinary levels of trimethylamine oxide (TMAO) [233], a gut microbiota metabolite mostly deriving from animal proteins, and recognized as a potential risk factor for cardiovascular disease as well as for brain aging and cognitive impairment [234,235].

Compared with a Western diet, a 2-week intervention with a Mediterranean diet was associated with a higher abundance of butyrate-producing bacteria and a remarkable change in the metabolic activity of gut microbiota [236]. The ability of the Mediterranean diet to restore the gut microbiome dysbiosis has been also demonstrated in a randomized clinical trial in which obese subjects adhering to the Mediterranean diet for two years presented an improved insulin sensitivity and a concomitant increase in *Bacteroides*, *Prevotella*, and *Faecalibacterium* genera, and, most importantly, of the *Roseburia* and *Ruminococcus* genera and *Parabacteroides distasonis* and *Faecalibacterium prausnitzii* bacterial species, known for their saccharolytic activity and the ability to convert carbohydrates in SCFA [237]. Features of the Mediterranean diet that can be associated with these positive effects on the gut bacteria ecosystem include the high content in unsaturated fat, such as MUFA and n-3 PUFA, which have been shown to increase the abundance of beneficial bacteria including *Akkermansia* and *Bifidobacterium* and to reduce detrimental bacteria such as *Streptococcus* and *Escherichia* spp. [238]. The high fiber content of the Mediterranean diet also plays an important role, with a known prebiotic effect and being key bacteria fermentable substrates to generate SCFAs, increasing the abundance of *Bifidobacterium* and reducing the ratio of *Firmicutes*/*Bacteroidetes*, and promoting the growth and diversity of the gut microbiota taxa in humans and experimental animals [238].

Besides macronutrients, polyphenols of the Mediterranean diet are important modulators of gut bacteria ecosystems, thus influencing the gut microbiota–brain axis. On the one hand, ingested polyphenols can be utilized by gut bacteria and transformed into phenolic metabolites with increased bioavailability and bioactivity compared with the parent compounds [239]. These metabolites can exert health effects locally in the intestine and systemically on target organs. The gut microbiota composition and metabolic activity, which may depend on several external and host factors, may therefore influence the bioavailability and biological effects of polyphenols. Obesity can promote overgrowth of pathogenic microorganisms and is associated with perturbations in the composition and metabolic function of the gut microbiota, with some studies reporting an increased *Firmicutes* to *Bacteroidetes* ratio in obese subjects compared with normal-weight controls [240], which may impinge on the biotransformation of polyphenols and hence on their bioactivity [178]. On the other hand, ingested polyphenols, including resveratrol, quercetin, hesperidin, tannins, catechins, phenolic acids, and secoiridoids, can shape the gut microbiota by exerting a direct antimicrobial effect, selectively inhibiting the development of potential pathogenic species, such as LPS producers (*Escherichia coli* and *Enterobacter cloacae*) [241]; they also exert a direct stimulatory/modulatory effect on gut bacteria favoring an increased abundance of bacteria with health benefits to the host, as manifested, for example, by decreased *Firmicutes* to *Bacteroidetes* ratio, and an increase in *Akkermansia muciniphila*,

Bacteroides thetaiotaomicrom, *Faecalibacterium prausnitzii*, *Bifidobacteria*, and *Lactobacilli*, thus modifying gut-derived metabolite profile, promoting a higher production of SCFA and serotonin, with parallel improvements of oxidative stress, inflammation, as well as metabolic, neurologic, and cognitive functions [239,242,243]. Small clinical trials and animal studies have found that olive oil and its main phenolic compounds favorably changed gut microbiota composition and metabolic function, so that an increase in *Bacteroidetes* or a reduction in *Firmicutes/Bacteroidetes*, and an increase in beneficial bacteria such as *Bifidobacteria*, and *Lactobacillus*, have been observed after olive oil polyphenol administration, in association with an increase in SCFA production [239]. A similar beneficial modulation of gut microbiota has been shown for red grape and red wine, or some vegetables [244].

Whether the sleep effects of the Mediterranean diet and its main components are mediated by the modulation of the gut microbiota, though plausible, remains unexplained and deserves further research in experimental animals and in humans.

5. Future Perspective and Conclusions

The epidemiological evidence gathered in the present review concordantly suggests a positive influence of the Mediterranean diet on sleep pattern, improving sleep quality and duration in otherwise healthy subjects. The underlying mechanisms are still partially explained and generally supposed, based on the biological plausibility of the multiple beneficial effects—and underlying mechanisms—exerted by this dietary pattern on (patho)physiological processes, including inflammation, oxidative stress, metabolism, neuroendocrine system, or cell signaling. Considering the role of physiological sleep in health maintenance and, contrarily, of sleep curtailment in chronic disease development, beneficial sleep modulation by the Mediterranean diet seems to represent an additional pathway through which the Mediterranean diet may prevent the risk for major diseases, including cardiometabolic, neurodegenerative and psychiatric diseases, and cancer.

Similarly to other health parameters modulated by the Mediterranean diet [245], though the Mediterranean diet is a rich source of potential sleep-promoting foods and nutrients, no single unifying factor exists in the relation between the Mediterranean diet and sleep, but the additive and synergistic actions of the multiple and pleiotropic nutrients combined in this dietary pattern seems to be more important than the effect of individual nutrients in providing the final effects on sleep. The whole diet reflects real life conditions where foods or nutrients are eaten in combinations and interact among them, so that it can be difficult to disentangle their independent effects. This comprehensive approach is emphasized by dietary guidelines and has been used in several clinical settings, including cardiovascular disease, cancer, and type 2 diabetes [246].

A bidirectional direct relationship links diet and sleep, with poor and/or short sleep negatively influencing dietary intakes, and in turn unhealthy diet unfavorably affecting sleep quantity and/or quality. Thus, a possible (vicious or virtuous) cycle can emerge between diet and sleep, which are both modifiable determinants of health outcomes. As such, the directionality of the associations found in the available epidemiological studies cannot be established.

As outlined in the earlier observations made by Ancel Keys in the Seven Countries' Study and in the recently updated Mediterranean diet concept [9], the Mediterranean diet is a part of a healthy lifestyle pattern, which includes preference for seasonal, local, fresh and raw foods, eating with moderation and frugality, conviviality and social interactions, lengthy meals and post-lunch naps, leisure activities such as moderate physical activity, adequate and regular nocturnal sleep according to the circadian sleep–wake rhythm, and a less stressful way of life [245]. This implies that the health benefit against the risk of chronic diseases associated with adherence to the Mediterranean diet, as observed in epidemiological studies including those presented here, may depend on the clustering of healthy lifestyle features, which also include a better sleep.

Although many non-Mediterranean countries are increasingly adopting the Mediterranean dietary pattern due to its evidence-based international recognition as one of the healthiest dietary patterns, there has been a progressive worldwide abandonment of the traditional Mediterranean-style diet mostly in countries of the Mediterranean region [247]. This trend highlights the need of individual and public health policies prioritizing the preservation and promotion of the Mediterranean diet approach and lifestyle habits to counteract risk factors, including impaired sleep, and the increasing rates of chronic disease, mostly in times of economic crisis and pandemic outbreak. In parallel, a progressive nocturnal sleep curtailment has been documented over time, thus exposing individuals starting from early ages to the negative health consequences of poor or short sleep [16].

As an integral part of homeostasis, essential physiological functions, such as sleep–wake cycle, metabolism, immune system, and cardiorespiratory functions, are controlled by the circadian clock system, which coordinates internal time with the external environment resulting in 24-h day/night cycles. Disorders of the circadian rhythms have severe consequences on human health. In the modern 24/7 society, the school and work schedules and other social demands, the exposure to artificial light at night and the use of electronic devices can cause delayed sleep onset, poor sleep quality and sleep deprivation, which people attempt to compensate for during weekends, resulting in a misalignment between the internal circadian clock and the actual sleep–wake cycle, described as social jet lag [248]. This phenomenon is mostly pronounced among individuals who prefer late bedtimes and later awakening (evening chronotypes), and is associated with an increased risk of obesity and metabolic disorders. A recently discovered feature of social jet lag is its association with a lower adherence to the Mediterranean diet and irregular eating timing, which may be linked to sleep derangements and may contribute to explain part of the negative metabolic effects of social jet lag [249]. A similar circadian misalignment with poor sleep as well as poor food choices and irregular meals [250] can be observed in night- and shift-workers, which are frequently occurring in the current 24-h economy and are at an increased risk of chronic diseases. Dietary interventions based on Mediterranean diet might represent an effective strategy to counteract the detrimental effects of this work schedule, but studies are warranted in this context.

The following tasks might be implemented in future research and policy agenda:

- large prospective cohort studies are needed to assess the preventive effect of the Mediterranean diet on sleep and identify the potential qualitative and quantitative contribution of typical foods and nutrients;
- clinical intervention studies with the Mediterranean diet as exposure and sleep as outcome in large samples are needed to confirm associations and provide causality;
- preclinical studies in animal models could provide insights into mechanisms and pathways mediating the benefit of the Mediterranean diet on sleep features;
- objective neurophysiological tools for sleep assessment (actigraphy, polysomnography) should be used in the scientific studies;
- the effects of meal timing and frequency, and the influence of individuals' chronotype in the relation of the Mediterranean diet and sleep should be investigated;
- more studies on both genders and in different age groups, including during pregnancy, are needed to substantiate the influence of the Mediterranean diet on sleep as a life-long exposome;
- the influence of risk factors, such as overweight/obesity, on the relation between the Mediterranean diet and sleep requires investigation;
- the role of gut microbiota in the Mediterranean diet–brain–sleep axis should be further studied, using strategies such as fecal microbiota transplantation, in order to assess whether and how the modulation of gut bacteria ecology by the Mediterranean diet could mediate the effects on sleep, potentially providing new therapeutic pathways and biomarkers.
- public health policy and health promotion programs should include focused attention on diet, especially the Mediterranean diet, and sleep.
- knowledge and eventually education about the diet–sleep relation should be improved among healthcare professionals including nutrition professionals, in order to imple-

ment the discussion of this topic with patients during routine healthcare practices in promoting the adoption of healthy and protective lifestyles.

These new directions for future actions would contribute to substantiate the clinical relevance of the Mediterranean diet for promoting adequate sleep and preventing sleep disturbances, ultimately leading to evidence-based dietary recommendations focused on Mediterranean-style diet to adopt early in life as part of a healthy lifestyle.

In conclusion, taking account of the limitations of the available evidence and pending future research, the Mediterranean diet can be considered an easy-to-implement and safe lifestyle intervention to promote proper sleep hygiene, favorably nourishing the connection between these two allies along the path towards wellbeing and health.

Supplementary Materials: The following is available online at https://www.mdpi.com/article/10.3390/nu14142998/s1, Table S1: Characteristics and main results of included studies (by publication date).

Author Contributions: Conceptualization, E.S. and S.G.; methodology, E.S. and M.R.T.; writing—original draft preparation, E.S.; writing—review and editing, E.S. and S.G.; visualization, E.S. and M.R.T.; supervision, E.S. and S.G. All authors have read and agreed to the published version of the manuscript.

Funding: This research received no external funding.

Institutional Review Board Statement: Not applicable.

Informed Consent Statement: Not applicable.

Data Availability Statement: No new data were created or analyzed in this study. Data sharing is not applicable to this article.

Conflicts of Interest: The authors declare no conflict of interest.

References

1. Santos, L. The impact of nutrition and lifestyle modification on health. *Eur. J. Intern. Med.* **2021**, *97*, 18–25. [CrossRef] [PubMed]
2. Koehler, K.; Drenowatz, C. Integrated Role of Nutrition and Physical Activity for Lifelong Health. *Nutrients* **2019**, *11*, 1437. [CrossRef] [PubMed]
3. GBD 2017 Diet Collaborators. Health effects of dietary risks in 195 countries, 1990–2017: A systematic analysis for the Global Burden of Disease Study 2017. *Lancet* **2019**, *393*, 1958–1972. [CrossRef]
4. English, L.K.; Ard, J.D.; Bailey, R.L.; Bates, M.; Bazzano, L.A.; Boushey, C.J.; Brown, C.; Butera, G.; Callahan, E.H.; de Jesus, J.; et al. Evaluation of Dietary Patterns and All-Cause Mortality: A Systematic Review. *JAMA Netw. Open* **2021**, *4*, e2122277. [CrossRef]
5. Lichtenstein, A.H.; Appel, L.J.; Vadiveloo, M.; Hu, F.B.; Kris-Etherton, P.M.; Rebholz, C.M.; Sacks, F.M.; Thorndike, A.N.; Van Horn, L.; Wylie-Rosett, J.; et al. 2021 Dietary Guidance to Improve Cardiovascular Health: A Scientific Statement from the American Heart Association. *Circulation* **2021**, *144*, e472–e487. [CrossRef]
6. Dinu, M.; Pagliai, G.; Casini, A.; Sofi, F. Mediterranean diet and multiple health outcomes: An umbrella review of meta-analyses of observational studies and randomised trials. *Eur. J. Clin. Nutr.* **2018**, *72*, 30–43. [CrossRef]
7. Guasch-Ferre, M.; Willett, W.C. The Mediterranean diet and health: A comprehensive overview. *J. Intern. Med.* **2021**, *290*, 549–566. [CrossRef]
8. Schwingshackl, L.; Morze, J.; Hoffmann, G. Mediterranean diet and health status: Active ingredients and pharmacological mechanisms. *Br. J. Pharmacol.* **2020**, *177*, 1241–1257. [CrossRef]
9. Bach-Faig, A.; Berry, E.M.; Lairon, D.; Reguant, J.; Trichopoulou, A.; Dernini, S.; Medina, F.X.; Battino, M.; Belahsen, R.; Miranda, G.; et al. Mediterranean diet pyramid today. Science and cultural updates. *Public Health Nutr.* **2011**, *14*, 2274–2284. [CrossRef]
10. Trichopoulou, A.; Martinez-Gonzalez, M.A.; Tong, T.Y.; Forouhi, N.G.; Khandelwal, S.; Prabhakaran, D.; Mozaffarian, D.; de Lorgeril, M. Definitions and potential health benefits of the Mediterranean diet: Views from experts around the world. *BMC Med.* **2014**, *12*, 112. [CrossRef]
11. Luyster, F.S.; Strollo, P.J., Jr.; Zee, P.C.; Walsh, J.K.; Boards of Directors of the American Academy of Sleep Medicine; Sleep Research Society. Sleep: A health imperative. *Sleep* **2012**, *35*, 727–734. [CrossRef] [PubMed]
12. Grigg-Damberger, M.M. The AASM Scoring Manual four years later. *J. Clin. Sleep Med.* **2012**, *8*, 323–332. [CrossRef] [PubMed]
13. Grandner, M.A. Sleep, Health, and Society. *Sleep Med. Clin.* **2017**, *12*, 1–22. [CrossRef] [PubMed]
14. Consensus Conference Panel; Watson, N.F.; Badr, M.S.; Belenky, G.; Bliwise, D.L.; Buxton, O.M.; Buysse, D.; Dinges, D.F.; Gangwisch, J.; Grandner, M.A.; et al. Joint Consensus Statement of the American Academy of Sleep Medicine and Sleep Research Society on the Recommended Amount of Sleep for a Healthy Adult: Methodology and Discussion. *J. Clin. Sleep Med.* **2015**, *11*, 931–952. [CrossRef] [PubMed]

15. Consensus Conference Panel; Watson, N.F.; Badr, M.S.; Belenky, G.; Bliwise, D.L.; Buxton, O.M.; Buysse, D.; Dinges, D.F.; Gangwisch, J.; Grandner, M.A.; et al. Recommended Amount of Sleep for a Healthy Adult: A Joint Consensus Statement of the American Academy of Sleep Medicine and Sleep Research Society. *J. Clin. Sleep Med.* **2015**, *11*, 591–592. [CrossRef] [PubMed]
16. Ford, E.S.; Cunningham, T.J.; Croft, J.B. Trends in Self-Reported Sleep Duration among US Adults from 1985 to 2012. *Sleep* **2015**, *38*, 829–832. [CrossRef]
17. Matricciani, L.; Olds, T.; Petkov, J. In search of lost sleep: Secular trends in the sleep time of school-aged children and adolescents. *Sleep Med. Rev.* **2012**, *16*, 203–211. [CrossRef]
18. Wheaton, A.G.; Jones, S.E.; Cooper, A.C.; Croft, J.B. Short Sleep Duration Among Middle School and High School Students—United States, 2015. *MMWR Morb. Mortal. Wkly. Rep.* **2018**, *67*, 85–90. [CrossRef]
19. Zomers, M.L.; Hulsegge, G.; van Oostrom, S.H.; Proper, K.I.; Verschuren, W.M.M.; Picavet, H.S.J. Characterizing Adult Sleep Behavior Over 20 Years-The Population-Based Doetinchem Cohort Study. *Sleep* **2017**, *40*, zsx085. [CrossRef]
20. Cappuccio, F.P.; Miller, M.A. Sleep and Cardio-Metabolic Disease. *Curr. Cardiol. Rep.* **2017**, *19*, 110. [CrossRef]
21. Smagula, S.F.; Stone, K.L.; Redline, S.; Ancoli-Israel, S.; Barrett-Connor, E.; Lane, N.E.; Orwoll, E.S.; Cauley, J.A.; Osteoporotic Fractures in Men (MrOS) Research Group. Actigraphy- and Polysomnography-Measured Sleep Disturbances, Inflammation, and Mortality Among Older Men. *Psychosom. Med.* **2016**, *78*, 686–696. [CrossRef] [PubMed]
22. Van Leeuwen, W.M.; Lehto, M.; Karisola, P.; Lindholm, H.; Luukkonen, R.; Sallinen, M.; Harma, M.; Porkka-Heiskanen, T.; Alenius, H. Sleep restriction increases the risk of developing cardiovascular diseases by augmenting proinflammatory responses through IL-17 and CRP. *PLoS ONE* **2009**, *4*, e4589. [CrossRef]
23. Vgontzas, A.N.; Liao, D.; Bixler, E.O.; Chrousos, G.P.; Vela-Bueno, A. Insomnia with objective short sleep duration is associated with a high risk for hypertension. *Sleep* **2009**, *32*, 491–497. [CrossRef] [PubMed]
24. Garbarino, S.; Lanteri, P.; Bragazzi, N.L.; Magnavita, N.; Scoditti, E. Role of sleep deprivation in immune-related disease risk and outcomes. *Commun. Biol.* **2021**, *4*, 1304. [CrossRef] [PubMed]
25. Aldabal, L.; Bahammam, A.S. Metabolic, endocrine, and immune consequences of sleep deprivation. *Open Respir. Med. J.* **2011**, *5*, 31–43. [CrossRef]
26. Irwin, M.R. Sleep and inflammation: Partners in sickness and in health. *Nat. Rev. Immunol.* **2019**, *19*, 702–715. [CrossRef]
27. St-Onge, M.P.; Grandner, M.A.; Brown, D.; Conroy, M.B.; Jean-Louis, G.; Coons, M.; Bhatt, D.L.; American Heart Association Obesity, Behavior Change, Diabetes; Nutrition Committees of the Council on Lifestyle and Cardiometabolic Health; Council on Cardiovascular Disease in the Young; et al. Sleep Duration and Quality: Impact on Lifestyle Behaviors and Cardiometabolic Health: A Scientific Statement from the American Heart Association. *Circulation* **2016**, *134*, e367–e386. [CrossRef]
28. Visseren, F.L.J.; Mach, F.; Smulders, Y.M.; Carballo, D.; Koskinas, K.C.; Back, M.; Benetos, A.; Biffi, A.; Boavida, J.M.; Capodanno, D.; et al. 2021 ESC Guidelines on cardiovascular disease prevention in clinical practice. *Eur. Heart J.* **2021**, *42*, 3227–3337. [CrossRef]
29. Hoevenaar-Blom, M.P.; Spijkerman, A.M.; Kromhout, D.; Verschuren, W.M. Sufficient sleep duration contributes to lower cardiovascular disease risk in addition to four traditional lifestyle factors: The MORGEN study. *Eur. J. Prev. Cardiol.* **2014**, *21*, 1367–1375. [CrossRef]
30. Theorell-Haglow, J.; Lemming, E.W.; Michaelsson, K.; Elmstahl, S.; Lind, L.; Lindberg, E. Sleep duration is associated with healthy diet scores and meal patterns: Results from the population-based EpiHealth study. *J. Clin. Sleep Med.* **2020**, *16*, 9–18. [CrossRef]
31. Zuraikat, F.M.; Makarem, N.; Liao, M.; St-Onge, M.P.; Aggarwal, B. Measures of Poor Sleep Quality Are Associated with Higher Energy Intake and Poor Diet Quality in a Diverse Sample of Women from the Go Red for Women Strategically Focused Research Network. *J. Am. Heart Assoc.* **2020**, *9*, e014587. [CrossRef] [PubMed]
32. Zuraikat, F.M.; Wood, R.A.; Barragan, R.; St-Onge, M.P. Sleep and Diet: Mounting Evidence of a Cyclical Relationship. *Annu. Rev. Nutr.* **2021**, *41*, 309–332. [CrossRef] [PubMed]
33. Binks, H.; Vincent, G.E.; Gupta, C.; Irwin, C.; Khalesi, S. Effects of Diet on Sleep: A Narrative Review. *Nutrients* **2020**, *12*, 936. [CrossRef]
34. Godos, J.; Grosso, G.; Castellano, S.; Galvano, F.; Caraci, F.; Ferri, R. Association between diet and sleep quality: A systematic review. *Sleep Med. Rev.* **2021**, *57*, 101430. [CrossRef] [PubMed]
35. Gupta, C.C.; Irwin, C.; Vincent, G.E.; Khalesi, S. The Relationship Between Diet and Sleep in Older Adults: A Narrative Review. *Curr. Nutr. Rep.* **2021**, *10*, 166–178. [CrossRef] [PubMed]
36. Peuhkuri, K.; Sihvola, N.; Korpela, R. Diet promotes sleep duration and quality. *Nutr. Res.* **2012**, *32*, 309–319. [CrossRef] [PubMed]
37. St-Onge, M.P.; Crawford, A.; Aggarwal, B. Plant-based diets: Reducing cardiovascular risk by improving sleep quality? *Curr. Sleep Med. Rep.* **2018**, *4*, 74–78. [CrossRef]
38. St-Onge, M.P.; Mikic, A.; Pietrolungo, C.E. Effects of Diet on Sleep Quality. *Adv. Nutr.* **2016**, *7*, 938–949. [CrossRef]
39. Sutanto, C.N.; Wang, M.X.; Tan, D.; Kim, J.E. Association of Sleep Quality and Macronutrient Distribution: A Systematic Review and Meta-Regression. *Nutrients* **2020**, *12*, 126. [CrossRef]
40. Hu, F.B. Dietary pattern analysis: A new direction in nutritional epidemiology. *Curr. Opin. Lipidol.* **2002**, *13*, 3–9. [CrossRef]
41. Cao, Y.; Taylor, A.W.; Wittert, G.; Adams, R.; Shi, Z. Dietary patterns and sleep parameters in a cohort of community dwelling Australian men. *Asia Pac. J. Clin. Nutr.* **2017**, *26*, 1158–1169. [CrossRef] [PubMed]

42. Jansen, E.C.; Baylin, A.; Cantoral, A.; Tellez Rojo, M.M.; Burgess, H.J.; O'Brien, L.M.; Torres Olascoaga, L.; Peterson, K.E. Dietary Patterns in Relation to Prospective Sleep Duration and Timing among Mexico City Adolescents. *Nutrients* **2020**, *12*, 2305. [CrossRef] [PubMed]
43. Mousavi, S.A.; Mirzababaei, A.; Shiraseb, F.; Clark, C.C.T.; Mirzaei, K. The association between modified Nordic diet with sleep quality and circadian rhythm in overweight and obese woman: A cross-sectional study. *Eat Weight Disord.* **2021**, *27*, 1835–1845. [CrossRef]
44. Rostami, H.; Khayyatzadeh, S.S.; Tavakoli, H.; Bagherniya, M.; Mirmousavi, S.J.; Farahmand, S.K.; Tayefi, M.; Ferns, G.A.; Ghayour-Mobarhan, M. The relationship between adherence to a Dietary Approach to Stop Hypertension (DASH) dietary pattern and insomnia. *BMC Psychiatry* **2019**, *19*, 234. [CrossRef] [PubMed]
45. Adelantado-Renau, M.; Beltran-Valls, M.R.; Esteban-Cornejo, I.; Martinez-Vizcaino, V.; Santaliestra-Pasias, A.M.; Moliner-Urdiales, D. The influence of adherence to the Mediterranean diet on academic performance is mediated by sleep quality in adolescents. *Acta Paediatr.* **2019**, *108*, 339–346. [CrossRef]
46. Arriscado, D.; Knox, E.; Zabala, M.; Zurita-Ortega, F.; Dalmau, J.M.; Muros, J.J. Different healthy habits between northern and southern Spanish school children. *Z. Gesundh.* **2017**, *25*, 653–660. [CrossRef]
47. Campanini, M.Z.; Guallar-Castillon, P.; Rodriguez-Artalejo, F.; Lopez-Garcia, E. Mediterranean Diet and Changes in Sleep Duration and Indicators of Sleep Quality in Older Adults. *Sleep* **2017**, *40*, zsw083. [CrossRef]
48. Castro-Diehl, C.; Wood, A.C.; Redline, S.; Reid, M.; Johnson, D.A.; Maras, J.E.; Jacobs, D.R., Jr.; Shea, S.; Crawford, A.; St-Onge, M.P. Mediterranean diet pattern and sleep duration and insomnia symptoms in the Multi-Ethnic Study of Atherosclerosis. *Sleep* **2018**, *41*, zsy158. [CrossRef]
49. Ferranti, R.; Marventano, S.; Castellano, S.; Giogianni, G.; Nolfo, F.; Rametta, S.; Matalone, M.; Mistretta, A. Sleep quality and duration is related with diet and obesity in young adolescent living in Sicily, Southern Italy. *Sleep Sci.* **2016**, *9*, 117–122. [CrossRef]
50. Flor-Alemany, M.; Nestares, T.; Alemany-Arrebola, I.; Marin-Jimenez, N.; Borges-Cosic, M.; Aparicio, V.A. Influence of Dietary Habits and Mediterranean Diet Adherence on Sleep Quality during Pregnancy. The GESTAFIT Project. *Nutrients* **2020**, *12*, 3569. [CrossRef]
51. Godos, J.; Ferri, R.; Caraci, F.; Cosentino, F.I.I.; Castellano, S.; Galvano, F.; Grosso, G. Adherence to the Mediterranean Diet is Associated with Better Sleep Quality in Italian Adults. *Nutrients* **2019**, *11*, 976. [CrossRef] [PubMed]
52. Gupta, K.; Jansen, E.C.; Campos, H.; Baylin, A. Associations between sleep duration and Mediterranean diet score in Costa Rican adults. *Appetite* **2022**, *170*, 105881. [CrossRef] [PubMed]
53. Jaussent, I.; Dauvilliers, Y.; Ancelin, M.L.; Dartigues, J.F.; Tavernier, B.; Touchon, J.; Ritchie, K.; Besset, A. Insomnia symptoms in older adults: Associated factors and gender differences. *Am. J. Geriatr. Psychiatry* **2011**, *19*, 88–97. [CrossRef]
54. Mamalaki, E.; Anastasiou, C.A.; Ntanasi, E.; Tsapanou, A.; Kosmidis, M.H.; Dardiotis, E.; Hadjigeorgiou, G.M.; Sakka, P.; Scarmeas, N.; Yannakoulia, M. Associations between the mediterranean diet and sleep in older adults: Results from the hellenic longitudinal investigation of aging and diet study. *Geriatr. Gerontol. Int.* **2018**, *18*, 1543–1548. [CrossRef] [PubMed]
55. Muscogiuri, G.; Barrea, L.; Aprano, S.; Framondi, L.; Di Matteo, R.; Laudisio, D.; Pugliese, G.; Savastano, S.; Colao, A.; OPERA PREVENTION Project. Sleep Quality in Obesity: Does Adherence to the Mediterranean Diet Matter? *Nutrients* **2020**, *12*, 1364. [CrossRef] [PubMed]
56. Naja, F.; Hasan, H.; Khadem, S.H.; Buanq, M.A.; Al-Mulla, H.K.; Aljassmi, A.K.; Faris, M.E. Adherence to the Mediterranean Diet and Its Association with Sleep Quality and Chronotype Among Youth: A Cross-Sectional Study. *Front. Nutr.* **2021**, *8*, 805955. [CrossRef]
57. O'Connor, L.E.; Biberstine, S.L.; Paddon-Jones, D.; Schwichtenberg, A.J.; Campbell, W.W. Adopting a Mediterranean-Style Eating Pattern with Different Amounts of Lean Unprocessed Red Meat Does Not Influence Short-Term Subjective Personal Well-Being in Adults with Overweight or Obesity. *J. Nutr.* **2018**, *148*, 1917–1923. [CrossRef]
58. Rosi, A.; Giopp, F.; Milioli, G.; Melegari, G.; Goldoni, M.; Parrino, L.; Scazzina, F. Weight Status, Adherence to the Mediterranean Diet, Physical Activity Level, and Sleep Behavior of Italian Junior High School Adolescents. *Nutrients* **2020**, *12*, 478. [CrossRef]
59. Van Egmond, L.; Tan, X.; Sjogren, P.; Cederholm, T.; Benedict, C. Association between Healthy Dietary Patterns and Self-Reported Sleep Disturbances in Older Men: The ULSAM Study. *Nutrients* **2019**, *11*, 1029. [CrossRef]
60. Zaidalkilani, A.T.; Alhaj, O.A.; Serag El-Dine, M.F.; Fekih-Romdhane, F.; AlRasheed, M.M.; Jahrami, H.A.; Bragazzi, N.L. Arab Women Adherence to the Mediterranean Diet and Insomnia. *Medicina* **2021**, *58*, 17. [CrossRef]
61. Zuraikat, F.M.; Makarem, N.; St-Onge, M.P.; Xi, H.; Akkapeddi, A.; Aggarwal, B. A Mediterranean Dietary Pattern Predicts Better Sleep Quality in US Women from the American Heart Association Go Red for Women Strategically Focused Research Network. *Nutrients* **2020**, *12*, 2830. [CrossRef] [PubMed]
62. Hutchins-Wiese, H.L.; Bales, C.W.; Porter Starr, K.N. Mediterranean diet scoring systems: Understanding the evolution and applications for Mediterranean and non-Mediterranean countries. *Br. J. Nutr.* **2021**, 1–22. [CrossRef] [PubMed]
63. Trichopoulou, A.; Costacou, T.; Bamia, C.; Trichopoulos, D. Adherence to a Mediterranean diet and survival in a Greek population. *N. Engl. J. Med.* **2003**, *348*, 2599–2608. [CrossRef] [PubMed]
64. Schroder, H.; Fito, M.; Estruch, R.; Martinez-Gonzalez, M.A.; Corella, D.; Salas-Salvado, J.; Lamuela-Raventos, R.; Ros, E.; Salaverria, I.; Fiol, M.; et al. A short screener is valid for assessing Mediterranean diet adherence among older Spanish men and women. *J. Nutr.* **2011**, *141*, 1140–1145. [CrossRef]

65. Sofi, F.; Dinu, M.; Pagliai, G.; Marcucci, R.; Casini, A. Validation of a literature-based adherence score to Mediterranean diet: The MEDI-LITE score. *Int. J. Food Sci. Nutr.* **2017**, *68*, 757–762. [CrossRef]
66. Pengo, M.F.; Won, C.H.; Bourjeily, G. Sleep in Women Across the Life Span. *Chest* **2018**, *154*, 196–206. [CrossRef]
67. Okun, M.L.; Schetter, C.D.; Glynn, L.M. Poor sleep quality is associated with preterm birth. *Sleep* **2011**, *34*, 1493–1498. [CrossRef]
68. Sedov, I.D.; Cameron, E.E.; Madigan, S.; Tomfohr-Madsen, L.M. Sleep quality during pregnancy: A meta-analysis. *Sleep Med. Rev.* **2018**, *38*, 168–176. [CrossRef]
69. Benetou, V.; Kanellopoulou, A.; Kanavou, E.; Fotiou, A.; Stavrou, M.; Richardson, C.; Orfanos, P.; Kokkevi, A. Diet-Related Behaviors and Diet Quality among School-Aged Adolescents Living in Greece. *Nutrients* **2020**, *12*, 3804. [CrossRef]
70. Kechribari, I.; Kontogianni, M.D.; Georgoulis, M.; Lamprou, K.; Critselis, E.; Vagiakis, E.; Yiannakouris, N. Association of adherence to the Mediterranean diet and physical activity habits with the presence of insomnia in patients with obstructive sleep apnea. *Sleep Breath.* **2021**, *26*, 89–97. [CrossRef]
71. Bertisch, S.M.; Pollock, B.D.; Mittleman, M.A.; Buysse, D.J.; Bazzano, L.A.; Gottlieb, D.J.; Redline, S. Insomnia with objective short sleep duration and risk of incident cardiovascular disease and all-cause mortality: Sleep Heart Health Study. *Sleep* **2018**, *41*, zsy047. [CrossRef] [PubMed]
72. Estruch, R.; Ros, E. The role of the Mediterranean diet on weight loss and obesity-related diseases. *Rev. Endocr. Metab. Discrd.* **2020**, *21*, 315–327. [CrossRef] [PubMed]
73. Lotfi, K.; Saneei, P.; Hajhashemy, Z.; Esmaillzadeh, A. Adherence to the Mediterranean Diet, Five-Year Weight Change, and Risk of Overweight and Obesity: A Systematic Review and Dose-Response Meta-Analysis of Prospective Cohort Studies. *Adv. Nutr.* **2021**, *13*, 152–166. [CrossRef] [PubMed]
74. Mancini, J.G.; Filion, K.B.; Atallah, R.; Eisenberg, M.J. Systematic Review of the Mediterranean Diet for Long-Term Weight Loss. *Am. J. Med.* **2016**, *129*, 407–415.e404. [CrossRef] [PubMed]
75. Katagiri, R.; Asakura, K.; Kobayashi, S.; Suga, H.; Sasaki, S. Low intake of vegetables, high intake of confectionary, and unhealthy eating habits are associated with poor sleep quality among middle-aged female Japanese workers. *J. Occup. Health* **2014**, *56*, 359–368. [CrossRef]
76. St-Onge, M.P.; Roberts, A.; Shechter, A.; Choudhury, A.R. Fiber and Saturated Fat Are Associated with Sleep Arousals and Slow Wave Sleep. *J. Clin. Sleep Med.* **2016**, *12*, 19–24. [CrossRef]
77. Kurotani, K.; Kochi, T.; Nanri, A.; Eguchi, M.; Kuwahara, K.; Tsuruoka, H.; Akter, S.; Ito, R.; Pham, N.M.; Kabe, I.; et al. Dietary patterns and sleep symptoms in Japanese workers: The Furukawa Nutrition and Health Study. *Sleep Med.* **2015**, *16*, 298–304. [CrossRef]
78. Noorwali, E.; Hardie, L.; Cade, J. Bridging the Reciprocal Gap between Sleep and Fruit and Vegetable Consumption: A Review of the Evidence, Potential Mechanisms, Implications, and Directions for Future Work. *Nutrients* **2019**, *11*, 1382. [CrossRef]
79. Hysing, M.; Kvestad, I.; Kjellevold, M.; Kolden Midtbo, L.; Graff, I.E.; Lie, O.; Hurum, H.; Stormark, K.M.; Oyen, J. Fatty Fish Intake and the Effect on Mental Health and Sleep in Preschool Children in FINS-KIDS, a Randomized Controlled Trial. *Nutrients* **2018**, *10*, 1478. [CrossRef]
80. Del Brutto, O.H.; Mera, R.M.; Ha, J.E.; Gillman, J.; Zambrano, M.; Castillo, P.R. Dietary fish intake and sleep quality: A population-based study. *Sleep Med.* **2016**, *17*, 126–128. [CrossRef]
81. Hansen, A.L.; Dahl, L.; Olson, G.; Thornton, D.; Graff, I.E.; Froyland, L.; Thayer, J.F.; Pallesen, S. Fish consumption, sleep, daily functioning, and heart rate variability. *J. Clin. Sleep Med.* **2014**, *10*, 567–575. [CrossRef] [PubMed]
82. Montgomery, P.; Burton, J.R.; Sewell, R.P.; Spreckelsen, T.F.; Richardson, A.J. Fatty acids and sleep in UK children: Subjective and pilot objective sleep results from the DOLAB study—A randomized controlled trial. *J. Sleep Res.* **2014**, *23*, 364–388. [CrossRef]
83. Jansen, E.C.; Conroy, D.A.; Burgess, H.J.; O'Brien, L.M.; Cantoral, A.; Tellez-Rojo, M.M.; Peterson, K.E.; Baylin, A. Plasma DHA Is Related to Sleep Timing and Duration in a Cohort of Mexican Adolescents. *J. Nutr.* **2020**, *150*, 592–598. [CrossRef]
84. Fraga, C.G.; Croft, K.D.; Kennedy, D.O.; Tomas-Barberan, F.A. The effects of polyphenols and other bioactives on human health. *Food Funct.* **2019**, *10*, 514–528. [CrossRef]
85. Manach, C.; Williamson, G.; Morand, C.; Scalbert, A.; Remesy, C. Bioavailability and bioefficacy of polyphenols in humans. I. Review of 97 bioavailability studies. *Am. J. Clin. Nutr.* **2005**, *81*, 230S–242S. [CrossRef]
86. Tresserra-Rimbau, A.; Medina-Remon, A.; Perez-Jimenez, J.; Martinez-Gonzalez, M.A.; Covas, M.I.; Corella, D.; Salas-Salvado, J.; Gomez-Gracia, E.; Lapetra, J.; Aros, F.; et al. Dietary intake and major food sources of polyphenols in a Spanish population at high cardiovascular risk: The PREDIMED study. *Nutr. Metab. Cardiovasc. Dis.* **2013**, *23*, 953–959. [CrossRef]
87. Di Lorenzo, C.; Colombo, F.; Biella, S.; Stockley, C.; Restani, P. Polyphenols and Human Health: The Role of Bioavailability. *Nutrients* **2021**, *13*, 273. [CrossRef]
88. Cao, Y.; Taylor, A.W.; Zhen, S.; Adams, R.; Appleton, S.; Shi, Z. Soy Isoflavone Intake and Sleep Parameters over 5 Years among Chinese Adults: Longitudinal Analysis from the Jiangsu Nutrition Study. *J. Acad. Nutr. Diet.* **2017**, *117*, 536–544.e532. [CrossRef]
89. Cui, Y.; Niu, K.; Huang, C.; Momma, H.; Guan, L.; Kobayashi, Y.; Guo, H.; Chujo, M.; Otomo, A.; Nagatomi, R. Relationship between daily isoflavone intake and sleep in Japanese adults: A cross-sectional study. *Nutr. J.* **2015**, *14*, 127. [CrossRef]
90. Godos, J.; Ferri, R.; Castellano, S.; Angelino, D.; Mena, P.; Del Rio, D.; Caraci, F.; Galvano, F.; Grosso, G. Specific Dietary (Poly)phenols Are Associated with Sleep Quality in a Cohort of Italian Adults. *Nutrients* **2020**, *12*, 1226. [CrossRef]
91. Tuttolomondo, A.; Simonetta, I.; Daidone, M.; Mogavero, A.; Ortello, A.; Pinto, A. Metabolic and Vascular Effect of the Mediterranean Diet. *Int. J. Mol. Sci.* **2019**, *20*, 4716. [CrossRef]

92. Esposito, K.; Maiorino, M.I.; Ceriello, A.; Giugliano, D. Prevention and control of type 2 diabetes by Mediterranean diet: A systematic review. *Diabetes Res. Clin. Pract.* **2010**, *89*, 97–102. [CrossRef]
93. Estruch, R.; Martinez-Gonzalez, M.A.; Corella, D.; Salas-Salvado, J.; Ruiz-Gutierrez, V.; Covas, M.I.; Fiol, M.; Gomez-Gracia, E.; Lopez-Sabater, M.C.; Vinyoles, E.; et al. Effects of a Mediterranean-style diet on cardiovascular risk factors: A randomized trial. *Ann. Intern. Med.* **2006**, *145*, 1–11. [CrossRef]
94. Storniolo, C.E.; Casillas, R.; Bullo, M.; Castaner, O.; Ros, E.; Saez, G.T.; Toledo, E.; Estruch, R.; Ruiz-Gutierrez, V.; Fito, M.; et al. A Mediterranean diet supplemented with extra virgin olive oil or nuts improves endothelial markers involved in blood pressure control in hypertensive women. *Eur. J. Nutr.* **2017**, *56*, 89–97. [CrossRef]
95. Carroll, C.M.; Macauley, S.L. The Interaction Between Sleep and Metabolism in Alzheimer's Disease: Cause or Consequence of Disease? *Front. Aging Neurosci.* **2019**, *11*, 258. [CrossRef]
96. MacIntosh, B.J.; Shirzadi, Z.; Atwi, S.; Detre, J.A.; Dolui, S.; Bryan, R.N.; Launer, L.J.; Swardfager, W. Metabolic and vascular risk factors are associated with reduced cerebral blood flow and poorer midlife memory performance. *Hum. Brain Mapp.* **2020**, *41*, 855–864. [CrossRef]
97. Yu, G.X.; Zhang, T.; Hou, X.H.; Ou, Y.N.; Hu, H.; Wang, Z.T.; Guo, Y.; Xu, W.; Tan, L.; Yu, J.T.; et al. Associations of Vascular Risk with Cognition, Brain Glucose Metabolism, and Clinical Progression in Cognitively Intact Elders. *J. Alzheimers Dis.* **2021**, *80*, 321–330. [CrossRef]
98. Boccellino, M.; D'Angelo, S. Anti-Obesity Effects of Polyphenol Intake: Current Status and Future Possibilities. *Int. J. Mol. Sci.* **2020**, *21*, 5642. [CrossRef]
99. Vgontzas, A.N.; Bixler, E.O.; Chrousos, G.P. Sleep apnea is a manifestation of the metabolic syndrome. *Sleep Med. Rev.* **2005**, *9*, 211–224. [CrossRef]
100. Jordan, A.S.; McSharry, D.G.; Malhotra, A. Adult obstructive sleep apnoea. *Lancet* **2014**, *383*, 736–747. [CrossRef]
101. Papandreou, C.; Schiza, S.E.; Bouloukaki, I.; Hatzis, C.M.; Kafatos, A.G.; Siafakas, N.M.; Tzanakis, N.E. Effect of Mediterranean diet versus prudent diet combined with physical activity on OSAS: A randomised trial. *Eur. Respir. J.* **2012**, *39*, 1398–1404. [CrossRef] [PubMed]
102. Georgoulis, M.; Yiannakouris, N.; Tenta, R.; Fragopoulou, E.; Kechribari, I.; Lamprou, K.; Perraki, E.; Vagiakis, E.; Kontogianni, M.D. A weight-loss Mediterranean diet/lifestyle intervention ameliorates inflammation and oxidative stress in patients with obstructive sleep apnea: Results of the "MIMOSA" randomized clinical trial. *Eur. J. Nutr.* **2021**, *60*, 3799–3810. [CrossRef] [PubMed]
103. Schwingshackl, L.; Hoffmann, G. Mediterranean dietary pattern, inflammation and endothelial function: A systematic review and meta-analysis of intervention trials. *Nutr. Metab. Cardiovasc. Dis.* **2014**, *24*, 929–939. [CrossRef]
104. Schwingshackl, L.; Christoph, M.; Hoffmann, G. Effects of Olive Oil on Markers of Inflammation and Endothelial Function—A Systematic Review and Meta-Analysis. *Nutrients* **2015**, *7*, 7651–7675. [CrossRef] [PubMed]
105. Scoditti, E.; Capurso, C.; Capurso, A.; Massaro, M. Vascular effects of the Mediterranean diet-part II: Role of omega-3 fatty acids and olive oil polyphenols. *Vasc. Pharmacol.* **2014**, *63*, 127–134. [CrossRef] [PubMed]
106. Gardener, H.; Wright, C.B.; Rundek, T.; Sacco, R.L. Brain health and shared risk factors for dementia and stroke. *Nat. Rev. Neurol.* **2015**, *11*, 651–657. [CrossRef]
107. Klohs, J. An Integrated View on Vascular Dysfunction in Alzheimer's Disease. *Neurodegener. Dis.* **2019**, *19*, 109–127. [CrossRef]
108. Naiberg, M.R.; Newton, D.F.; Goldstein, B.I. Flow-Mediated Dilation and Neurocognition: Systematic Review and Future Directions. *Psychosom. Med.* **2016**, *78*, 192–207. [CrossRef]
109. Rajendran, P.; Rengarajan, T.; Thangavel, J.; Nishigaki, Y.; Sakthisekaran, D.; Sethi, G.; Nishigaki, I. The vascular endothelium and human diseases. *Int. J. Biol. Sci.* **2013**, *9*, 1057–1069. [CrossRef]
110. Cherubini, J.M.; Cheng, J.L.; Williams, J.S.; MacDonald, M.J. Sleep deprivation and endothelial function: Reconciling seminal evidence with recent perspectives. *Am. J. Physiol. Heart Circ. Physiol.* **2021**, *320*, H29–H35. [CrossRef]
111. Quick, S.; Moss, J.; Rajani, R.M.; Williams, A. A Vessel for Change: Endothelial Dysfunction in Cerebral Small Vessel Disease. *Trends Neurosci.* **2021**, *44*, 289–305. [CrossRef] [PubMed]
112. Faraguna, U.; Vyazovskiy, V.V.; Nelson, A.B.; Tononi, G.; Cirelli, C. A causal role for brain-derived neurotrophic factor in the homeostatic regulation of sleep. *J. Neurosci.* **2008**, *28*, 4088–4095. [CrossRef] [PubMed]
113. Giese, M.; Unternahrer, E.; Huttig, H.; Beck, J.; Brand, S.; Calabrese, P.; Holsboer-Trachsler, E.; Eckert, A. BDNF: An indicator of insomnia? *Mol. Psychiatry* **2014**, *19*, 151–152. [CrossRef] [PubMed]
114. Marie, C.; Pedard, M.; Quirie, A.; Tessier, A.; Garnier, P.; Totoson, P.; Demougeot, C. Brain-derived neurotrophic factor secreted by the cerebral endothelium: A new actor of brain function? *J. Cereb. Blood Flow Metab.* **2018**, *38*, 935–949. [CrossRef] [PubMed]
115. Sanchez-Villegas, A.; Galbete, C.; Martinez-Gonzalez, M.A.; Martinez, J.A.; Razquin, C.; Salas-Salvado, J.; Estruch, R.; Buil-Cosiales, P.; Marti, A. The effect of the Mediterranean diet on plasma brain-derived neurotrophic factor (BDNF) levels: The PREDIMED-NAVARRA randomized trial. *Nutr. Neurosci.* **2011**, *14*, 195–201. [CrossRef] [PubMed]
116. Khansari, N.; Shakiba, Y.; Mahmoudi, M. Chronic inflammation and oxidative stress as a major cause of age-related diseases and cancer. *Recent Pat. Inflamm. Allergy Drug Discov.* **2009**, *3*, 73–80. [CrossRef]
117. Agostinho, P.; Cunha, R.A.; Oliveira, C. Neuroinflammation, oxidative stress and the pathogenesis of Alzheimer's disease. *Curr. Pharm. Des.* **2010**, *16*, 2766–2778. [CrossRef]

118. Leonard, B.E. Inflammation and depression: A causal or coincidental link to the pathophysiology? *Acta Neuropsychiatr.* **2018**, *30*, 1–16. [CrossRef]
119. Clark, I.A.; Vissel, B. Inflammation-sleep interface in brain disease: TNF, insulin, orexin. *J. Neuroinflamm.* **2014**, *11*, 51. [CrossRef]
120. Kanagasabai, T.; Ardern, C.I. Contribution of Inflammation, Oxidative Stress, and Antioxidants to the Relationship between Sleep Duration and Cardiometabolic Health. *Sleep* **2015**, *38*, 1905–1912. [CrossRef]
121. Pourreza, S.; Khademi, Z.; Mirzababaei, A.; Yekaninejad, M.S.; Sadeghniiat-Haghighi, K.; Naghshi, S.; Mirzaei, K. Association of plant-based diet index with inflammatory markers and sleep quality in overweight and obese female adults: A cross-sectional study. *Int. J. Clin. Pract.* **2021**, *75*, e14429. [CrossRef]
122. Kanagasabai, T.; Ardern, C.I. Inflammation, Oxidative Stress, and Antioxidants Contribute to Selected Sleep Quality and Cardiometabolic Health Relationships: A Cross-Sectional Study. *Mediat. Inflamm.* **2015**, *2015*, 824589. [CrossRef]
123. Shivappa, N.; Steck, S.E.; Hurley, T.G.; Hussey, J.R.; Hebert, J.R. Designing and developing a literature-derived, population-based dietary inflammatory index. *Public Health Nutr.* **2014**, *17*, 1689–1696. [CrossRef]
124. Godos, J.; Ferri, R.; Caraci, F.; Cosentino, F.I.I.; Castellano, S.; Shivappa, N.; Hebert, J.R.; Galvano, F.; Grosso, G. Dietary Inflammatory Index and Sleep Quality in Southern Italian Adults. *Nutrients* **2019**, *11*, 1324. [CrossRef]
125. Bazyar, H.; Zare Javid, A.; Bavi Behbahani, H.; Shivappa, N.; Hebert, J.R.; Khodaramhpour, S.; Khaje Zadeh, S.; Aghamohammadi, V. The association between dietary inflammatory index with sleep quality and obesity amongst iranian female students: A cross-sectional study. *Int. J. Clin. Pract.* **2021**, *75*, e14061. [CrossRef]
126. Masaad, A.A.; Yusuf, A.M.; Shakir, A.Z.; Khan, M.S.; Khaleel, S.; Cheikh Ismail, L.; Faris, M.A.E.; Jahrami, H.A.; Shivappa, N.; Hebert, J.R.; et al. Sleep quality and Dietary Inflammatory Index among university students: A cross-sectional study. *Sleep Breath.* **2021**, *25*, 2221–2229. [CrossRef]
127. Lopes, T.V.C.; Borba, M.E.S.; Lopes, R.V.C.; Fisberg, R.M.; Paim, S.L.; Teodoro, V.V.; Zimberg, I.Z.; Araujo, L.B.; Shivappa, N.; Hebert, J.R.; et al. Association between inflammatory potential of the diet and sleep parameters in sleep apnea patients. *Nutrition* **2019**, *66*, 5–10. [CrossRef]
128. Kase, B.E.; Liu, J.; Wirth, M.D.; Shivappa, N.; Hebert, J.R. Associations between dietary inflammatory index and sleep problems among adults in the United States, NHANES 2005–2016. *Sleep Health* **2021**, *7*, 273–280. [CrossRef]
129. Wirth, M.D.; Jessup, A.; Turner-McGrievy, G.; Shivappa, N.; Hurley, T.G.; Hebert, J.R. Changes in dietary inflammatory potential predict changes in sleep quality metrics, but not sleep duration. *Sleep* **2020**, *43*, zsaa093. [CrossRef]
130. Gordon-Dseagu, V.L.Z.; Derkach, A.; Xiao, Q.; Williams, I.; Sampson, J.; Stolzenberg-Solomon, R.Z. The association of sleep with metabolic pathways and metabolites: Evidence from the Dietary Approaches to Stop Hypertension (DASH)-sodium feeding study. *Metabolomics* **2019**, *15*, 48. [CrossRef]
131. Abshirini, M.; Siassi, F.; Koohdani, F.; Qorbani, M.; Khosravi, S.; Hedayati, M.; Aslani, Z.; Soleymani, M.; Sotoudeh, G. Dietary total antioxidant capacity is inversely related to menopausal symptoms: A cross-sectional study among Iranian postmenopausal women. *Nutrition* **2018**, *55–56*, 161–167. [CrossRef]
132. Daneshzad, E.; Keshavarz, S.A.; Qorbani, M.; Larijani, B.; Azadbakht, L. Dietary total antioxidant capacity and its association with sleep, stress, anxiety, and depression score: A cross-sectional study among diabetic women. *Clin. Nutr. ESPEN* **2020**, *37*, 187–194. [CrossRef]
133. Billingsley, H.E.; Carbone, S. The antioxidant potential of the Mediterranean diet in patients at high cardiovascular risk: An in-depth review of the PREDIMED. *Nutr. Diabetes* **2018**, *8*, 13. [CrossRef]
134. Gantenbein, K.V.; Kanaka-Gantenbein, C. Mediterranean Diet as an Antioxidant: The Impact on Metabolic Health and Overall Wellbeing. *Nutrients* **2021**, *13*, 1951. [CrossRef]
135. Pitsavos, C.; Panagiotakos, D.B.; Tzima, N.; Chrysohoou, C.; Economou, M.; Zampelas, A.; Stefanadis, C. Adherence to the Mediterranean diet is associated with total antioxidant capacity in healthy adults: The ATTICA study. *Am. J. Clin. Nutr.* **2005**, *82*, 694–699. [CrossRef]
136. Casas, R.; Urpi-Sarda, M.; Sacanella, E.; Arranz, S.; Corella, D.; Castaner, O.; Lamuela-Raventos, R.M.; Salas-Salvado, J.; Lapetra, J.; Portillo, M.P.; et al. Anti-Inflammatory Effects of the Mediterranean Diet in the Early and Late Stages of Atheroma Plaque Development. *Mediat. Inflamm.* **2017**, *2017*, 3674390. [CrossRef]
137. Centritto, F.; Iacoviello, L.; di Giuseppe, R.; De Curtis, A.; Costanzo, S.; Zito, F.; Grioni, S.; Sieri, S.; Donati, M.B.; de Gaetano, G.; et al. Dietary patterns, cardiovascular risk factors and C-reactive protein in a healthy Italian population. *Nutr. Metab. Cardiovasc. Dis.* **2009**, *19*, 697–706. [CrossRef]
138. Esposito, K.; Marfella, R.; Ciotola, M.; Di Palo, C.; Giugliano, F.; Giugliano, G.; D'Armiento, M.; D'Andrea, F.; Giugliano, D. Effect of a mediterranean-style diet on endothelial dysfunction and markers of vascular inflammation in the metabolic syndrome: A randomized trial. *JAMA* **2004**, *292*, 1440–1446. [CrossRef]
139. Estruch, R. Anti-inflammatory effects of the Mediterranean diet: The experience of the PREDIMED study. *Proc. Nutr. Soc.* **2010**, *69*, 333–340. [CrossRef]
140. Castaner, O.; Covas, M.I.; Khymenets, O.; Nyyssonen, K.; Konstantinidou, V.; Zunft, H.F.; de la Torre, R.; Munoz-Aguayo, D.; Vila, J.; Fito, M. Protection of LDL from oxidation by olive oil polyphenols is associated with a downregulation of CD40-ligand expression and its downstream products in vivo in humans. *Am. J. Clin. Nutr.* **2012**, *95*, 1238–1244. [CrossRef]

141. Pounis, G.; Bonaccio, M.; Di Castelnuovo, A.; Costanzo, S.; de Curtis, A.; Persichillo, M.; Sieri, S.; Donati, M.B.; Cerletti, C.; de Gaetano, G.; et al. Polyphenol intake is associated with low-grade inflammation, using a novel data analysis from the Moli-sani study. *Thromb. Haemost.* **2016**, *115*, 344–352. [CrossRef]
142. Chiva-Blanch, G.; Arranz, S.; Lamuela-Raventos, R.M.; Estruch, R. Effects of wine, alcohol and polyphenols on cardiovascular disease risk factors: Evidences from human studies. *Alcohol Alcohol.* **2013**, *48*, 270–277. [CrossRef]
143. Kelley, D.S.; Siegel, D.; Fedor, D.M.; Adkins, Y.; Mackey, B.E. DHA supplementation decreases serum C-reactive protein and other markers of inflammation in hypertriglyceridemic men. *J. Nutr.* **2009**, *139*, 495–501. [CrossRef]
144. Landberg, R.; Sun, Q.; Rimm, E.B.; Cassidy, A.; Scalbert, A.; Mantzoros, C.S.; Hu, F.B.; van Dam, R.M. Selected dietary flavonoids are associated with markers of inflammation and endothelial dysfunction in U.S. women. *J. Nutr.* **2011**, *141*, 618–625. [CrossRef]
145. Medina-Remon, A.; Casas, R.; Tressserra-Rimbau, A.; Ros, E.; Martinez-Gonzalez, M.A.; Fito, M.; Corella, D.; Salas-Salvado, J.; Lamuela-Raventos, R.M.; Estruch, R.; et al. Polyphenol intake from a Mediterranean diet decreases inflammatory biomarkers related to atherosclerosis: A substudy of the PREDIMED trial. *Br. J. Clin. Pharmacol.* **2017**, *83*, 114–128. [CrossRef]
146. Urpi-Sarda, M.; Casas, R.; Chiva-Blanch, G.; Romero-Mamani, E.S.; Valderas-Martinez, P.; Salas-Salvado, J.; Covas, M.I.; Toledo, E.; Andres-Lacueva, C.; Llorach, R.; et al. The Mediterranean diet pattern and its main components are associated with lower plasma concentrations of tumor necrosis factor receptor 60 in patients at high risk for cardiovascular disease. *J. Nutr.* **2012**, *142*, 1019–1025. [CrossRef]
147. Carluccio, M.A.; Martinelli, R.; Massaro, M.; Calabriso, N.; Scoditti, E.; Maffia, M.; Verri, T.; Gatta, V.; De Caterina, R. Nutrigenomic Effect of Hydroxytyrosol in Vascular Endothelial Cells: A Transcriptomic Profile Analysis. *Nutrients* **2021**, *13*, 3990. [CrossRef]
148. Konstantinidou, V.; Covas, M.I.; Sola, R.; Fito, M. Up-to date knowledge on the in vivo transcriptomic effect of the Mediterranean diet in humans. *Mol. Nutr. Food Res.* **2013**, *57*, 772–783. [CrossRef]
149. Scoditti, E.; Nestola, A.; Massaro, M.; Calabriso, N.; Storelli, C.; De Caterina, R.; Carluccio, M.A. Hydroxytyrosol suppresses MMP-9 and COX-2 activity and expression in activated human monocytes via PKCalpha and PKCbeta1 inhibition. *Atherosclerosis* **2014**, *232*, 17–24. [CrossRef]
150. Garcia-Casares, N.; Gallego Fuentes, P.; Barbancho, M.A.; Lopez-Gigosos, R.; Garcia-Rodriguez, A.; Gutierrez-Bedmar, M. Alzheimer's Disease, Mild Cognitive Impairment and Mediterranean Diet. A Systematic Review and Dose-Response Meta-Analysis. *J. Clin. Med.* **2021**, *10*, 4642. [CrossRef]
151. Gardener, H.; Caunca, M.R. Mediterranean Diet in Preventing Neurodegenerative Diseases. *Curr. Nutr. Rep.* **2018**, *7*, 10–20. [CrossRef]
152. Siervo, M.; Shannon, O.M.; Llewellyn, D.J.; Stephan, B.C.; Fontana, L. Mediterranean diet and cognitive function: From methodology to mechanisms of action. *Free Radic. Biol. Med.* **2021**, *176*, 105–117. [CrossRef]
153. Lassale, C.; Batty, G.D.; Baghdadli, A.; Jacka, F.; Sanchez-Villegas, A.; Kivimaki, M.; Akbaraly, T. Healthy dietary indices and risk of depressive outcomes: A systematic review and meta-analysis of observational studies. *Mol. Psychiatry* **2019**, *24*, 965–986. [CrossRef]
154. Sadeghi, O.; Keshteli, A.H.; Afshar, H.; Esmaillzadeh, A.; Adibi, P. Adherence to Mediterranean dietary pattern is inversely associated with depression, anxiety and psychological distress. *Nutr. Neurosci.* **2021**, *24*, 248–259. [CrossRef]
155. Fang, H.; Tu, S.; Sheng, J.; Shao, A. Depression in sleep disturbance: A review on a bidirectional relationship, mechanisms and treatment. *J. Cell. Mol. Med.* **2019**, *23*, 2324–2332. [CrossRef]
156. Peterson, M.J.; Benca, R.M. Sleep in mood disorders. *Psychiatr. Clin. N. Am.* **2006**, *29*, 1009–1032. [CrossRef]
157. Tahmasian, M.; Samea, F.; Khazaie, H.; Zarei, M.; Kharabian Masouleh, S.; Hoffstaedter, F.; Camilleri, J.; Kochunov, P.; Yeo, B.T.T.; Eickhoff, S.B.; et al. The interrelation of sleep and mental and physical health is anchored in grey-matter neuroanatomy and under genetic control. *Commun. Biol.* **2020**, *3*, 171. [CrossRef]
158. Luchtman, D.W.; Song, C. Cognitive enhancement by omega-3 fatty acids from child-hood to old age: Findings from animal and clinical studies. *Neuropharmacology* **2013**, *64*, 550–565. [CrossRef]
159. Hornedo-Ortega, R.; Cerezo, A.B.; de Pablos, R.M.; Krisa, S.; Richard, T.; Garcia-Parrilla, M.C.; Troncoso, A.M. Phenolic Compounds Characteristic of the Mediterranean Diet in Mitigating Microglia-Mediated Neuroinflammation. *Front. Cell. Neurosci.* **2018**, *12*, 373. [CrossRef]
160. Ruxton, C.H.; Calder, P.C.; Reed, S.C.; Simpson, M.J. The impact of long-chain n-3 polyunsaturated fatty acids on human health. *Nutr. Res. Rev.* **2005**, *18*, 113–129. [CrossRef]
161. Janssen, C.I.; Kiliaan, A.J. Long-chain polyunsaturated fatty acids (LCPUFA) from genesis to senescence: The influence of LCPUFA on neural development, aging, and neurodegeneration. *Prog. Lipid Res.* **2014**, *53*, 1–17. [CrossRef] [PubMed]
162. Horman, T.; Fernandes, M.F.; Tache, M.C.; Hucik, B.; Mutch, D.M.; Leri, F. Dietary n-6/n-3 Ratio Influences Brain Fatty Acid Composition in Adult Rats. *Nutrients* **2020**, *12*, 1847. [CrossRef] [PubMed]
163. Simopoulos, A.P. The importance of the omega-6/omega-3 fatty acid ratio in cardiovascular disease and other chronic diseases. *Exp. Biol. Med. (Maywood)* **2008**, *233*, 674–688. [CrossRef]
164. Calder, P.C. n-3 fatty acids, inflammation and immunity: New mechanisms to explain old actions. *Proc. Nutr. Soc.* **2013**, *72*, 326–336. [CrossRef]
165. Joffre, C.; Rey, C.; Laye, S. N-3 Polyunsaturated Fatty Acids and the Resolution of Neuroinflammation. *Front. Pharmacol.* **2019**, *10*, 1022. [CrossRef] [PubMed]

166. Chalon, S.; Vancassel, S.; Zimmer, L.; Guilloteau, D.; Durand, G. Polyunsaturated fatty acids and cerebral function: Focus on monoaminergic neurotransmission. *Lipids* **2001**, *36*, 937–944. [CrossRef] [PubMed]
167. Delion, S.; Chalon, S.; Guilloteau, D.; Besnard, J.C.; Durand, G. alpha-Linolenic acid dietary deficiency alters age-related changes of dopaminergic and serotoninergic neurotransmission in the rat frontal cortex. *J. Neurochem.* **1996**, *66*, 1582–1591. [CrossRef]
168. Grosso, G.; Galvano, F.; Marventano, S.; Malaguarnera, M.; Bucolo, C.; Drago, F.; Caraci, F. Omega-3 fatty acids and depression: Scientific evidence and biological mechanisms. *Oxid. Med. Cell. Longev.* **2014**, *2014*, 313570. [CrossRef]
169. Vines, A.; Delattre, A.M.; Lima, M.M.; Rodrigues, L.S.; Suchecki, D.; Machado, R.B.; Tufik, S.; Pereira, S.I.; Zanata, S.M.; Ferraz, A.C. The role of 5-HT(1)A receptors in fish oil-mediated increased BDNF expression in the rat hippocampus and cortex: A possible antidepressant mechanism. *Neuropharmacology* **2012**, *62*, 184–191. [CrossRef]
170. Zhang, W.; Li, P.; Hu, X.; Zhang, F.; Chen, J.; Gao, Y. Omega-3 polyunsaturated fatty acids in the brain: Metabolism and neuroprotection. *Front. Biosci. (Landmark Ed.)* **2011**, *16*, 2653–2670. [CrossRef]
171. Zimmer, L.; Vancassel, S.; Cantagrel, S.; Breton, P.; Delamanche, S.; Guilloteau, D.; Durand, G.; Chalon, S. The dopamine mesocorticolimbic pathway is affected by deficiency in n-3 polyunsaturated fatty acids. *Am. J. Clin. Nutr.* **2002**, *75*, 662–667. [CrossRef] [PubMed]
172. Cutuli, D.; Landolfo, E.; Decandia, D.; Nobili, A.; Viscomi, M.T.; La Barbera, L.; Sacchetti, S.; De Bartolo, P.; Curci, A.; D'Amelio, M.; et al. Neuroprotective Role of Dietary Supplementation with Omega-3 Fatty Acids in the Presence of Basal Forebrain Cholinergic Neurons Degeneration in Aged Mice. *Int. J. Mol. Sci.* **2020**, *21*, 1741. [CrossRef] [PubMed]
173. Titova, O.E.; Sjogren, P.; Brooks, S.J.; Kullberg, J.; Ax, E.; Kilander, L.; Riserus, U.; Cederholm, T.; Larsson, E.M.; Johansson, L.; et al. Dietary intake of eicosapentaenoic and docosahexaenoic acids is linked to gray matter volume and cognitive function in elderly. *Age (Dordr)* **2013**, *35*, 1495–1505. [CrossRef] [PubMed]
174. Araya-Quintanilla, F.; Gutierrez-Espinoza, H.; Sanchez-Montoya, U.; Munoz-Yanez, M.J.; Baeza-Vergara, A.; Petersen-Yanjari, M.; Fernandez-Lecaros, L. Effectiveness of omega-3 fatty acid supplementation in patients with Alzheimer disease: A systematic review and meta-analysis. *Neurologia (Engl. Ed.)* **2020**, *35*, 105–114. [CrossRef] [PubMed]
175. Parletta, N.; Zarnowiecki, D.; Cho, J.; Wilson, A.; Bogomolova, S.; Villani, A.; Itsiopoulos, C.; Niyonsenga, T.; Blunden, S.; Meyer, B.; et al. A Mediterranean-style dietary intervention supplemented with fish oil improves diet quality and mental health in people with depression: A randomized controlled trial (HELFIMED). *Nutr. Neurosci.* **2019**, *22*, 474–487. [CrossRef]
176. Patan, M.J.; Kennedy, D.O.; Husberg, C.; Hustvedt, S.O.; Calder, P.C.; Khan, J.; Forster, J.; Jackson, P.A. Supplementation with oil rich in eicosapentaenoic acid, but not in docosahexaenoic acid, improves global cognitive function in healthy, young adults: Results from randomized controlled trials. *Am. J. Clin. Nutr.* **2021**, *114*, 914–924. [CrossRef]
177. Figueira, I.; Garcia, G.; Pimpao, R.C.; Terrasso, A.P.; Costa, I.; Almeida, A.F.; Tavares, L.; Pais, T.F.; Pinto, P.; Ventura, M.R.; et al. Polyphenols journey through blood-brain barrier towards neuronal protection. *Sci. Rep.* **2017**, *7*, 11456. [CrossRef]
178. Frolinger, T.; Sims, S.; Smith, C.; Wang, J.; Cheng, H.; Faith, J.; Ho, L.; Hao, K.; Pasinetti, G.M. The gut microbiota composition affects dietary polyphenols-mediated cognitive resilience in mice by modulating the bioavailability of phenolic acids. *Sci. Rep.* **2019**, *9*, 3546. [CrossRef]
179. Turner, R.S.; Thomas, R.G.; Craft, S.; van Dyck, C.H.; Mintzer, J.; Reynolds, B.A.; Brewer, J.B.; Rissman, R.A.; Raman, R.; Aisen, P.S.; et al. A randomized, double-blind, placebo-controlled trial of resveratrol for Alzheimer disease. *Neurology* **2015**, *85*, 1383–1391. [CrossRef]
180. Youdim, K.A.; Qaiser, M.Z.; Begley, D.J.; Rice-Evans, C.A.; Abbott, N.J. Flavonoid permeability across an in situ model of the blood-brain barrier. *Free Radic. Biol. Med.* **2004**, *36*, 592–604. [CrossRef]
181. Quincozes-Santos, A.; Gottfried, C. Resveratrol modulates astroglial functions: Neuroprotective hypothesis. *Ann. N. Y. Acad. Sci.* **2011**, *1215*, 72–78. [CrossRef] [PubMed]
182. Scoditti, E. Neuroinflammation and Neurodegeneration: The Promising Protective Role of the Citrus Flavanone Hesperetin. *Nutrients* **2020**, *12*, 2336. [CrossRef] [PubMed]
183. Valls-Pedret, C.; Lamuela-Raventos, R.M.; Medina-Remon, A.; Quintana, M.; Corella, D.; Pinto, X.; Martinez-Gonzalez, M.A.; Estruch, R.; Ros, E. Polyphenol-rich foods in the Mediterranean diet are associated with better cognitive function in elderly subjects at high cardiovascular risk. *J. Alzheimers Dis.* **2012**, *29*, 773–782. [CrossRef] [PubMed]
184. Batarseh, Y.S.; Mohamed, L.A.; Al Rihani, S.B.; Mousa, Y.M.; Siddique, A.B.; El Sayed, K.A.; Kaddoumi, A. Oleocanthal ameliorates amyloid-beta oligomers' toxicity on astrocytes and neuronal cells: In vitro studies. *Neuroscience* **2017**, *352*, 204–215. [CrossRef]
185. Crespo, M.C.; Tome-Carneiro, J.; Pintado, C.; Davalos, A.; Visioli, F.; Burgos-Ramos, E. Hydroxytyrosol restores proper insulin signaling in an astrocytic model of Alzheimer's disease. *Biofactors* **2017**, *43*, 540–548. [CrossRef]
186. Feng, X.; Liang, N.; Zhu, D.; Gao, Q.; Peng, L.; Dong, H.; Yue, Q.; Liu, H.; Bao, L.; Zhang, J.; et al. Resveratrol inhibits beta-amyloid-induced neuronal apoptosis through regulation of SIRT1-ROCK1 signaling pathway. *PLoS ONE* **2013**, *8*, e59888. [CrossRef]
187. Feng, Y.; Cui, Y.; Gao, J.L.; Li, R.; Jiang, X.H.; Tian, Y.X.; Wang, K.J.; Li, M.H.; Zhang, H.A.; Cui, J.Z. Neuroprotective effects of resveratrol against traumatic brain injury in rats: Involvement of synaptic proteins and neuronal autophagy. *Mol. Med. Rep.* **2016**, *13*, 5248–5254. [CrossRef]
188. Muhammad, T.; Ikram, M.; Ullah, R.; Rehman, S.U.; Kim, M.O. Hesperetin, a Citrus Flavonoid, Attenuates LPS-Induced Neuroinflammation, Apoptosis and Memory Impairments by Modulating TLR4/NF-kappaB Signaling. *Nutrients* **2019**, *11*, 648. [CrossRef]

189. Rehman, S.U.; Ali, T.; Alam, S.I.; Ullah, R.; Zeb, A.; Lee, K.W.; Rutten, B.P.F.; Kim, M.O. Ferulic Acid Rescues LPS-Induced Neurotoxicity via Modulation of the TLR4 Receptor in the Mouse Hippocampus. *Mol. Neurobiol.* **2019**, *56*, 2774–2790. [CrossRef]
190. Arus, B.A.; Souza, D.G.; Bellaver, B.; Souza, D.O.; Goncalves, C.A.; Quincozes-Santos, A.; Bobermin, L.D. Resveratrol modulates GSH system in C6 astroglial cells through heme oxygenase 1 pathway. *Mol. Cell. Biochem.* **2017**, *428*, 67–77. [CrossRef]
191. Yao, Y.; Li, J.; Niu, Y.; Yu, J.Q.; Yan, L.; Miao, Z.H.; Zhao, X.X.; Li, Y.J.; Yao, W.X.; Zheng, P.; et al. Resveratrol inhibits oligomeric Abeta-induced microglial activation via NADPH oxidase. *Mol. Med. Rep.* **2015**, *12*, 6133–6139. [CrossRef] [PubMed]
192. Ikram, M.; Muhammad, T.; Rehman, S.U.; Khan, A.; Jo, M.G.; Ali, T.; Kim, M.O. Hesperetin Confers Neuroprotection by Regulating Nrf2/TLR4/NF-kappaB Signaling in an Abeta Mouse Model. *Mol. Neurobiol.* **2019**, *56*, 6293–6309. [CrossRef] [PubMed]
193. Gravesteijn, E.; Mensink, R.P.; Plat, J. Effects of nutritional interventions on BDNF concentrations in humans: A systematic review. *Nutr. Neurosci.* **2021**, *25*, 1–12. [CrossRef] [PubMed]
194. Kurauchi, Y.; Hisatsune, A.; Isohama, Y.; Mishima, S.; Katsuki, H. Caffeic acid phenethyl ester protects nigral dopaminergic neurons via dual mechanisms involving haem oxygenase-1 and brain-derived neurotrophic factor. *Br. J. Pharmacol.* **2012**, *166*, 1151–1168. [CrossRef] [PubMed]
195. Dias, G.P.; Cavegn, N.; Nix, A.; do Nascimento Bevilaqua, M.C.; Stangl, D.; Zainuddin, M.S.; Nardi, A.E.; Gardino, P.F.; Thuret, S. The role of dietary polyphenols on adult hippocampal neurogenesis: Molecular mechanisms and behavioural effects on depression and anxiety. *Oxid. Med. Cell. Longev.* **2012**, *2012*, 541971. [CrossRef]
196. Hsieh, C.P.; Chang, W.T.; Chen, L.; Chen, H.H.; Chan, M.H. Differential inhibitory effects of resveratrol on excitotoxicity and synaptic plasticity: Involvement of NMDA receptor subtypes. *Nutr. Neurosci.* **2021**, *24*, 443–458. [CrossRef]
197. Abraham, J.; Johnson, R.W. Consuming a diet supplemented with resveratrol reduced infection-related neuroinflammation and deficits in working memory in aged mice. *Rejuvenation Res.* **2009**, *12*, 445–453. [CrossRef]
198. Tu, Y.; Cheng, S.X.; Sun, H.T.; Ma, T.Z.; Zhang, S. Ferulic acid potentiates pentobarbital-induced sleep via the serotonergic system. *Neurosci. Lett.* **2012**, *525*, 95–99. [CrossRef]
199. Fernandez, S.P.; Wasowski, C.; Paladini, A.C.; Marder, M. Synergistic interaction between hesperidin, a natural flavonoid, and diazepam. *Eur. J. Pharmacol.* **2005**, *512*, 189–198. [CrossRef]
200. Kim, J.W.; Kim, C.S.; Hu, Z.; Han, J.Y.; Kim, S.K.; Yoo, S.K.; Yeo, Y.M.; Chong, M.S.; Lee, K.; Hong, J.T.; et al. Enhancement of pentobarbital-induced sleep by apigenin through chloride ion channel activation. *Arch. Pharm. Res.* **2012**, *35*, 367–373. [CrossRef]
201. Park, K.S.; Han, J.Y.; Moon, D.C.; Hong, J.T.; Oh, K.W. (-)-epigallocatechin-3-O-gallate augments pentobarbital-induced sleeping behaviors through Cl- channel activation. *J. Med. Food* **2011**, *14*, 1456–1462. [CrossRef] [PubMed]
202. Yang, H.; Xie, J.; Mu, W.; Ruan, X.; Zhang, J.; Yao, L.; Diao, Z.; Wu, M.; Li, Y.; Ren, W.; et al. Tea polyphenols protect learning and memory in sleep-deprived mice by promoting AMPA receptor internalization. *NeuroReport* **2020**, *31*, 857–864. [CrossRef] [PubMed]
203. Frolinger, T.; Smith, C.; Cobo, C.F.; Sims, S.; Brathwaite, J.; de Boer, S.; Huang, J.; Pasinetti, G.M. Dietary polyphenols promote resilience against sleep deprivation-induced cognitive impairment by activating protein translation. *FASEB J.* **2018**, *32*, 5390–5404. [CrossRef] [PubMed]
204. Zhao, W.; Wang, J.; Bi, W.; Ferruzzi, M.; Yemul, S.; Freire, D.; Mazzola, P.; Ho, L.; Dubner, L.; Pasinetti, G.M. Novel application of brain-targeting polyphenol compounds in sleep deprivation-induced cognitive dysfunction. *Neurochem. Int.* **2015**, *89*, 191–197. [CrossRef]
205. Masters, A.; Pandi-Perumal, S.R.; Seixas, A.; Girardin, J.L.; McFarlane, S.I. Melatonin, the Hormone of Darkness: From Sleep Promotion to Ebola Treatment. *Brain Disord. Ther.* **2014**, *4*, 1000151. [CrossRef]
206. Iriti, M.; Varoni, E.M.; Vitalini, S. Melatonin in traditional Mediterranean diets. *J. Pineal Res.* **2010**, *49*, 101–105. [CrossRef]
207. Yu, E.; Ruiz-Canela, M.; Guasch-Ferre, M.; Zheng, Y.; Toledo, E.; Clish, C.B.; Salas-Salvado, J.; Liang, L.; Wang, D.D.; Corella, D.; et al. Increases in Plasma Tryptophan Are Inversely Associated with Incident Cardiovascular Disease in the Prevencion con Dieta Mediterranea (PREDIMED) Study. *J. Nutr.* **2017**, *147*, 314–322. [CrossRef]
208. Razquin, C.; Ruiz-Canela, M.; Toledo, E.; Hernandez-Alonso, P.; Clish, C.B.; Guasch-Ferre, M.; Li, J.; Wittenbecher, C.; Dennis, C.; Alonso-Gomez, A.; et al. Metabolomics of the tryptophan-kynurenine degradation pathway and risk of atrial fibrillation and heart failure: Potential modification effect of Mediterranean diet. *Am. J. Clin. Nutr.* **2021**, *114*, 1646–1654. [CrossRef]
209. Chen, Y.; Guillemin, G.J. Kynurenine pathway metabolites in humans: Disease and healthy States. *Int. J. Tryptophan Res.* **2009**, *2*, IJTR-S2097. [CrossRef]
210. Pocivavsek, A.; Baratta, A.M.; Mong, J.A.; Viechweg, S.S. Acute Kynurenine Challenge Disrupts Sleep-Wake Architecture and Impairs Contextual Memory in Adult Rats. *Sleep* **2017**, *40*, zsx141. [CrossRef]
211. Bravo, R.; Matito, S.; Cubero, J.; Paredes, S.D.; Franco, L.; Rivero, M.; Rodriguez, A.B.; Barriga, C. Tryptophan-enriched cereal intake improves nocturnal sleep, melatonin, serotonin, and total antioxidant capacity levels and mood in elderly humans. *Age (Dordr)* **2013**, *35*, 1277–1285. [CrossRef] [PubMed]
212. Martinez-Rodriguez, A.; Rubio-Arias, J.A.; Ramos-Campo, D.J.; Reche-Garcia, C.; Leyva-Vela, B.; Nadal-Nicolas, Y. Psychological and Sleep Effects of Tryptophan and Magnesium-Enriched Mediterranean Diet in Women with Fibromyalgia. *Int. J. Environ. Res. Public Health* **2020**, *17*, 2227. [CrossRef] [PubMed]
213. Afaghi, A.; O'Connor, H.; Chow, C.M. High-glycemic-index carbohydrate meals shorten sleep onset. *Am. J. Clin. Nutr.* **2007**, *85*, 426–430. [CrossRef] [PubMed]

214. Lavialle, M.; Champeil-Potokar, G.; Alessandri, J.M.; Balasse, L.; Guesnet, P.; Papillon, C.; Pevet, P.; Vancassel, S.; Vivien-Roels, B.; Denis, I. An (n-3) polyunsaturated fatty acid-deficient diet disturbs daily locomotor activity, melatonin rhythm, and striatal dopamine in Syrian hamsters. *J. Nutr.* **2008**, *138*, 1719–1724. [CrossRef] [PubMed]
215. Sato-Mito, N.; Sasaki, S.; Murakami, K.; Okubo, H.; Takahashi, Y.; Shibata, S.; Yamada, K.; Sato, K.; Freshmen in Dietetic Courses Study II Group. The midpoint of sleep is associated with dietary intake and dietary behavior among young Japanese women. *Sleep Med.* **2011**, *12*, 289–294. [CrossRef]
216. Sarubbo, F.; Ramis, M.R.; Aparicio, S.; Ruiz, L.; Esteban, S.; Miralles, A.; Moranta, D. Improving effect of chronic resveratrol treatment on central monoamine synthesis and cognition in aged rats. *Age (Dordr)* **2015**, *37*, 9777. [CrossRef]
217. Ramis, M.R.; Sarubbo, F.; Moranta, D.; Tejada, S.; Llado, J.; Miralles, A.; Esteban, S. Cognitive and Neurochemical Changes Following Polyphenol-Enriched Diet in Rats. *Nutrients* **2020**, *13*, 59. [CrossRef]
218. Malhotra, R.K.; Desai, A.K. Healthy brain aging: What has sleep got to do with it? *Clin. Geriatr. Med.* **2010**, *26*, 45–56. [CrossRef]
219. Fan, Y.; Pedersen, O. Gut microbiota in human metabolic health and disease. *Nat. Rev. Microbiol.* **2021**, *19*, 55–71. [CrossRef]
220. Sen, P.; Molinero-Perez, A.; O'Riordan, K.J.; McCafferty, C.P.; O'Halloran, K.D.; Cryan, J.F. Microbiota and sleep: Awakening the gut feeling. *Trends Mol. Med.* **2021**, *27*, 935–945. [CrossRef]
221. Smith, R.P.; Easson, C.; Lyle, S.M.; Kapoor, R.; Donnelly, C.P.; Davidson, E.J.; Parikh, E.; Lopez, J.V.; Tartar, J.L. Gut microbiome diversity is associated with sleep physiology in humans. *PLoS ONE* **2019**, *14*, e0222394. [CrossRef] [PubMed]
222. Han, M.; Yuan, S.; Zhang, J. The interplay between sleep and gut microbiota. *Brain Res. Bull.* **2022**, *180*, 131–146. [CrossRef] [PubMed]
223. Szentirmai, E.; Millican, N.S.; Massie, A.R.; Kapas, L. Butyrate, a metabolite of intestinal bacteria, enhances sleep. *Sci. Rep.* **2019**, *9*, 7035. [CrossRef] [PubMed]
224. Lu, J.; Synowiec, S.; Lu, L.; Yu, Y.; Bretherick, T.; Takada, S.; Yarnykh, V.; Caplan, J.; Caplan, M.; Claud, E.C.; et al. Microbiota influence the development of the brain and behaviors in C57BL/6J mice. *PLoS ONE* **2018**, *13*, e0201829. [CrossRef]
225. Tran, S.M.; Mohajeri, M.H. The Role of Gut Bacterial Metabolites in Brain Development, Aging and Disease. *Nutrients* **2021**, *13*, 732. [CrossRef]
226. Agrawal, R.; Ajami, N.J.; Malhotra, S.; Chen, L.; White, D.L.; Sharafkhaneh, A.; Hoffman, K.L.; Graham, D.Y.; El-Serag, H.B.; Petrosino, J.F.; et al. Habitual Sleep Duration and the Colonic Mucosa-Associated Gut Microbiota in Humans—A Pilot Study. *Clocks Sleep* **2021**, *3*, 387–397. [CrossRef]
227. Ogawa, Y.; Miyoshi, C.; Obana, N.; Yajima, K.; Hotta-Hirashima, N.; Ikkyu, A.; Kanno, S.; Soga, T.; Fukuda, S.; Yanagisawa, M. Gut microbiota depletion by chronic antibiotic treatment alters the sleep/wake architecture and sleep EEG power spectra in mice. *Sci. Rep.* **2020**, *10*, 19554. [CrossRef]
228. Grosicki, G.J.; Riemann, B.L.; Flatt, A.A.; Valentino, T.; Lustgarten, M.S. Self-reported sleep quality is associated with gut microbiome composition in young, healthy individuals: A pilot study. *Sleep Med.* **2020**, *73*, 76–81. [CrossRef]
229. Badran, M.; Khalyfa, A.; Ericsson, A.; Gozal, D. Fecal microbiota transplantation from mice exposed to chronic intermittent hypoxia elicits sleep disturbances in naive mice. *Exp. Neurol.* **2020**, *334*, 113439. [CrossRef]
230. Kurokawa, S.; Kishimoto, T.; Mizuno, S.; Masaoka, T.; Naganuma, M.; Liang, K.C.; Kitazawa, M.; Nakashima, M.; Shindo, C.; Suda, W.; et al. The effect of fecal microbiota transplantation on psychiatric symptoms among patients with irritable bowel syndrome, functional diarrhea and functional constipation: An open-label observational study. *J. Affect. Disord* **2018**, *235*, 506–512. [CrossRef]
231. De Filippo, C.; Cavalieri, D.; Di Paola, M.; Ramazzotti, M.; Poullet, J.B.; Massart, S.; Collini, S.; Pieraccini, G.; Lionetti, P. Impact of diet in shaping gut microbiota revealed by a comparative study in children from Europe and rural Africa. *Proc. Natl. Acad. Sci. USA* **2010**, *107*, 14691–14696. [CrossRef] [PubMed]
232. Requena, T.; Martinez-Cuesta, M.C.; Pelaez, C. Diet and microbiota linked in health and disease. *Food Funct.* **2018**, *9*, 688–704. [CrossRef] [PubMed]
233. De Filippis, F.; Pellegrini, N.; Vannini, L.; Jeffery, I.B.; La Storia, A.; Laghi, L.; Serrazanetti, D.I.; Di Cagno, R.; Ferrocino, I.; Lazzi, C.; et al. High-level adherence to a Mediterranean diet beneficially impacts the gut microbiota and associated metabolome. *Gut* **2016**, *65*, 1812–1821. [CrossRef] [PubMed]
234. Govindarajulu, M.; Pinky, P.D.; Steinke, I.; Bloemer, J.; Ramesh, S.; Kariharan, T.; Rella, R.T.; Bhattacharya, S.; Dhanasekaran, M.; Suppiramaniam, V.; et al. Gut Metabolite TMAO Induces Synaptic Plasticity Deficits by Promoting Endoplasmic Reticulum Stress. *Front. Mol. Neurosci.* **2020**, *13*, 138. [CrossRef] [PubMed]
235. Li, D.; Ke, Y.; Zhan, R.; Liu, C.; Zhao, M.; Zeng, A.; Shi, X.; Ji, L.; Cheng, S.; Pan, B.; et al. Trimethylamine-N-oxide promotes brain aging and cognitive impairment in mice. *Aging Cell* **2018**, *17*, e12768. [CrossRef] [PubMed]
236. Barber, C.; Mego, M.; Sabater, C.; Vallejo, F.; Bendezu, R.A.; Masihy, M.; Guarner, F.; Espin, J.C.; Margolles, A.; Azpiroz, F. Differential Effects of Western and Mediterranean-Type Diets on Gut Microbiota: A Metagenomics and Metabolomics Approach. *Nutrients* **2021**, *13*, 2638. [CrossRef] [PubMed]
237. Haro, C.; Garcia-Carpintero, S.; Rangel-Zuniga, O.A.; Alcala-Diaz, J.F.; Landa, B.B.; Clemente, J.C.; Perez-Martinez, P.; Lopez-Miranda, J.; Perez-Jimenez, F.; Camargo, A. Consumption of Two Healthy Dietary Patterns Restored Microbiota Dysbiosis in Obese Patients with Metabolic Dysfunction. *Mol. Nutr. Food Res.* **2017**, *61*, 1700300. [CrossRef]
238. Yang, Q.; Liang, Q.; Balakrishnan, B.; Belobrajdic, D.P.; Feng, Q.J.; Zhang, W. Role of Dietary Nutrients in the Modulation of Gut Microbiota: A Narrative Review. *Nutrients* **2020**, *12*, 381. [CrossRef]

239. Rodriguez-Daza, M.C.; Pulido-Mateos, E.C.; Lupien-Meilleur, J.; Guyonnet, D.; Desjardins, Y.; Roy, D. Polyphenol-Mediated Gut Microbiota Modulation: Toward Prebiotics and Further. *Front. Nutr.* **2021**, *8*, 689456. [CrossRef]
240. Aoun, A.; Darwish, F.; Hamod, N. The Influence of the Gut Microbiome on Obesity in Adults and the Role of Probiotics, Prebiotics, and Synbiotics for Weight Loss. *Prev. Nutr. Food Sci.* **2020**, *25*, 113–123. [CrossRef]
241. Correa, T.A.F.; Rogero, M.M.; Hassimotto, N.M.A.; Lajolo, F.M. The Two-Way Polyphenols-Microbiota Interactions and Their Effects on Obesity and Related Metabolic Diseases. *Front. Nutr.* **2019**, *6*, 188. [CrossRef] [PubMed]
242. Gonzalez-Sarrias, A.; Romo-Vaquero, M.; Garcia-Villalba, R.; Cortes-Martin, A.; Selma, M.V.; Espin, J.C. The Endotoxemia Marker Lipopolysaccharide-Binding Protein is Reduced in Overweight-Obese Subjects Consuming Pomegranate Extract by Modulating the Gut Microbiota: A Randomized Clinical Trial. *Mol. Nutr. Food Res.* **2018**, *62*, e1800160. [CrossRef] [PubMed]
243. Yu, Y.C.; Li, J.; Zhang, M.; Pan, J.C.; Yu, Y.; Zhang, J.B.; Zheng, L.; Si, J.M.; Xu, Y. Resveratrol Improves Brain-Gut Axis by Regulation of 5-HT-Dependent Signaling in the Rat Model of Irritable Bowel Syndrome. *Front. Cell. Neurosci.* **2019**, *13*, 30. [CrossRef] [PubMed]
244. Plamada, D.; Vodnar, D.C. Polyphenols-Gut Microbiota Interrelationship: A Transition to a New Generation of Prebiotics. *Nutrients* **2021**, *14*, 137. [CrossRef] [PubMed]
245. Yannakoulia, M.; Kontogianni, M.; Scarmeas, N. Cognitive health and Mediterranean diet: Just diet or lifestyle pattern? *Ageing Res. Rev.* **2015**, *20*, 74–78. [CrossRef]
246. Schulze, M.B.; Martinez-Gonzalez, M.A.; Fung, T.T.; Lichtenstein, A.H.; Forouhi, N.G. Food based dietary patterns and chronic disease prevention. *BMJ* **2018**, *361*, k2396. [CrossRef]
247. Vilarnau, C.; Stracker, D.M.; Funtikov, A.; da Silva, R.; Estruch, R.; Bach-Faig, A. Worldwide adherence to Mediterranean Diet between 1960 and 2011. *Eur. J. Clin. Nutr.* **2019**, *72*, 83–91. [CrossRef]
248. Wittmann, M.; Dinich, J.; Merrow, M.; Roenneberg, T. Social jetlag: Misalignment of biological and social time. *Chronobiol. Int.* **2006**, *23*, 497–509. [CrossRef]
249. Zeron-Rugerio, M.F.; Cambras, T.; Izquierdo-Pulido, M. Social Jet Lag Associates Negatively with the Adherence to the Mediterranean Diet and Body Mass Index among Young Adults. *Nutrients* **2019**, *11*, 1756. [CrossRef]
250. Heath, G.; Dorrian, J.; Coates, A. Associations between shift type, sleep, mood, and diet in a group of shift working nurses. *Scand. J. Work Environ. Health* **2019**, *45*, 402–412. [CrossRef]

nutrients

MDPI

Review

Cyrcadian Rhythm, Mood, and Temporal Patterns of Eating Chocolate: A Scoping Review of Physiology, Findings, and Future Directions

Sergio Garbarino [1], Emanuela Garbarino [2] and Paola Lanteri [3,*]

[1] Department of Neuroscience, Rehabilitation, Ophthalmology, Genetics and Maternal/Child Sciences (DINOGMI), University of Genoa, 16132 Genoa, Italy; sgarbarino.neuro@gmail.com
[2] Department of Psychology, University Milan-Bicocca, 20126 Milan, Italy; e.garbarino@campus.unimib.it
[3] Neurophysiopathology Center, Fondazione IRCCS Istituto Neurologico Carlo Besta, 20133 Milan, Italy
* Correspondence: paola.lanteri@istituto-besta.it

Abstract: This paper discusses the effect of chrononutrition on the regulation of circadian rhythms; in particular, that of chocolate on the resynchronization of the human internal biological central and peripheral clocks with the main external synchronizers, light–dark cycle and nutrition-fasting cycle. The desynchronization of internal clocks with external synchronizers, which is so frequent in our modern society due to the tight rhythms imposed by work, social life, and technology, has a negative impact on our psycho-physical performance, well-being, and health. Taking small amounts of chocolate, in the morning at breakfast at the onset of the active phase, helps speed up resynchronization time. The high flavonoid contents in chocolate promote cardioprotection, metabolic regulation, neuroprotection, and neuromodulation with direct actions on brain function, neurogenesis, angiogenesis, and mood. Although the mechanisms of action of chocolate compounds on brain function and mood as well as on the regulation of circadian rhythms have yet to be fully understood, data from the literature currently available seem to agree in suggesting that chocolate intake, in compliance with chrononutrition, could be a strategy to reduce the negative effects of desynchronization. This strategy appears to be easily implemented in different age groups to improve work ability and daily life.

Keywords: chrononutrition; circadian rhythms; chocolate; flavonoid; brain function; mood

Citation: Garbarino, S.; Garbarino, E.; Lanteri, P. Cyrcadian Rhythm, Mood, and Temporal Patterns of Eating Chocolate: A Scoping Review of Physiology, Findings, and Future Directions. *Nutrients* **2022**, *14*, 3113. https://doi.org/10.3390/nu14153113

Academic Editor: Giovanna Muscogiuri

Received: 22 June 2022
Accepted: 25 July 2022
Published: 28 July 2022

1. Introduction: Life, Health, and Circadian Rhythms

The Earth's rotation around its axis causes periodic light and dark variations in the environment over a 24-h period. These predictable variations in the light environment allow organisms to optimally organize their physiology and activity-rest rhythms to specific periods of the day-night cycle [1]. Indeed, there is a clear selective advantage in anticipating changes in the environment, particularly light–dark, by adapting their activities. Therefore, evolved organisms have an internal biological clock that functions with a period close to 24 h, which is circadian, in the absence of environmental influences such as light [2]. Anticipation of daily events organized in this way allows optimal management of time and available energy, giving those organisms a significant advantage [3]. The correct alignment between the light–dark rhythm, the circadian clock, and behavior determines a temporal order in organisms that is of fundamental importance for ensuring survival [4].

The circadian clock drives many outputs, including sleep–wakefulness, hormonal, and metabolic rhythm. Locomotor activity and sleep are segregated into specific phases of the light–dark cycle [5]. Endocrine processes are also regulated by a rhythmic periodicity: the cortisol concentration in circulation reaches a peak level in the morning, and unlike many hormones involved in metabolism, such as leptin and ghrelin, reach their peak at night [6]. Urine production has a circadian nature [7], as well as perception–nociception [8], temperature [9], vigilance [10], and cognitive abilities such as memory and mathematical performance [11,12]. Mood also appears to have a circadian cycle [13,14], with more evidence supporting a role in positive affect [15].

The light–dark cycles drive the central clock located in the hypothalamic suprachiasmatic nucleus (SCN), where it mainly dominates the rhythms related to activity–rest oscillations. SCN favors the active phase by acting on the sympathetic nervous system, anticipating the rise in body temperature and blood pressure. The same model also applies to the feeding–fasting systems that prepare/activate before the activation phase, breakfast, according to the time of local clocks and guide the peripheral clocks present in most tissues and, to some extent, in the brain [16]. Indeed, peripheral clocks dominate local physiological processes, such as glucose and lipid homeostasis, hormone secretion, xenobiotics, immune response, and the digestive system [16]. Since the central clock organizes local clocks through neuronal and humoral signals, the mismatch between the central and peripheral clocks results in a condition of desynchronization [17]. The desynchronization condition contributes to the sleep disorders, adverse metabolic effects of circadian misalignment, such as increased risk of type 2 diabetes among shift workers [18], cancer, and psychological/psychiatric disorders [19].

In this paper, we aim to review the available data in the literature on the role of nutrition as a synchronizer of biological clocks with regard to chocolate and its mechanisms of action on brain function and mood.

2. Nutrition: Not Only Light Acts as (De)Synchronizer

While it has long been recognized that the light–dark cycle is the main zeitgeber for the central clock in the brain, providing input to central clock genes (CLOCK, BMAL, PER1, PER2, PER3, CRY1, CRY2, Tim) in central nervous system cell populations, food intake may act as an important time synchronizer for tissue metabolic clocks, but not for the brain [20]. Temporal signals from the SCN drive circadian rhythms of feeding–fasting and activity–rest, which in turn act by synchronizing clock genes in peripheral clocks [21], such as liver cells and skeletal muscle, the heart, white adipose tissue, and other metabolic tissues.

The timing of food intake is a key factor in determining the phase state of peripheral circadian clocks. As such, it is a powerful signal of activation of peripheral oscillators involved in metabolic and digestive functions, such as liver, kidney, intestine, and adipose tissue [22].

Moreover, although food exerts a strong synchronizing effect on peripheral clocks, it also exerts a function at the brain level by acting mainly on brain areas involved in energy balance and motivation/reward for food intake [23–25]. It is important to note that the two main synchronizers, feeding–fasting and light–dark, act differently on circadian rhythms in different systems and organ tissues, as the signals provided by the light–dark cycle act predominantly on central systems and certain brain regions, including the SCN, while feeding–fasting acts mainly on peripheral systems [26,27]. In fact, restricted food intake at certain times during the day for one week is able to completely alter the expression phase of genes controlled by the circadian clock in the peripheral tissues of nocturnal rodents, whereas it does not affect the central clock, which is dominated by light–dark cycles [27–29].

Among peripheral clocks, the liver adapts most rapidly to changes in feeding–fasting rhythms, within about 3 days, whereas the kidney, heart, pancreas, and lung take longer [28,29].

Among different feeding times, studies in mice mimicking human dietary patterns have shown that breakfast is usually the most effective meal in determining liver clock phase because breakfast is taken after the longest fasting phase of the day [30]. Therefore, late meals or midnight snacks distort the hunger period and significantly modify the phase of peripheral clocks [22].

3. Chrononutrition and Food Components Affecting Circadian Clocks

Both central and peripheral circadian clocks can be influenced by the consumption of different food components by different mechanisms.

The reciprocal influences of circadian clocks and energy metabolism suggest that feeding time has a critical impact on metabolism. The quality and quantity as well as the timing of food intake are all important for nutrition: this is referred to as 'chrononutrition'. Indeed, food intake has a knock-on effect on peripheral clocks, mainly the liver, but not on the SCN, so that food intake outside the physiologically designated 'time' window induces desynchronization between peripheral clocks and SCN [29].

Tissue-specific gene expression of peripheral clocks is regulated by food intake, whereas the central gene expression of the SCN itself is protected by nutrient deregulation [20]. Timing of food intake also influences body weight and obesity risk in humans [31,32]: a two-group study evaluating weight loss during an isocaloric diet showed improvements in several metabolic indicators such as body weight, fasting blood glucose, insulin, ghrelin, average hunger, and satiety scores in the group eating a larger breakfast and lighter dinner compared to the other way around [33]. In both humans and experimental animals, eating dinner late at night and skipping breakfast leads to increased body weight and obesity [3].

With regard to individual food components associated differently with different eating times, it has been shown that eating a diet rich in fat *ad libitum* attenuates the amplitude of the clocks, while eating a diet limited in time restores their amplitude by regulating the circadian cycle. A combination of carbohydrates and protein also appears to be essential for proper resetting of peripheral clocks such as the liver clock [34].

In contrast, the intake of protein nutrients at times of phase-shifting hepatic cell clocks correlates with blood glucose absorption, and readily digestible starches with a high glycemic index have a powerful entrainment effect on peripheral liver clocks [35].

The action of specific nutrients occurs through the different expression of clock-controlled genes [36]. Glucose is able to regulate BMAL1 and period expression [37] as demonstrated, for example, in a study of mice fed a high-fat diet in which an alteration of lipid metabolism genes was documented [38]. Indeed, this finding is further corroborated by a study conducted by Eckel-Mahn et al., which illustrated that most hepatic metabolites also follow changes according to a circadian rhythm, and that these variations are regulated by both the clock transcriptome and the feeding–fasting cycles that favor the maintenance of hepatic homeostasis [39].

A high-fat diet, such as the ketogenic diet with a high fat content and low carbohydrate content, has the greatest effect on the central clock and/or eating behavior: a high fat content reduces the duration of the circadian rhythm of locomotor activity under conditions of constant darkness, whereas under normal conditions of alternating light–dark, there is an advancement in the rhythm of clock expression and clock-controlled genes in peripheral tissues [40]. In recent years, the field of nutraceuticals and functional foods has developed alongside interest in the potential modulatory effects of food constituents on human health. In this context, mention should be made of caffeine/theophylline, which could influence our circadian system, and flavonoids. Caffeine in food and drink has been shown to prolong circadian locomotor rhythms in Drosophila and mice [41,42]. The dose of caffeine needed to influence circadian rhythms appears to be low, at 0.05%, equivalent to the dose contained in coffee. Indeed, consumption of non-decaffeinated coffee prolongs the rhythm of circadian activity in mice under constant dark conditions [42]. Interestingly, both caffeine and theophylline also prolong the circadian period in cultured cells, mouse tissues, Neurospora, Chlamydomonas, Drosophila, and even higher plants [37,43–45].

Among the dietary constituents, flavonoids, a class of polyphenolic compounds, have received particular interest for their various beneficial biological actions, such as neurological and cardiological protection and neuromodulation [46]. In particular, a rich source of flavonoids has been detected in the cocoa bean, especially the subclass of flavanols in the form of epicatechin and catechin [47].

3.1. Chocolate

Cocoa products and especially chocolate are foodstuffs originating from South America. Cocoa is obtained from the seeds of the Theobroma cacao tree, which are then dried, shelled, fermented, and ground with other substances such as sugar, fat, and other flavorings to produce the wide variety of chocolate available on the market, from dark to milk variations.

Chocolate consists of a combination of several ingredients, the main ones being cocoa, cocoa butter, and sugar, which make up the solid food product. Cocoa beans, along with foods such as tea, red wine, and fruit, are a rich source of flavanols, a subgroup of natural flavonoids that are bioactive plant compounds.

There have been reports of the health benefits of chocolate since ancient times, the earliest dating back to Aztec and Mayan medical practice [48]. It was not until the end of the 20th century, however, that claims about the alleged health benefits of chocolate attracted more and more scientific interest. Flavonoids, contained in cocoa, have been approved by the European Food Safety Agency as being beneficial to health [49].

Due to dark chocolate's high concentrations of flavanols, a subgroup of flavonoids, especially epicatechin, [50] known as powerful antioxidant agents [51], have received a health claim in relation to their impact on 'maintaining normal endothelium-dependent vasodilation'.

Many studies conducted so far focus on the effects of chocolate on the cardiovascular system, skin, cholesterol levels, the release of the neurotransmitters anandamide and serotonin, and on the properties of specific stimulating constituents such as theobromine and caffeine [52].

3.1.1. Biochemical Components and Neurobiological Impact

Chocolate contains more than 300–500 known chemicals, some of which also act on brain cells and modulate mood [53]. Of these 300 to 500 chemicals in chocolate, some play an important role in humans, influencing neurocognitive functions.

The main psychoactive components of chocolate are [54,55] as follows.

- Carbohydrates, which have known behavioral effects.
- Flavanols, which are ubiquitous in the plant kingdom. In foods normally consumed in the diet, high levels of flavonoids can be found in green and black tea, grapes, red wine, apples, and especially in cocoa and cocoa-containing products. In fact, cocoa is particularly rich in flavonoids and contains a distinct complement of flavanols (a subclass of flavonoids), flavan-3-ols, mainly present in the form of epicatechin and catechin [50], and their derivatives in high concentrations [56]. Flavan-3-ols are the building blocks for polymeric procyanidin type B-2.
- Methylxanthines (MX), such as caffeine and its highly fat-soluble derivative and metabolite theobromine, which have peak plasma levels 60–120 min after ingestion. Like caffeine, theobromine binds to adenosine receptors, exhibiting its psychoactive potential similar to that of caffeine. However, these two MX have distinct functional binding properties.
- Biogenic amines, such as serotonin, tryptophan, phenylethylamine, tyrosine, tryptamine, and tyramine, have a concentration that increases during fermentation and decreases during roasting and alkalinization.
- Anandamide, an endogenous ligand for the cannabinoid receptor that is found in low quantities, such as 0.5 mg g^{-1}, salsolinol, and tetrahydro-b-carboline.

The most studied substances are flavanols and their metabolites: it has been shown in animal studies that these products can pass through the blood-brain barrier (BBB), with positive effects on brain tissue, vessels, and function (angio- and neurogenesis, changes in neuron morphology), stimulating cerebral blood circulation [57]. Epicatechin [50], the most common flavanol found in cocoa, is rapidly absorbed by humans. Thirty min after intake it is detectable in blood plasma, and its peak is reached after 2–3 h, showing a concentration directly related to the dose of chocolate ingested [58], and then it returns to basal level within 6–8 h after consumption.

3.1.2. Chocolate and Brain Functions

In order to exert effects in the brain, flavanols cross the BBB by a process that is not only time-dependent but also stereoselective, which, thus, favors the passage of epicatechin more than catechin, as has been demonstrated in two cell lines, one of rat and one of human origin [59]; the permeability is proportional to the degree of lipophilicity and inversely proportional to the degree of polarity.

In humans, the main polyphenols increase cerebral blood flow (CBF) and are derived from nutrients such as cocoa, wine, grape seeds, berries, tea, tomatoes, and soy [60].

Optimal brain function requires an adequate CBF that promotes the correct and constant supply of oxygen and glucose to neurons and the excretion of deposited waste products. Increasing CBF is also a potential tool for enhancing brain function. Flavanols and their metabolites have the ability to reach and accumulate in the brain regions mainly involved in learning processes and memory, and are therefore thought to exert a direct brain action on cognitive function and neuroprotection [55]. After chronic administration of chocolate, high concentrations of tangeretin were found in the striatum nucleus, hypothalamus, and hippocampus of the rat [61].

The neurobiological impact of flavanols on the brain in the areas of learning, memory, cognitive function, and mood is thought to occur mainly in two ways. The effects of flavonoids on the brain are mediated by their function of neuroprotecting vulnerable neurons in particular, improving neuronal function and stimulating regeneration (neurogenesis) [62] by interacting with intracellular neuronal signaling pathways that control neuronal survival and differentiation, long-term potentiation (LTP), and memory.

First, flavonoids interact with a number of cellular signaling pathways by activating gene expression and protein synthesis for the maintenance of LTP, and the stabilization of long-term memories [63] are critical for neurogenesis, synaptic growth, and survival of neurons, mainly in the brain hippocampus and subventricular area related to learning and memory [64,65].

Secondly, flavonoids induce vasodilation through nitric oxide at both cardiovascular and peripheral levels through the production of nitric oxide (NO), a key regulator of vascular function, which acts as a signaling molecule by inhibiting the action of adhesion molecules in atheromatous plaque that cause inflammation [66], and, most importantly, promote and improve the function of the vascular endothelium by acting, with dilating action, on the smooth muscle tissue of blood vessels [67]. This effect on the vascular system with endothelium-dependent vasodilation, contributes to the maintenance of normal blood flow and improvement of blood pressure; it also induces a reduction in platelet aggregation [68–73]. This. in turn. results in increased CBF and blood perfusion throughout the central and peripheral nervous system [74], allowing better oxygen and glucose delivery to neurons and removal of waste metabolites in the nervous systems [75]. Among flavanols, epicatechin has the greatest ability to increase nitric oxide (NO) bioavailability, leading to improvements in vascular tone and blood pressure regulation [76]. These vascular changes occurring at the peripheral level may also extend to cerebral perfusion, leading to optimized cerebrovascular integration during neuronal activation phases, a mechanism considered crucial for the functional and structural integrity of the brain and for promoting adult neurogenesis in the hippocampus [77].

Administration of cocoa flavonoids, therefore, also stimulates angiogenesis in the hippocampus [51,78], as demonstrated by administering epicatechin, given to mice at a dose of 500 mg g^{-1} (daily supply of 2.5 mg). Combining epicatechin with exercise also improved the consolidation of spatial memory and the density of dendritic spines in the dentate gyrus of the hippocampus. In the same study, epicatechin treatment was shown to increase learning-associated gene expression in the hippocampus, while it did not appear to influence neurogenesis in the adult hippocampus [78].

Thus, flavonoids could also exert their neurocognitive effects both directly and indirectly by interacting with the cellular network and molecular system deputed to memory acquisition, storage, retrieval, and learning [62], also through long-term potentiation, synaptic plasticity [79], enhanced neuronal connection, and communication.

Epidemiological studies suggest that regular flavonoid intake may be associated with better cognitive function [80], to decreased risk of dementia and cognitive decline [81–83], better cognitive development over a 10-year period [84], and improved dose-dependent cognitive performance in physiological aging [85].

3.1.3. Chocolate and Mood

A lot of data from the literature support the hypothesis of the influence of theobromine and caffeine on mood and cognitive function [86–89], but the impact and mechanism by which flavanols affect mood remains unclear.

It is commonly believed that eating chocolate improves mood and rapidly induces a sense of well-being in people [89]. An initial rapid effect of chocolate on emotional comfort appears to be related to the ability of the carbohydrates it contains, to promote such positive feelings through the release of several gut and brain peptides [90].

In rats, the intake of cocoa-extracted polyphenols, while significantly reducing the duration of immobility in a forced swimming test, had no effect on locomotor activity in the open field, confirming its specific antidepressant effect [91]. The most likely basis for this effect may be attributable to endorphin release [92]. Indeed, sweet food intake is increased by opioid agonists and decreased by opioid antagonists [93,94]. The effect of chocolate is also exerted through interaction with neurotransmitters such as dopamine (tyrosine contained in chocolate is the precursor to dopamine), serotonin, and endorphins, which contribute to appetite, reward, and mood regulation. The dopaminergic system contributes to the desire to consume chocolate, probably by acting mainly non-specifically towards food. After carbohydrate ingestion, only when the protein component of the meal is less than 2% does it induce an increase in serotonin concentrations in the brain [90]. It should be noted that chocolate contains 5% of its caloric content in the form of protein, which would cancel out any effect of serotonin. Moreover, manipulations of tryptophan, the precursor of serotonin, also cause physiological changes that are too slow to explain the mood effects described during or immediately after eating chocolate [95]. Another area where chocolate might act could be in the area of opioids, which are known to play a role in the palatability of preferred foods [96], releasing endorphins during food intake, and, thus, justifying the increase in pleasure during food intake [97]. The mood effects of cocoa may also be partly due to opioids released in response to the ingestion of sweets and other pleasantly palatable foods [98,99]. The increase in central opioidergic activity, in turn, stimulates the immediate release of beta-endorphin in the hypothalamus, which exerts an analgesic effect. Bad mood stimulates consumption of comfort foods such as chocolate in two different ways [100]. The former is called craving and is associated with an impulsive desire for chocolate, and its compulsive consumption occurs especially when under high emotional stress, showing a clear link between the perception of a negative mood and the intense desire to consume chocolate [101]. The association between chocolate craving and consumption under emotional stress was demonstrated in a study in which subjects had to listen to music that induced a happy or sad mood. Chocolate consumption was increased by listening to sad music [100].

The second modality to be considered is the palatability of the food. The pleasure induced by palatable food is regulated by endogenous opioids that stimulate food intake in rats. The pleasure induced by palatable foods is regulated by endogenous opioids that stimulate their intake in rats. In humans, however, the critical factor in satisfying chocolate craving appears to be taste and mouthfeel [102]. Females, mainly in the perimenstrual period, seems more sensitive to chocolate. The response to satiety seems to vary by gender [103].

It is more conceivable that an important role in liking or craving chocolate is due more to the composite sensory properties of chocolate than to its role in appetite and satiety [104]. During the consumption of chocolate, different brain areas are also activated depending on the motivation to eat chocolate, based on positive/appetitive stimuli or associated with negative/adverse stimuli. Modulation of brain activity has been observed in chemosensory cortical areas such as the insula, prefrontal regions, and caudomedial and caudolateral orbitofrontal cortex, with overlapping and co-activation under contrasting motivational conditions [105]. The ability to activate images of appetizing foods involved in food motivation and hedonism in a fronto–striatal–amygdala–midbrain network appears to be dependent on individual variability in reward sensitivity. If this same neuronal circuit is stimulated in the animal, it may result in the cancellation of the sense of satiety and cause overeating of highly palatable foods [106].

The smell of chocolate itself is sufficient to modulate brain activity recorded on the electroencephalography (EEG). The smell of chocolate induces a significant reduction in theta activity compared to any other stimulus. Theta activity is considered to be closely related to attentional level, cognitive load in general, and, in this specific case, to olfactory perception, so a reduction in theta activity could be indicative of a reduced level of attention and an increased propensity to distraction [107]. In addition to olfaction, the sight of chocolate also evokes activations in the brain and especially in the medial orbitofrontal cortex and ventral striatum, particularly in subjects who crave chocolate compared to those who do not: the combination of an image of chocolate with chocolate in the mouth evoked greater brain activation than the sight of the sum of the different components in the medial orbitofrontal cortex and cingulate cortex [108].

The motivation for chocolate preference appears to be primarily, if not entirely, sensory. The origin of the liking of its sensory properties is unclear; it could be innate or acquired based on the sweetness, texture, and aroma characteristics of chocolate, or it could depend on the interaction between a person's state and the post-gastronomic effects of chocolate. Surprisingly, there is little evidence of a relationship between chocolate addiction and chocolate liking [102]. However, chocolate consumption fails to activate the key structure for drug addiction, the nucleus accumbens [109–111].

The effect of chocolate on mood may be attributed to the affinity for adenosine and benzodiazepine (GABAa) receptors of polyphenolic compounds, which means that their ingestion may have a soothing effect [112]. Some polyphenolic compounds indeed have anxiolytic properties [113]. A small randomized controlled pilot study in humans with chronic fatigue reported a reduction in anxiety-related symptoms after eating polyphenol-rich chocolate, compared to polyphenol-poor chocolate [114].

Pase et al. [115] investigated the acute and subchronic effects of polyphenol supplementation on mood and cognitive performance in a randomized, placebo-controlled, double-blind study. Thirty days of treatment with a high dose of cocoa polyphenols reduced self-rated anxiety and contentment. No significant effect on cognitive performance was recorded with either the high or low dose in either the acute or chronic phase.

The optimal dosage of cocoa polyphenols needed to improve cognitive function and mood remains unclear.

3.1.4. Chocolate, Sleep and Circadian Rhythms

Our modern society functions at a hectic pace of activity 24/7, which leads individuals to sacrifice sleep hours and disregard daily sleep–wake rhythms. As experienced daily by shift workers, jet-lagged travelers, or those with so-called social jet-lag syndrome, disturbed sleep–wake rhythms create a conflict, a temporal mismatch between the circadian system and temporal signals derived from the cyclical environmental changes, such as the light–dark cycle, [116] or a desynchronization [117]. This condition, if prolonged over time, leads to chronic sleep disorders (CSD) that result in deficits to health and psychophysical well-being. The most frequent symptoms belong to the neurobehavioral sphere and are often associated with mood changes such as a tendency to depression and impairment of cognitive functions, especially executive functions, as well as cardio- and cerebrovascular disorders, stroke, hypertension, obesity, and diabetes [118]. CSD are a socio-economic and public health issue due to their high prevalence in the general and working population, their impact on health, and work output given the higher incidence of absenteeism, and increased rates of errors and accidents at work [119]. In addition, CSD often have a bidirectional relationship with stress [120].

Jet lag results from a sudden change in the light–dark cycle due to trans-south travel or social life (in the case of social jet lag), which leads to a misalignment between internal circadian rhythms, mainly but not limited to SCN, and the external rhythmic time-cues (Zeitgeber), mainly light, for the day–night cycle. The days required for the process of coordinating the internal circadian clock to external rhythmic are often associated with behavioral and physiological discomfort.

Escobar C. et al. [121] observed that chocolate administration resulted in a faster rate of realignment and synchronization between activity–rest cycle and circadian temperature rhythms. Programmed access to chocolate activates brain areas involved in motivation and metabolic response to food [23] as well as the circadian system by improving neuronal activation in the SCN [122]. Other studies have also reported a similar effect on the speed of realignment to feeding schedules in peripheral oscillatory clocks [26,27,123].

However, it is not only the type of food that determines the ability to realign and re-synchronize internal clocks with external signals but also the time of intake of food, especially for palatable food.

Due to this complex interaction between external Zeitgeber and internal circadian rhythms, the greatest beneficial effects of entrainment on circadian function are seen when food intake coincides with the activity phase [124,125], whereas an inhibitory effect occurs when food is taken during the rest phase [126,127] (Figure 1).

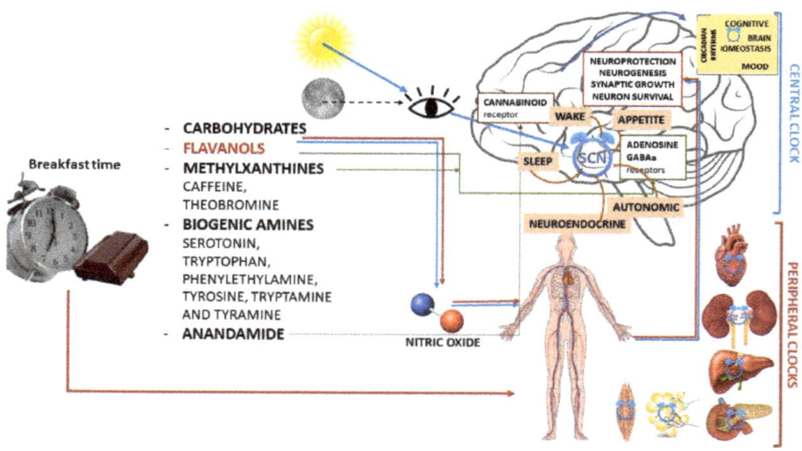

Figure 1. Schematic illustration of chrononutrition with chocolate for breakfast and components of chocolate with the main mechanisms of action in both the central nervous system and peripheral organs and their actions.

Furthermore, to maintain a coordinated and synchronized circadian function, food intake must be phased with the light–dark cycle. The main effect seems to be due to a direct synchronization action on brain oscillators and central and peripheral clocks [26,128].

Under normal light and dark (LD) conditions, programmed food intake does not shift the SCN phase [27,129]. Other studies indicate that SCN is inhibited during food anticipation and fasting as observed with c-Fos, a major early gene that is activated by external signals [130–132] or electrophysiological recordings, whereas the ventral SCN is activated both after re-feeding and with light [133].

Recent results indicate that the SCN may also respond to palatable food construed as hedonic information, via dopaminergic projections from the ventral tegmental area [134]. The rapid achievement of synchronization with limited daily chocolate intake may also be partly due to the increase in arousal induced by chocolate intake as a hedonic effect.

When planned for breakfast, an appetizing food, such as chocolate, can influence activation in the SCN, at the level of the dorsomedial region [121]. This rhythmic pattern in the dorsal SCN may promote faster re-entrainment [135] when bounded by a time window in which chocolate was administered during the active phase, whereas chocolate did not promote re-entrainment when administered during the resting phase.

Scheduled feeding has been shown to be a strong entrainment signal for circadian rhythm; especially when food intake is in phase with the period of activity. This exerts beneficial effects on the circadian system by favoring its synchronization and activation of the metabolism [136,137], as demonstrated in experimental studies on shift-worker models. Time-limited access to food accelerates resynchronization in a jet-lag model, prevents circadian desynchronization in a shift-work model, and induces positive effects in metabolism [138,139]. In contrast, food scheduled in the sleep–rest phase slows circadian synchronization and metabolism and alters behavior [138–140].

Recently, Oishi et al. [141] confirmed the positive action of cocoa on sleep disturbance induced by psychophysiological stress in mice using EEG. Cocoa intake attenuated the alteration of circadian sleep–wake rhythms. The EEG revealed that cocoa significantly improved both the increase in the level of alertness during the first half of the light period and the increase in NREM sleep during the first half of the dark period in mice with CSD. Under non-CSD conditions, cocoa does not appear to influence either the rhythms of run-rest activity or sleep–wake cycles. It is hypothesized that this positive action may be attributable to high concentrations of flavonoids in cocoa (epicatechin, catechin, and procyanidins), which improve blood flow and have antioxidant and neuroprotective properties. Indeed, in experimental animals, sleep deprivation and CSDs in general have been shown to increase oxidative stress levels in specific brain regions such as the hypothalamus, hippocampus, and thalamus [142].

Flavanols acting on endothelial function could also play a role in insomnia, as it is just endothelial dysfunction that appears to be responsible for some insomnia-related symptoms and the association between insomnia and cardiovascular disease [143]. Cocoa flavanols, by facilitating nitric oxide production, improve vascular endothelial function due to their vasodilatory effect [51]. CBF also appears to play an important role in sleep regulation [144] as well as cognitive and emotional processes, although it is not known how cerebral blood flow varies during alternating sleep–wake cycles [145]. Flavanol-rich cocoa significantly increases cerebral blood flow in humans [46,51] and attenuates the CSD-induced disturbance of circadian activity rhythm, sleep–wake cycles, cognitive functions by improving cerebral endothelial cell function, and blood flow, as demonstrated by Grassi et al. [76].

In addition, another pathway through which cocoa has a protective effect on the synchronous maintenance of the sleep–wake rhythm in subjects with CSD is the modulated neurotransmission of serotonin [146,147]. In fact, regular cocoa consumption has been shown to increase serotonin concentrations in the brain [148].

Acute administration of flavanol-rich chocolate can counteract the negative effects of total sleep deprivation both on working memory performance in healthy young people [76] and on endothelial and arterial function and, thus, on blood pressure (Figure 1). Natural cocoa seems to be an ideal nutrient for ameliorating stress-induced psychophysiological sleep disturbance without distorting behavioral or sleep regulation under normal conditions.

4. Conclusions

Modern society imposes increasingly stressful rhythms of life and work, and technological innovations have led us to live an active life 24/7. The immediate effect is on the quality and quantity of sleep, which in turn has repercussions on our lower tolerance to stress, alterations in mood, and greater susceptibility to infectious, metabolic, and cardiovascular diseases. Lifestyles are undoubtedly the main targets to work on to maintain a harmonious synchronization between our central and peripheral endogenous clocks and external synchronizers. Of course, the circadian rhythm of light and dark is the most powerful synchronizer, but we must not forget the other fundamental synchronizer: food and fasting.

Chocolate is a food that, due to its high flavonoid content, when taken in the morning, can have positive effects on our mental and cognitive well-being, our cardiovascular system, and our metabolism. Chocolate also induces positive effects on mood and is, therefore, often spontaneously consumed under conditions of emotional stress. In addition to the beneficial effects on the vascular system and cerebral blood flow, flavonoids have a protective function on the neuronal cell by inhibiting neuronal death by apoptosis induced by neurotoxicants, such as oxygen radicals through interaction with signaling cascades involving proteins and lipid kinases. They also promote neuronal survival, neurogenesis, and synaptic plasticity [89], preserving cognitive abilities during the stages of aging. Crossing the BBB, they act on both the vascular and neurocellular sides of the brain, on one hand, by stimulating cerebral perfusion and promoting angiogenesis, and, on the other, by modulating changes in the morphology of the neurons participating in learning and memory processes. All these properties are of great interest, but it is currently unclear when consumption of flavonoid-rich cocoa and chocolate should be initiated to achieve beneficial and protective effects against the mechanisms underlying cognitive decline and age-dependent neurodegenerative diseases [52,149]. Many studies are still needed to understand the mechanisms of action and the timing of the neuroprotective activity of cocoa and chocolate.

Although the dosages to be taken in order to achieve the beneficial effects are still unclear, the literature data support the suggestion that the beneficial effect is obtained by eating cocoa only in modest amounts, preferably dark chocolate, during the activity hours of the day and in the early part of the day to avoid problems of weight gain. Frequent consumption of chocolate managed in this way might actually be associated with a lower body mass index [150].

Time of food intake is now suggested as a chronotherapeutic strategy that may help speed up the time needed to resynchronize biological clocks and, thus, reduce circadian disruption caused by shift work, jet lag, and social jet lag [135,137,138]. Palatable food, such as chocolate, scheduled for breakfast is a valuable aid in maintaining circadian synchrony and improving body weight. The present data agree with the previous literature and indicate that the thermogenic effect of a high-calorie food such as chocolate is actually different depending on the time of intake in relation to the activity–rest phase: high postprandial thermogenic response takes place during the active phase, not during the rest phase. Breakfast induces a strong postprandial thermogenesis, which induces an increase in energy expenditure [151–153].

A piece of chocolate a day may also modulate circadian oscillations in brain areas involved in the reward system [154,155]. It may also have effects on behavior, mood, and cognitive functions.

In a socio-historic period such as the current one, when psychophysical distress is significantly increasing, due in part to the pandemic that has led to social distancing, significant repercussions on economic stability and the ability to envision future prospects are to be expected. The intake of chocolate rich in flavanols at breakfast and during the active phase, in accordance with the rules of chrononutrition, could provide a new tool that is economical, quick, and easily applicable for all ages to support cognitive performance during periods of sleep deprivation and psychophysical stress as well as during phases of desynchronization of circadian rhythms, with a positive effect on mood, while also helping to protect our cardiovascular system and regulate our metabolism.

Future studies in selected populations will be essential to establish the correct ways of taking chocolate by defining dosages as well as timing during the day and over a lifetime.

Author Contributions: Conceptualization, S.G. and P.L.; methodology, S.G. and P.L.; data curation, S.G., E.G. and P.L.; writing—original draft preparation, S.G., E.G. and P.L.; writing—review and editing, S.G., E.G. and P.L.; supervision, S.G., E.G. and P.L. All authors have read and agreed to the published version of the manuscript.

Funding: This research received no external funding.

Institutional Review Board Statement: Not applicable.

Informed Consent Statement: Not applicable.

Data Availability Statement: Not applicable.

Conflicts of Interest: The authors declare no conflict of interest.

References

1. LeGates, T.; Fernandez, D.C.; Hattar, S. Light as a central modulator of circadian rhythms, sleep and affect. *Nat. Rev. Neurosci.* **2014**, *15*, 443–454. [CrossRef] [PubMed]
2. Reppert, S.M.; Weaver, D.R. Coordination of circadian timing in mammals. *Nature* **2002**, *418*, 935–941. [CrossRef] [PubMed]
3. Oike, H.; Oishi, K.; Kobori, M. Nutrients, Clock Genes, and Chrononutrition. *Curr. Nutr. Rep.* **2014**, *3*, 204–212. [CrossRef] [PubMed]
4. Hastings, M.H.; Reddy, A.B.; Maywood, E.S. A clockwork web: Circadian timing in brain and periphery, in health and disease. *Nat. Rev. Neurosci.* **2003**, *4*, 649–661. [CrossRef]
5. Ospri, L.L.; Prusky, G.; Hattar, S. Mood, the Circadian System, and Melanopsin Retinal Ganglion Cells. *Annu. Rev. Neurosci.* **2017**, *40*, 539–556. [CrossRef]
6. Huang, W.; Ramsey, K.M.; Marcheva, B.; Bass, J. Circadian rhythms, sleep, and metabolism. *J. Clin. Investig.* **2011**, *121*, 2133–2141. [CrossRef]
7. Noh, J.-Y.; Han, D.-H.; Yoon, J.-A.; Kim, M.-H.; Kim, S.-E.; Ko, I.-G.; Kim, K.-H.; Kim, C.-J.; Cho, S. Circadian Rhythms in Urinary Functions: Possible Roles of Circadian Clocks? *Int. Neurourol. J.* **2011**, *15*, 64–73. [CrossRef]
8. Pickard, G.E. Circadian rhythm of nociception in the golden hamster. *Brain Res.* **1987**, *425*, 395–400. [CrossRef]
9. Refinetti, R. Rhythms of temperature selection and body temperature are out of phase in the golden hamster. *Behav. Neurosci.* **1995**, *109*, 523–527. [CrossRef]
10. Lockley, S.W.; Evans, E.E.; Scheer, F.; Brainard, G.C.; Czeisler, C.A.; Aeschbach, D. Short-Wavelength Sensitivity for the Direct Effects of Light on Alertness, Vigilance, and the Waking Electroencephalogram in Humans. *Sleep* **2006**, *29*, 161–168. [CrossRef]
11. Johnson, M.P.; Duffy, J.F.; Dijk, D.J.; Ronda, J.M.; Dyal, C.M.; Czeisler, C.A. Short-term memory, alertness and performance: A reappraisal of their relationship to body temperature. *J. Sleep Res.* **1992**, *1*, 24–29. [CrossRef]
12. Foster, R.G.; Kreitzman, L. The rhythms of life: What your body clock means to you! *Exp. Physiol.* **2013**, *99*, 599–606. [CrossRef]
13. Boivin, D.B.; Czeisler, C.A.; Dijk, D.-J.; Duffy, J.F.; Folkard, S.; Minors, D.S.; Totterdell, P.; Waterhouse, J.M. Complex Interaction of the Sleep-Wake Cycle and Circadian Phase Modulates Mood in Healthy Subjects. *Arch. Gen. Psychiatry* **1997**, *54*, 145–152. [CrossRef]
14. Birchler-Pedross, A.; Schröder, C.M.; Münch, M.; Knoblauch, V.; Blatter, K.; Schnitzler-Sack, C.; Wirz-Justice, A.; Cajochen, C. Subjective Well-Being Is Modulated by Circadian Phase, Sleep Pressure, Age, and Gender. *J. Biol. Rhythm.* **2009**, *24*, 232–242. [CrossRef]
15. Murray, G.; Allen, N.B.; Trinder, J. Mood and the circadian system: Investigation of a circadian component in positive affect. *Chronobiol. Int.* **2002**, *19*, 1151–1169. [CrossRef]
16. Tahara, Y.; Shibata, S. Chronobiology and nutrition. *Neuroscience* **2013**, *253*, 78–88. [CrossRef]
17. Asher, G.; Sassone-Corsi, P. Time for Food: The Intimate Interplay between Nutrition, Metabolism, and the Circadian Clock. *Cell* **2015**, *161*, 84–92. [CrossRef]
18. Gan, Y.; Yang, C.; Tong, X.; Sun, H.; Cong, Y.; Yin, X.; Li, L.; Cao, S.; Dong, X.; Gong, Y.; et al. Shift work and diabetes mellitus: A meta-analysis of observational studies. *Occup. Environ. Med.* **2014**, *72*, 72–78. [CrossRef]
19. Albrecht, U. Timing to Perfection: The Biology of Central and Peripheral Circadian Clocks. *Neuron* **2012**, *74*, 246–260. [CrossRef]
20. Gupta, N.J. Lifestyle and Circadian Health: Where the Challenges Lie? *Nutr. Metab. Insights* **2019**, *12*. [CrossRef]
21. Hughes, M.E.; DiTacchio, L.; Hayes, K.R.; Vollmers, C.; Pulivarthy, S.; Baggs, J.E.; Panda, S.; HogenEsch, J.B. Harmonics of Circadian Gene Transcription in Mammals. *PLoS Genet.* **2009**, *5*, e1000442. [CrossRef]
22. Kuroda, H.; Tahara, Y.; Saito, K.; Ohnishi, N.; Kubo, Y.; Seo, Y.; Otsuka, M.; Fuse, Y.; Ohura, Y.; Hirao, A.; et al. Meal frequency patterns determine the phase of mouse peripheral circadian clocks. *Sci. Rep.* **2012**, *2*, 711. [CrossRef]

23. Ángeles-Castellanos, M.; Salgado-Delgado, R.; Rodríguez, K.; Buijs, R.; Escobar, C. Expectancy for food or expectancy for chocolate reveals timing systems for metabolism and reward. *Neuroscience* **2008**, *155*, 297–307. [CrossRef]
24. Feillet, C.A.; Albrecht, U.; Challet, E. "Feeding time" for the brain: A matter of clocks. *J. Physiol. Paris* **2006**, *100*, 252–260. [CrossRef]
25. Verwey, M.; Amir, S. Food-entrainable circadian oscillators in the brain. *Eur. J. Neurosci.* **2009**, *30*, 1650–1657. [CrossRef]
26. Wu, T.; Jin, Y.; Ni, Y.; Zhang, D.; Kato, H.; Fu, Z. Effects of light cues on re-entrainment of the food-dominated peripheral clocks in mammals. *Gene* **2008**, *419*, 27–34. [CrossRef]
27. Hara, R.; Wan, K.; Wakamatsu, H.; Aida, R.; Moriya, T.; Akiyama, M.; Shibata, S. Restricted feeding entrains liver clock without participation of the suprachiasmatic nucleus. *Genes Cells* **2001**, *6*, 269–278. [CrossRef]
28. Stokkan, K.-A.; Yamazaki, S.; Tei, H.; Sakaki, Y.; Menaker, M. Entrainment of the Circadian Clock in the Liver by Feeding. *Science* **2001**, *291*, 490–493. [CrossRef]
29. Damiola, F.; Minh, N.L.; Preitner, N.; Kornmann, B.; Fleury-Olela, F.; Schibler, U. Restricted feeding uncouples circadian oscillators in peripheral tissues from the central pacemaker in the suprachiasmatic nucleus. *Genes Dev.* **2000**, *14*, 2950–2961. [CrossRef]
30. Hirao, A.; Nagahama, H.; Tsuboi, T.; Hirao, M.; Tahara, Y.; Shibata, S. Combination of starvation interval and food volume determines the phase of liver circadian rhythm in *Per2::Luc* knock-in mice under two meals per day feeding. *Am. J. Physiol. Liver Physiol.* **2010**, *299*, G1045–G1053. [CrossRef]
31. Garaulet, M.; Gomez-Abellan, P.; Alburquerque-Bejar, J.J.; Lee, Y.C.; Ordovas, J.M.; Scheer, F.A. Timing of food intake predicts weight loss effectiveness. *Int. J. Obes.* **2013**, *37*, 604–611. [CrossRef] [PubMed]
32. Wang, J.B.; Patterson, R.E.; Ang, A.; Emond, J.A.; Shetty, N.; Arab, L. Timing of energy intake during the day is associated with the risk of obesity in adults. *J. Hum. Nutr. Diet.* **2013**, *27*, 255–262. [CrossRef] [PubMed]
33. Jakubowicz, D.; Barnea, M.; Wainstein, J.; Froy, O. High Caloric intake at breakfast vs. dinner differentially influences weight loss of overweight and obese women. *Obesity* **2013**, *21*, 2504–2512. [CrossRef] [PubMed]
34. Hirao, A.; Tahara, Y.; Kimura, I.; Shibata, S. A Balanced Diet Is Necessary for Proper Entrainment Signals of the Mouse Liver Clock. *PLoS ONE* **2009**, *4*, e6909. [CrossRef]
35. Itokawa, M.; Hirao, A.; Nagahama, H.; Otsuka, M.; Ohtsu, T.; Furutani, N.; Hirao, K.; Hatta, T.; Shibata, S. Time-restricted feeding of rapidly digested starches causes stronger entrainment of the liver clock in PER2::LUCIFERASE knock-in mice. *Nutr. Res.* **2013**, *33*, 109–119. [CrossRef]
36. Micó, V.; Díez-Ricote, L.; Daimiel, L. Nutrigenetics and Nutrimiromics of the Circadian System: The Time for Human Health. *Int. J. Mol. Sci.* **2016**, *17*, 299. [CrossRef]
37. Wu, T.; Yao, C.; Huang, L.; Mao, Y.; Zhang, W.; Jiang, J.; Fu, Z. Nutrients and Circadian Rhythms in Mammals. *J. Nutr. Sci. Vitaminol.* **2015**, *61*, S89–S91. [CrossRef]
38. Oosterman, J.E.; Kalsbeek, A.; la Fleur, S.E.; Belsham, D.D. Impact of nutrients on circadian rhythmicity. *Am. J. Physiol. Integr. Comp. Physiol.* **2015**, *308*, R337–R350. [CrossRef]
39. Eckel-Mahan, K.L.; Patel, V.R.; Mohney, R.P.; Vignola, K.S.; Baldi, P.; Sassone-Corsi, P. Coordination of the transcriptome and metabolome by the circadian clock. *Proc. Natl. Acad. Sci. USA* **2012**, *109*, 5541–5546. [CrossRef]
40. Oishi, K.; Uchida, D.; Ohkura, N.; Doi, R.; Ishida, N.; Kadota, K.; Horie, S. Ketogenic Diet Disrupts the Circadian Clock and Increases Hypofibrinolytic Risk by Inducing Expression of Plasminogen Activator Inhibitor-1. *Arter. Thromb. Vasc. Biol.* **2009**, *29*, 1571–1577. [CrossRef]
41. Wu, M.N.; Ho, K.; Crocker, A.; Yue, Z.; Koh, K.; Sehgal, A. The Effects of Caffeine on Sleep in Drosophila Require PKA Activity, But Not the Adenosine Receptor. *J. Neurosci.* **2009**, *29*, 11029–11037. [CrossRef]
42. Oike, H.; Kobori, M.; Suzuki, T.; Ishida, N. Caffeine lengthens circadian rhythms in mice. *Biochem. Biophys. Res. Commun.* **2011**, *410*, 654–658. [CrossRef]
43. Feldman, J.F. Circadian Periodicity in Neurospora: Alteration by Inhibitors of Cyclic AMP Phosphodiesterase. *Science* **1975**, *190*, 789–790. [CrossRef]
44. Goodenough, J.E.; Bruce, V.G. The effects of caffeine and theophylline on the phototactic rhythm of Chlamydomonas Reinhardtii. *Biol. Bull.* **1980**, *159*, 649–655. [CrossRef]
45. Bollig, I.; Mayer, K.; Mayer, W.-E.; Engelmann, W. Effects of cAMP, theophylline, imidazole, and 4-(3,4-dimethoxybenzyl)-2-imidazolidone on the leaf movement rhythm of Trifolium repens?a test of the cAMP-hypothesis of circadian rhythms. *Planta* **1978**, *141*, 225–230. [CrossRef]
46. Socci, V.; Tempesta, D.; Desideri, G.; De Gennaro, L.; Ferrara, M. Enhancing Human Cognition with Cocoa Flavonoids. *Front. Nutr.* **2017**, *4*, 19. [CrossRef]
47. Williams, R.J.; Spencer, J.P. Flavonoids, cognition, and dementia: Actions, mechanisms, and potential therapeutic utility for Alzheimer disease. *Free Radic. Biol. Med.* **2012**, *52*, 35–45. [CrossRef]
48. Hurst, W.J.; Jr, S.M.T.; Powis, T.G.; Valdez, F.; Hester, T.R. Cacao usage by the earliest Maya civilization. *Nature* **2002**, *418*, 289–290. [CrossRef]
49. EFSA Panel on Dietetic Products, Nutrition and Allergies (NDA). Scientific opinion on a health claim related to cocoa flavanols and maintenance of normal endothelium-dependent vasodilation as referred to in Article 13(5) of Regulation (EC) No 1924/2006. *EFSA J.* **2014**, *12*, 3654.

50. Whiting, D.A. Natural phenolic compounds 1900–2000: A bird's eye view of a century's chemistry. *Nat. Prod. Rep.* **2001**, *18*, 583–606. [CrossRef]
51. Sokolov, A.N.; Pavlova, M.A.; Klosterhalfen, S.; Enck, P. Chocolate and the brain: Neurobiological impact of cocoa flavanols on cognition and behavior. *Neurosci. Biobehav. Rev.* **2013**, *37 Pt 2*, 2445–2453. [CrossRef]
52. Katz, D.L.; Doughty, K.; Ali, A. Cocoa and Chocolate in Human Health and Disease. *Antioxid. Redox Signal.* **2011**, *15*, 2779–2811. [CrossRef]
53. Chemistry Behind the Sweet Lure of Chocolate. Available online: https://www.worldofchemicals.com/113/chemistry-articles/chemistry-behind-the-sweet-lure-of-chocolate.html (accessed on 22 December 2021).
54. Scholey, A.; Owen, L. Effects of chocolate on cognitive function and mood: A systematic review. *Nutr. Rev.* **2013**, *71*, 665–681. [CrossRef]
55. Nehlig, A. The neuroprotective effects of cocoa flavanol and its influence on cognitive performance. *Br. J. Clin. Pharmacol.* **2013**, *75*, 716–727. [CrossRef]
56. Gu, L.; Kelm, M.A.; Hammerstone, J.F.; Beecher, G.; Holden, J.; Haytowitz, D.; Gebhardt, S.; Prior, R.L. Concentrations of Proanthocyanidins in Common Foods and Estimations of Normal Consumption. *J. Nutr.* **2004**, *134*, 613–617. [CrossRef]
57. Vauzour, D.; Vafeiadou, K.; Rodriguez-Mateos, A.; Rendeiro, C.; Spencer, J.P.E. The neuroprotective potential of flavonoids: A multiplicity of effects. *Genes Nutr.* **2008**, *3*, 115–126. [CrossRef]
58. Richelle, M.; Tavazzi, I.; Enslen, M.; Offord, E. Plasma kinetics in man of epicatechin from black chocolate. *Eur. J. Clin. Nutr.* **1999**, *53*, 22–26. [CrossRef]
59. Faria, A.; Pestana, D.; Teixeira, D.; Couraud, P.-O.; Romero, I.; Weksler, B.; de Freitas, V.; Mateus, N.; Calhau, C. Insights into the putative catechin and epicatechin transport across blood-brain barrier. *Food Funct.* **2010**, *2*, 39–44. [CrossRef]
60. Ghosh, D.; Scheepens, A. Vascular action of polyphenols. *Mol. Nutr. Food Res.* **2009**, *53*, 322–331. [CrossRef]
61. Datla, K.P.; Christidou, M.; Widmer, W.W.; Rooprai, H.K.; Dexter, D.T. Tissue distribution and neuroprotective effects of citrus flavonoid tangeretin in a rat model of Parkinson's disease. *NeuroReport* **2001**, *12*, 3871–3875. [CrossRef]
62. Spencer, J.P.E. Food for thought: The role of dietary flavonoids in enhancing human memory, learning and neuro-cognitive performance. *Proc. Nutr. Soc.* **2008**, *67*, 238–252. [CrossRef] [PubMed]
63. Kelleher, R.J.; Govindarajan, A.; Jung, H.-Y.; Kang, H.; Tonegawa, S. Translational Control by MAPK Signaling in Long-Term Synaptic Plasticity and Memory. *Cell* **2004**, *116*, 467–479. [CrossRef]
64. Kim, D.H.; Jeon, S.J.; Son, K.H.; Jung, J.W.; Lee, S.; Yoon, B.H.; Choi, J.W.; Cheong, J.H.; Ko, K.H.; Ryu, J.H. Effect of the flavonoid, oroxylin A, on transient cerebral hypoperfusion-induced memory impairment in mice. *Pharmacol. Biochem. Behav.* **2006**, *85*, 658–668. [CrossRef] [PubMed]
65. Valente, T.; Hidalgo, J.; Bolea, I.; Ramirez, B.; Anglés, N.; Reguant, J.; Morelló, J.R.; Gutiérrez, C.; Boada, M.; Unzeta, M. A Diet Enriched in Polyphenols and Polyunsaturated Fatty Acids, LMN Diet, Induces Neurogenesis in the Subventricular Zone and Hippocampus of Adult Mouse Brain. *J. Alzheimer's Dis.* **2009**, *18*, 849–865. [CrossRef]
66. González-Gallego, J.; Sánchez-Campos, S.; Tuñón, M.J. Anti-inflammatory properties of dietary flavonoids. *Nutr. Hosp.* **2007**, *22*, 287–293.
67. Schroeter, H.; Heiss, C.; Balzer, J.; Kleinbongard, P.; Keen, C.L.; Hollenberg, N.K.; Sies, H.; Kwik-Uribe, C.; Schmitz, H.H.; Kelm, M. (-)-Epicatechin mediates beneficial effects of flavanol-rich cocoa on vascular function in humans. *Proc. Natl. Acad. Sci. USA* **2006**, *103*, 1024–1029. [CrossRef]
68. Flammer, A.J.; Hermann, F.; Sudano, I.; Spieker, L.; Hermann, M.; Cooper, K.A.; Serafini, M.; Lüscher, T.F.; Ruschitzka, F.; Noll, G.; et al. Dark Chocolate Improves Coronary Vasomotion and Reduces Platelet Reactivity. *Circulation* **2007**, *116*, 2376–2382. [CrossRef]
69. Grassi, D.; Necozione, S.; Lippi, C.; Croce, G.; Valeri, L.; Pasqualetti, P.; Desideri, G.; Blumberg, J.B.; Ferri, C. Cocoa Reduces Blood Pressure and Insulin Resistance and Improves Endothelium-Dependent Vasodilation in Hypertensives. *Hypertension* **2005**, *46*, 398–405. [CrossRef]
70. Grassi, D.; Desideri, G.; Necozione, S.; Lippi, C.; Casale, R.; Properzi, G.; Blumberg, J.B.; Ferri, C. Blood Pressure Is Reduced and Insulin Sensitivity Increased in Glucose-Intolerant, Hypertensive Subjects after 15 Days of Consuming High-Polyphenol Dark Chocolate. *J. Nutr.* **2008**, *138*, 1671–1676. [CrossRef]
71. Grassi, D.; Desideri, G.; Ferri, C. Protective effects of dark chocolate on endothelial function and diabetes. *Curr. Opin. Clin. Nutr. Metab. Care* **2013**, *16*, 662–668. [CrossRef]
72. Grassi, D.; Desideri, G.; Necozione, S.; di Giosia, P.; Barnabei, R.; Allegaert, L.; Bernaert, H.; Ferri, C. Cocoa consumption dose-dependently improves flow-mediated dilation and arterial stiffness decreasing blood pressure in healthy individuals. *J. Hypertens.* **2015**, *33*, 294–303. [CrossRef]
73. Shrime, M.G.; Bauer, S.R.; McDonald, A.C.; Chowdhury, N.H.; Coltart, C.E.M.; Ding, E.L. Flavonoid-Rich Cocoa Consumption Affects Multiple Cardiovascular Risk Factors in a Meta-Analysis of Short-Term Studies. *J. Nutr.* **2011**, *141*, 1982–1988. [CrossRef]
74. Hollenberg, N.K.; Fisher, N.D.; McCullough, M.L. Flavanols, the Kuna, cocoa consumption, and nitric oxide. *J. Am. Soc. Hypertens.* **2009**, *3*, 105–112. [CrossRef]
75. Kalt, W.; Hanneken, A.; Milbury, P.; Tremblay, F. Recent Research on Polyphenolics in Vision and Eye Health. *J. Agric. Food Chem.* **2010**, *58*, 4001–4007. [CrossRef]

76. Grassi, D.; Ferri, C.; Desideri, G. Brain Protection and Cognitive Function: Cocoa Flavonoids as Nutraceuticals. *Curr. Pharm. Des.* **2015**, *22*, 145–151. [CrossRef]
77. Gage, F.H. Mammalian Neural Stem Cells. *Science* **2000**, *287*, 1433–1438. [CrossRef]
78. Van Praag, H.; Lucero, M.J.; Yeo, E.; Stecker, K.; Heivand, N.; Zhao, C.; Yip, E.; Afanador, M.; Schroeter, H.; Hammerstone, J.; et al. Plant-Derived Flavanol (-)Epicatechin Enhances Angiogenesis and Retention of Spatial Memory in Mice. *J. Neurosci.* **2007**, *27*, 5869–5878, Erratum in *J. Neurosci.* **2007**, *27*. [CrossRef]
79. Spencer, J.P.E. The interactions of flavonoids within neuronal signalling pathways. *Genes Nutr.* **2007**, *2*, 257–273. [CrossRef]
80. Crichton, G.E.; Elias, M.F.; Alkerwi, A. Chocolate intake is associated with better cognitive function: The Maine-Syracuse Longitudinal Study. *Appetite* **2016**, *100*, 126–132. [CrossRef]
81. Commenges, D.; Scotet, V.; Renaud, S.; Jacqmin-Gadda, H.; Barberger-Gateau, P.; Dartigues, J.-F. Intake of flavonoids and risk of dementia. *Eur. J. Epidemiol.* **2000**, *16*, 357–363. [CrossRef]
82. Moreira, A.; Diógenes, M.J.; de Mendonça, A.; Lunet, N.; Barros, H. Chocolate Consumption is Associated with a Lower Risk of Cognitive Decline. *J. Alzheimer's Dis.* **2016**, *53*, 85–93. [CrossRef]
83. Kuriyama, S.; Hozawa, A.; Ohmori, K.; Shimazu, T.; Matsui, T.; Ebihara, S.; Awata, S.; Nagatomi, R.; Arai, H.; Tsuji, I. Green tea consumption and cognitive function: A cross-sectional study from the Tsurugaya Project. *Am. J. Clin. Nutr.* **2006**, *83*, 355–361. [CrossRef]
84. Letenneur, L.; Proust-Lima, C.; Le Gouge, A.; Dartigues, J.F.; Barberger-Gateau, P. Flavonoid Intake and Cognitive Decline over a 10-Year Period. *Am. J. Epidemiol.* **2007**, *165*, 1364–1371. [CrossRef]
85. Nurk, E.; Refsum, H.; Drevon, C.A.; Tell, G.S.; Nygaard, H.A.; Engedal, K.; Smith, A.D. Intake of Flavonoid-Rich Wine, Tea, and Chocolate by Elderly Men and Women Is Associated with Better Cognitive Test Performance. *J. Nutr.* **2008**, *139*, 120–127. [CrossRef]
86. Smit, H.J.; Rogers, P.J. Effects of low doses of caffeine on cognitive performance, mood and thirst in low and higher caffeine consumers. *Psychopharmacology* **2000**, *152*, 167–173. [CrossRef]
87. Smit, H.J.; Gaffan, E.A.; Rogers, P.J. Methylxanthines are the psycho-pharmacologically active constituents of chocolate. *Psychopharmacology* **2004**, *176*, 412–419. [CrossRef]
88. Smit, H.J.; Blackburn, R.J. Reinforcing effects of caffeine and theobromine as found in chocolate. *Psychopharmacology* **2005**, *181*, 101–106. [CrossRef]
89. Nehlig, A. Is caffeine a cognitive enhancer? *J. Alzheimer's Dis.* **2010**, *20*, 85–94. [CrossRef]
90. Parker, G.; Roy, K.; Mitchell, P.; Wilhelm, K.; Malhi, G.; Hadzi-Pavlovic, D. Atypical Depression: A Reappraisal. *Am. J. Psychiatry* **2002**, *159*, 1470–1479. [CrossRef]
91. Messaoudi, M.; Bisson, J.-F.; Nejdi, A.; Rozan, P.; Javelot, H. Antidepressant-like effects of a cocoa polyphenolic extract in Wistar–Unilever rats. *Nutr. Neurosci.* **2008**, *11*, 269–276. [CrossRef]
92. Benton, D.; Donohoe, R.T. The effects of nutrients on mood. *Public Health Nutr.* **1999**, *2*, 403–409. [CrossRef] [PubMed]
93. Reid, L.D. Endogenous opioid peptides and regulation of drinking and feeding. *Am. J. Clin. Nutr.* **1985**, *42*, 1099–1132. [CrossRef] [PubMed]
94. Giraudo, S.; Grace, M.; Welch, C.; Billington, C.; Levine, A. Naloxone's anorectic effect is dependant upon the relative palatability of food. *Pharmacol. Biochem. Behav.* **1993**, *46*, 917–921. [CrossRef]
95. Young, S.N.; Smith, S.E.; Pihl, R.O.; Ervin, F.R. Tryptophan depletion causes a rapid lowering of mood in normal males. *Psychopharmacology* **1985**, *87*, 173–177. [CrossRef] [PubMed]
96. Si, E.; Bryant, H.; Yim, G. Opioid and non-opioid components of insulin-induced feeding. *Pharmacol. Biochem. Behav.* **1986**, *24*, 899–903. [CrossRef]
97. Ottley, C. Food and mood. *Nurs. Stand.* **2000**, *15*, 46–52, quiz 54–55.
98. Parker, G.; Parker, I.; Brotchie, H. Mood state effects of chocolate. *J. Affect. Disord.* **2006**, *92*, 149–159. [CrossRef]
99. Fullerton, D.T.; Getto, C.J.; Swift, W.J.; Carlson, I.H. Sugar, opioids and binge eating. *Brain Res. Bull.* **1985**, *14*, 673–680. [CrossRef]
100. Willner, P.; Benton, D.; Brown, E.; Cheeta, S.; Roderique-Davies, G.; Morgan, J.; Morgan, M. "Depression" increases "craving" for sweet rewards in animal and human models of depression and craving. *Psychopharmacology* **1998**, *136*, 272–283. [CrossRef]
101. Hetherington, M.M.; MacDiarmid, J.I. "Chocolate Addiction": A Preliminary Study of its Description and its Relationship to Problem Eating. *Appetite* **1993**, *21*, 233–246. [CrossRef]
102. Rozin, P.; Levine, E.; Stoess, C. Chocolate craving and liking. *Appetite* **1991**, *17*, 199–212. [CrossRef]
103. Smeets, P.A.M.; De Graaf, C.; Stafleu, A.; Van Osch, M.J.P.; Nievelstein, R.A.J.; Van Der Grond, J. Effect of satiety on brain activation during chocolate tasting in men and women. *Am. J. Clin. Nutr.* **2006**, *83*, 1297–1305. [CrossRef]
104. Michener, W.; Rozin, P. Pharmacological versus sensory factors in the satiation of chocolate craving. *Physiol. Behav.* **1994**, *56*, 419–422. [CrossRef]
105. Smits, M.; Peeters, R.R.; Van Hecke, P.; Sunaert, S. A 3 T event-related functional magnetic resonance imaging (fMRI) study of primary and secondary gustatory cortex localization using natural tastants. *Neuroradiology* **2006**, *49*, 61–71. [CrossRef]
106. Beaver, J.D.; Lawrence, A.D.; van Ditzhuijzen, J.; Davis, M.; Woods, A.; Calder, A.J. Individual Differences in Reward Drive Predict Neural Responses to Images of Food. *J. Neurosci.* **2006**, *26*, 5160–5166. [CrossRef]
107. Martin, G. Human electroencephalographic (EEG) response to olfactory stimulation: Two experiments using the aroma of food. *Int. J. Psychophysiol.* **1998**, *30*, 287–302. [CrossRef]

108. Rolls, E.T.; McCabe, C. Enhanced affective brain representations of chocolate in cravers vs. non-cravers. *Eur. J. Neurosci.* **2007**, *26*, 1067–1076. [CrossRef]
109. Nehlig, A. Are we dependent upon coffee and caffeine? A review on human and animal data. *Neurosci. Biobehav. Rev.* **1999**, *23*, 563–576. [CrossRef]
110. Di Chiara, G. Nucleus accumbens shell and core dopamine: Differential role in behavior and addiction. *Behav. Brain Res.* **2002**, *137*, 75–114. [CrossRef]
111. Schroeder, B.; Binzak, J.; Kelley, A. A common profile of prefrontal cortical activation following exposure to nicotine- or chocolate-associated contextual cues. *Neuroscience* **2001**, *105*, 535–545. [CrossRef]
112. Medina, J.H.; Viola, H.; Wolfman, C.; Marder, M.; Wasowski, C.; Calvo, D.; Paladini, A.C. Overview—Flavonoids: A New Family of Benzodiazepine Receptor Ligands. *Neurochem. Res.* **1997**, *22*, 419–425. [CrossRef]
113. Vignes, M.; Maurice, T.; Lanté, F.; Nedjar, M.; Thethi, K.; Guiramand, J.; Récasens, M. Anxiolytic properties of green tea polyphenol (−)-epigallocatechin gallate (EGCG). *Brain Res.* **2006**, *1110*, 102–115. [CrossRef]
114. Sathyapalan, T.; Beckett, S.; Rigby, A.S.; Mellor, D.D.; Atkin, S.L. High cocoa polyphenol rich chocolate may reduce the burden of the symptoms in chronic fatigue syndrome. *Nutr. J.* **2010**, *9*, 55. [CrossRef]
115. Pase, M.P.; Scholey, A.B.; Pipingas, A.; Kras, M.; Nolidin, K.; Gibbs, A.; Wesnes, K.; Stough, C. Cocoa polyphenols enhance positive mood states but not cognitive performance: A randomized, placebo-controlled trial. *J. Psychopharmacol.* **2013**, *27*, 451–458. [CrossRef]
116. Buijs, F.N.; Leon, L.; Guzmán-Ruiz, M.; Guerrero-Vargas, N.N.; Romo-Nava, F.; Buijs, R.M. The Circadian System: A Regulatory Feedback Network of Periphery and Brain. *Physiology* **2016**, *31*, 170–181. [CrossRef]
117. Bass, J.; Takahashi, J.S. Circadian Integration of Metabolism and Energetics. *Science* **2010**, *330*, 1349–1354. [CrossRef]
118. Knutson, K.L.; Spiegel, K.; Penev, P.; Van Cauter, E. The metabolic consequences of sleep deprivation. *Sleep Med. Rev.* **2007**, *11*, 163–178. [CrossRef]
119. Uehli, K.; Mehta, A.J.; Miedinger, D.; Hug, K.; Schindler, C.; Holsboer-Trachsler, E.; Leuppi, J.D.; Künzli, N. Sleep problems and work injuries: A systematic review and meta-analysis. *Sleep Med. Rev.* **2013**, *18*, 61–73. [CrossRef]
120. Kalmbach, D.A.; Anderson, J.R.; Drake, C.L. The impact of stress on sleep: Pathogenic sleep reactivity as a vulnerability to insomnia and circadian disorders. *J. Sleep Res.* **2018**, *27*, e12710. [CrossRef]
121. Escobar, C.; Espitia-Bautista, E.; Guzmán-Ruiz, M.A.; Vargas, N.N.G.; Hernández-Navarrete, M.; Ángeles-Castellanos, M.; Morales-Pérez, B.; Buijs, R.M. Chocolate for breakfast prevents circadian desynchrony in experimental models of jet-lag and shift-work. *Sci. Rep.* **2020**, *10*, 6243. [CrossRef] [PubMed]
122. Mendoza, J.; Angeles-Castellanos, M.; Escobar, C. A daily palatable meal without food deprivation entrains the suprachiasmatic nucleus of rats. *Eur. J. Neurosci.* **2005**, *22*, 2855–2862. [CrossRef] [PubMed]
123. Yamazaki, S.; Numano, R.; Abe, M.; Hida, A.; Takahashi, R.-I.; Ueda, M.; Block, G.D.; Sakaki, Y.; Menaker, M.; Tei, H. Resetting Central and Peripheral Circadian Oscillators in Transgenic Rats. *Science* **2000**, *288*, 682–685. [CrossRef] [PubMed]
124. Salgado-Delgado, R.C.; Saderi, N.; Basualdo, M.D.C.; Guerrero-Vargas, N.N.; Escobar, C.; Buijs, R.M. Shift Work or Food Intake during the Rest Phase Promotes Metabolic Disruption and Desynchrony of Liver Genes in Male Rats. *PLoS ONE* **2013**, *8*, e60052. [CrossRef]
125. Schilperoort, M.; van den Berg, R.; Dollé, M.E.T.; Van Oostrom, C.T.M.; Wagner, K.; Tambyrajah, L.L.; Wackers, P.; DeBoer, T.; Hulsegge, G.; Proper, K.I.; et al. Time-restricted feeding improves adaptation to chronically alternating light-dark cycles. *Sci. Rep.* **2019**, *9*, 7874. [CrossRef]
126. Mukherji, A.; Kobiita, A.; Damara, M.; Misra, N.; Meziane, H.; Champy, M.-F.; Chambon, P. Shifting eating to the circadian rest phase misaligns the peripheral clocks with the master SCN clock and leads to a metabolic syndrome. *Proc. Natl. Acad. Sci. USA* **2015**, *112*, E6691–E6698. [CrossRef]
127. Yasumoto, Y.; Hashimoto, C.; Nakao, R.; Yamazaki, H.; Hiroyama, H.; Nemoto, T.; Yamamoto, S.; Sakurai, M.; Oike, H.; Wada, N.; et al. Short-term feeding at the wrong time is sufficient to desynchronize peripheral clocks and induce obesity with hyperphagia, physical inactivity and metabolic disorders in mice. *Metabolism* **2016**, *65*, 714–727. [CrossRef]
128. Ekmekcioglu, C.; Touitou, Y. Chronobiological aspects of food intake and metabolism and their relevance on energy balance and weight regulation. *Obes. Rev.* **2010**, *12*, 14–25. [CrossRef]
129. Wakamatsu, H.; Yoshinobu, Y.; Aida, R.; Moriya, T.; Akiyama, M.; Shibata, S. Restricted-feeding-induced anticipatory activity rhythm is associated with a phase-shift of the expression of mPer1 and mPer2 mRNA in the cerebral cortex and hippocampus but not in the suprachiasmatic nucleus of mice. *Eur. J. Neurosci.* **2001**, *13*, 1190–1196. [CrossRef]
130. Herrera, D.; Robertson, H. Activation of c-fos in the brain. *Prog. Neurobiol.* **1996**, *50*, 83–107. [CrossRef]
131. Greco, C.; Sassone–Corsi, P. Circadian blueprint of metabolic pathways in the brain. *Nat. Rev. Neurosci.* **2018**, *20*, 71–82. [CrossRef]
132. Acosta-Galvan, G.; Yi, C.-X.; van der Vliet, J.; Jhamandas, J.H.; Panula, P.; Angeles-Castellanos, M.; Del Basualdo, M.C.; Escobar, C.; Buijs, R.M. Interaction between hypothalamic dorsomedial nucleus and the suprachiasmatic nucleus determines intensity of food anticipatory behavior. *Proc. Natl. Acad. Sci. USA* **2011**, *108*, 5813–5818. [CrossRef]
133. Saderi, N.; Cazarez-Márquez, F.; Buijs, F.; Salgado-Delgado, R.; Guzman-Ruiz, M.; Basualdo, M.D.C.; Escobar, C.; Buijs, R. The NPY intergeniculate leaflet projections to the suprachiasmatic nucleus transmit metabolic conditions. *Neuroscience* **2013**, *246*, 291–300. [CrossRef]

134. Grippo, R.M.; Purohit, A.M.; Zhang, Q.; Zweifel, L.S.; Güler, A.D. Direct Midbrain Dopamine Input to the Suprachiasmatic Nucleus Accelerates Circadian Entrainment. *Curr. Biol.* **2017**, *27*, 2465–2475.e3. [CrossRef]
135. Ubaldo, L.; Buijs, R.; Escobar, C.; Angeles-Castellanos, M. Scheduled meal accelerates entrainment to a 6-h phase advance by shifting central and peripheral oscillations in rats. *Eur. J. Neurosci.* **2017**, *46*, 1875–1886. [CrossRef]
136. Ángeles-Castellanos, M.; Amaya, J.M.; Salgado-Delgado, R.; Buijs, R.M.; Escobar, C. Scheduled Food Hastens Re-Entrainment More Than Melatonin Does after a 6-h Phase Advance of the Light-Dark Cycle in Rats. *J. Biol. Rhythm.* **2011**, *26*, 324–334. [CrossRef]
137. Salgado-Delgado, R.; Angeles-Castellanos, M.; Saderi, N.; Buijs, R.M.; Escobar, C. Food Intake during the Normal Activity Phase Prevents Obesity and Circadian Desynchrony in a Rat Model of Night Work. *Endocrinology* **2010**, *151*, 1019–1029. [CrossRef]
138. Guerrero-Vargas, N.N.; Espitia-Bautista, E.; Buijs, R.M.; Escobar, C. Shift-work: Is time of eating determining metabolic health? Evidence from animal models. *Proc. Nutr. Soc.* **2018**, *77*, 199–215. [CrossRef]
139. Hutchison, A.T.; Regmi, P.; Manoogian, E.N.; Fleischer, J.G.; Wittert, G.A.; Panda, S.; Heilbronn, L.K. Time-restricted feeding improves glucose tolerance in men at risk for type 2 diabetes: A randomized crossover trial. *Obesity* **2019**, *27*, 724–732. [CrossRef]
140. Chaix, A.; Manoogian, E.N.; Melkani, G.C.; Panda, S. Time-Restricted Eating to Prevent and Manage Chronic Metabolic Diseases. *Annu. Rev. Nutr.* **2019**, *39*, 291–315. [CrossRef]
141. Oishi, K.; Okauchi, H.; Yamamoto, S.; Higo-Yamamoto, S. Dietary natural cocoa ameliorates disrupted circadian rhythms in locomotor activity and sleep-wake cycles in mice with chronic sleep disorders caused by psychophysiological stress. *Nutrition* **2020**, *75–76*, 110751. [CrossRef]
142. Villafuerte, G.; Miguel-Puga, A.; Rodríguez, E.M.; Machado, S.; Manjarrez, E.; Arias-Carrión, O. Sleep Deprivation and Oxidative Stress in Animal Models: A Systematic Review. *Oxidative Med. Cell. Longev.* **2015**, *2015*, 234952. [CrossRef] [PubMed]
143. Routledge, F.S.; Dunbar, S.B.; Higgins, M.; Rogers, A.E.; Feeley, C.; Ioachimescu, O.; Euwer, K.; Eapen, D.; Quyyumi, A. Insomnia Symptoms Are Associated With Abnormal Endothelial Function. *J. Cardiovasc. Nurs.* **2017**, *32*, 78–85. [CrossRef] [PubMed]
144. Braun, A.R.; Balkin, T.J.; Wesenten, N.J.; E Carson, R.; Varga, M.; Baldwin, P.; Selbie, S.; Belenky, G.; Herscovitch, P. Regional cerebral blood flow throughout the sleep-wake cycle. An H2(15)O PET study. *Brain* **1997**, *120 Pt 7*, 1173–1197. [CrossRef] [PubMed]
145. Elvsåshagen, T.; Mutsaerts, H.J.; Zak, N.; Norbom, L.; Quraishi, S.H.; Pedersen, P.Ø.; Malt, U.F.; Westlye, L.T.; van Someren, E.J.; Bjørnerud, A.; et al. Cerebral blood flow changes after a day of wake, sleep, and sleep deprivation. *NeuroImage* **2018**, *186*, 497–509. [CrossRef]
146. Pich, E.M.; Solfrini, V.; Biagini, G.; Fuxe, K.; Agnati, L.F. Effects of indole-pyruvic acid on sleep and food intake in the rat. *Acta Physiol. Scand.* **1990**, *139*, 583–589. [CrossRef]
147. Biagini, G.; Pich, E.M.; Carani, C.; Marrama, P.; Gustafsson, J.; Fuxe, K.; Agnati, L.F. Indole-pyruvic acid, a tryptophan ketoanalogue, antagonizes the endocrine but not the behavioral effects of repeated stress in a model of depression. *Biol. Psychiatry* **1993**, *33*, 712–719. [CrossRef]
148. Yamada, T.; Yamada, Y.; Okano, Y.; Terashima, T.; Yokogoshi, H. Anxiolytic effects of short- and long-term administration of cacao mass on rat elevated T-maze test. *J. Nutr. Biochem.* **2009**, *20*, 948–955. [CrossRef]
149. McCarty, M. Vascular nitric oxide may lessen Alzheimer's risk. *Med. Hypotheses* **1998**, *51*, 465–476. [CrossRef]
150. Golomb, B.A.; Koperski, S.; White, H.L. Association between more frequent chocolate consumption and lower body mass index. *Arch. Intern. Med.* **2012**, *172*, 519–521. [CrossRef]
151. Jakubowicz, D.; Wainstein, J.; Landau, Z.; Ahren, B.; Barnea, M.; Bar-Dayan, Y.; Froy, O. High-energy breakfast based on whey protein reduces body weight, postprandial glycemia and HbA 1C in Type 2 diabetes. *J. Nutr. Biochem.* **2017**, *49*, 1–7. [CrossRef]
152. A Betts, J.; Richardson, J.D.; Chowdhury, E.; Holman, G.; Tsintzas, K.; Thompson, D. The causal role of breakfast in energy balance and health: A randomized controlled trial in lean adults. *Am. J. Clin. Nutr.* **2014**, *100*, 539–547. [CrossRef]
153. Romon, M.; Edme, J.L.; Boulenguez, C.; Lescroart, J.L.; Frimat, P. Circadian variation of diet-induced thermogenesis. *Am. J. Clin. Nutr.* **1993**, *57*, 476–480. [CrossRef]
154. Webb, I.C.; Baltazar, R.M.; Lehman, M.N.; Coolen, L.M. Bidirectional interactions between the circadian and reward systems: Is restricted food access a unique zeitgeber? *Eur. J. Neurosci.* **2009**, *30*, 1739–1748. [CrossRef]
155. Ángeles-Castellanos, M.; Mendoza, J.; Escobar, C. Restricted feeding schedules phase shift daily rhythms of c-Fos and protein Per1 immunoreactivity in corticolimbic regions in rats. *Neuroscience* **2007**, *144*, 344–355. [CrossRef]

nutrients

Review

Do Sleep Disorders and Western Diet Influence Psoriasis? A Scoping Review

Ilaria Controne [1,†], Egeria Scoditti [2,†], Alessandra Buja [3], Alessia Pacifico [4], Khalaf Kridin [5], Massimo Del Fabbro [1,6], Sergio Garbarino [7,†] and Giovanni Damiani [1,8,*,†]

1 Clinical Dermatology, IRCCS Istituto Ortopedico Galeazzi, 20161 Milan, Italy
2 Institute of Clinical Physiology, National Research Council (CNR), 73100 Lecce, Italy
3 Department of Cardiologic, Vascular and Thoracic Sciences, and Public Health, University of Padova, 35128 Padova, Italy
4 Clinical Dermatology Department, San Gallicano Dermatological Institute, IRCCS, 00144 Rome, Italy
5 Lübeck Institute of Experimental Dermatology, University of Lübeck, 23562 Lübeck, Germany
6 IRCCS Fondazione Ca' Granda IRCCS Ospedale Maggiore Policlinico, 20122 Milan, Italy
7 Department of Neuroscience, Rehabilitation, Ophthalmology, Genetics, and Maternal/Child Sciences (DINOGMI), University of Genoa, 16132 Genoa, Italy
8 Department of Biomedical, Surgical and Dental Sciences, University of Milan, 20122 Milan, Italy
* Correspondence: dr.giovanni.damiani@gmail.com; Tel.: +39-02-662141
† These authors contributed equally to this work.

Abstract: Western diet may trigger sleep disorders and vice versa, but their single and mutual effects on systemic inflammatory diseases (i.e., psoriasis) are far from being fully elucidated. At the same time, psoriatic patients display a great burden of sleep disorders and dysmetabolisms related to an unhealthy lifestyle (i.e., diet). These patients are also affected by a chronic disorder deeply modulated by environmental factors (i.e., sleep and diet) capable to influence drug-response and disease progression. Thus, we aimed to summarize the evidence in the literature that may highlight a potential link among psoriasis–diet–sleep in order to further promote a multidisciplinary approach to psoriatic patients in the scientific community.

Keywords: psoriasis; sleep; sleep disorders; western diet; diet; inflammation

Citation: Controne, I.; Scoditti, E.; Buja, A.; Pacifico, A.; Kridin, K.; Fabbro, M.D.; Garbarino, S.; Damiani, G. Do Sleep Disorders and Western Diet Influence Psoriasis? A Scoping Review. *Nutrients* **2022**, *14*, 4324. https://doi.org/10.3390/nu14204324

Academic Editor: Michael Wirth

Received: 24 August 2022
Accepted: 13 October 2022
Published: 16 October 2022

1. Introduction

Currently, the World Health Organization (WHO) reports psoriasis (PsO) as one of the five diseases that drastically influence patients' quality of life. PsO is a systemic, chronic inflammatory skin disease related to epidermal keratinocyte hyperplasia and epidermal immune cell over-activation via the interleukin (IL)-23/IL-17 axis [1]. It is one of the most common chronic inflammatory skin diseases, with a prevalence of 1–2% worldwide [2], and almost 6–11% of patients with PsO may have inflammatory arthropathy (psoriatic arthritis) [3]. The current pathogenetic hypotheses of PsO consider a trigger (i.e., trauma or infection) that destroys keratinocytes, activating both innate immunity (i.e., neutrophils) and plasmacytoid dendritic cells that promote naïve-T-cell differentiations into Th17. The pro-inflammatory microenvironment further activates keratinocytes, causing hyperproliferation (i.e., plaques) [4].

Recently, epidemiological studies claimed a link between PsO-related inflammation and a constellation of comorbidities affecting different systems, such as cardiovascular [5,6], respiratory [7,8], neurologic [9,10] or even gastrointestinal system [11]. To date, all the above-cited comorbidities may impair both the patient's quality of life and sleep; in fact, physicians also demonstrated that PsO patients are at a higher risk than the normal population to experience sleep disorders (i.e., insomnia or obstructive sleep apnea (OSA)) [11–14], but the real pathogenetic mechanism remains obscure.

At the same time, the pathological manifestations of PsO (phenotype) are the product of the intricate interaction between genetic susceptibility [15] and environmental factors (i.e., diet or circadian rhythm) [16] (Figure 1). Several studies pointed out that in western societies, unhealthy lifestyles may deeply condition the PsO march and flare frequency; in particular, unbalanced diets (i.e., western diet) with high fat and/or high sugar [17,18], inadequate sleep [19] and insufficient physical activity [20] may be triggers. Furthermore, all these PsO triggers may converge in creating gut dysbiosis [21], not necessarily identifiable as small intestinal bacterial overgrowth (SIBO) [22], but potentially critical for PsO and psoriatic arthritis (PsA) flares.

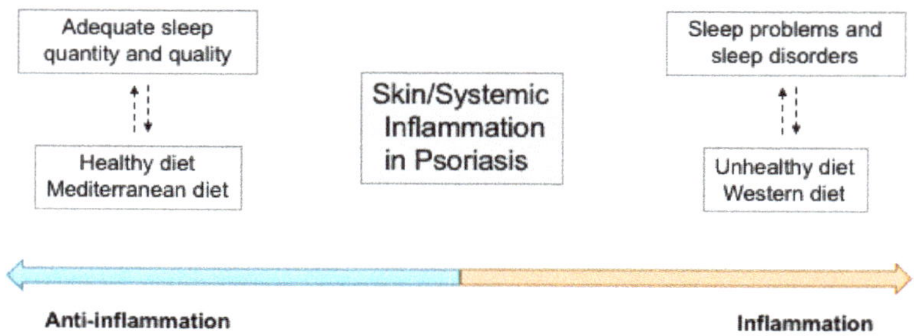

Figure 1. Psoriasis is associated with sleep disorders and unhealthy dietary patterns through the activation and/or regulation of systemic and skin inflammation. On the left, the scheme shows protective effects obtainable via adequate sleep quantity and quality together with healthy diet on psoriasis disease.

Thus, we decided to perform a scoping review to summarize the evidence toward PsO–western diet–sleep disorders.

2. Evidence Acquisition

For the present scoping review, we performed a comprehensive literature search on PubMed/MEDLINE from 1949 to 1 August 2022. No language restrictions were applied and we included only original articles, research letters and short reports; conversely, we excluded reviews, systematic reviews, letters without data and editorials.

The literature search method applied was this string: ("psoriasis" OR "psoriatic disease" OR "psoriatic arthritis" OR "inflammation") AND (("western diet") OR ("sleep" OR "sleep disorders" OR "sleep apnea" OR "insomnia")). Extensive cross-referencing and article references were carefully and manually evaluated. The literature search was performed independently by two authors (Giovanni Damiani and Ilaria Controne), and in case of disagreement, the article was discussed openly with a third experienced author (Egeria Scoditti for diet and Sergio Garbarino for sleep).

The literature search produced 839 papers; then, after eliminating duplicates and the screening phase, 81 papers were summarized into thematic paragraphs in the present scoping review (for more details, see the Preferred Reporting Items for Systematic Reviews and Meta-Analyses (PRISMA)) (Figure 2).

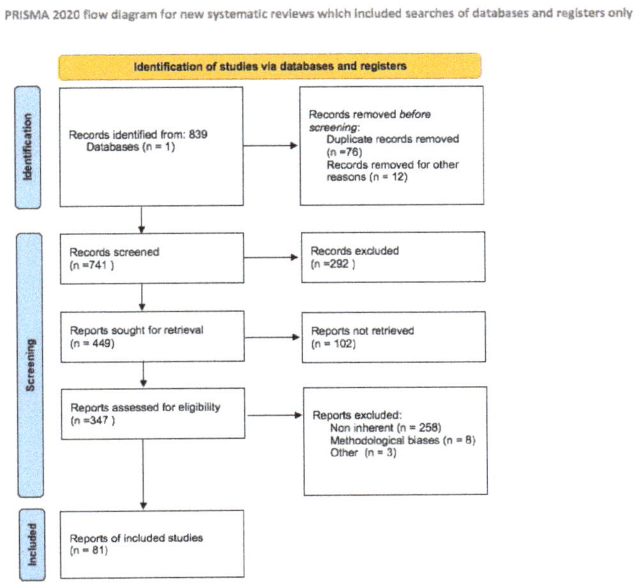

Figure 2. PRISMA flow diagram with a detailed report of the literature revision.

3. Sleep and Psoriasis

PsO seems to affect sleep quality, and this clinical finding is also confirmed by extensive epidemiological studies [23–47]. Recent studies displayed that obstructive sleep apnea (OSA) patients doubled the risk of PsO and PsA compared with heathy ones [33–35,38]. To further confirm this trend, PsO patients were found to be at higher risk of OSA than the general population using the apnea–hypopnea index (AHI) > 5 [23]. Researchers also observed an increase in the prevalence of resting leg syndrome (RLS) in PsO, with 15.1% and 18% in PsO versus 5% and 10% in the general populations in Europe and North America, respectively. Importantly, Chiu et al. [37] indicated that the concomitant presence of OSA and PsO exposed patients to a higher risk of Major Adverse Cardiovascular Events (MACE), especially ischemic heart attack and stroke.

Sleep disorders seem to be linked to PsO in a vicious circle in which one disease increases the risk of the other, with inflammation as a pathogenic basis bi-directionally linking sleep disorders and PsO [47–61]. In the PsO-to-sleep disorders directionality, there are various potential mechanisms through which PsO can act on sleep. These mechanisms operate through both direct and indirect effects. The direct effects are due to cutaneous symptoms present in PsO, such as pruritus, pain, burning sensations or, in the case of PsA, inability to move [28,39–43,62]. Pruritus in PsO often exhibits and/or increases during the evening, since it is regulated by circadian factors—for instance, lower cortisol levels and decreased epidermal barrier function—that lower the threshold for pruritus [30,40–46]. In addition, the skin plays an important role in sleep initiation, as it acts on thermoregulation and the control of the body core temperature, which normally decreases in the late evening. In contrast, these mechanisms result to be altered in PsO, where the skin diminishes the ability to dissipate heat and exhibits altered thermoregulation [52]. Instead, the indirect effects through which PsO contributes to the risk of sleep disorders may be attributable to psoriasis comorbidities that share with psoriasis a common underlying inflammatory basis [48–50,53]. In particular, diabetes and hypertension [57,58,60,61] were shown to be associated with insomnia, while cardiovascular disease, diabetes, obesity and psychiatric disorders were associated with OSA [59]. At the same time, systemic inflammation triggered by several pathological conditions such as OSA or even insomnia may trigger

PsO flares and vice versa, thus a multidisciplinary evaluation in patients with chronic inflammation should always be considered.

On the other hand, regarding psoriasis-related sleep disorders, recent studies suggested that a possible pathogenetic mechanism could be the overproduction of oxygen radicals due to the massive pro-inflammatory cytokines spill-over from the skin and other inflamed tissues [62]. Their inhibition with biologics seems to restore the oneiric dimension, by attenuating circadian rhythm dysregulation; from this perspective, also by properly treating sleep disorders, we can hypothesize that we can decrease the possibility to develop PsO, especially in patients with positive family history.

3.1. Pro-Inflammatory Cytokines Shared between Sleep Disorders and Psoriasis

Common inflammatory networks are involved in the pathomechanisms underlying both sleep disorders and psoriasis [56]. In particular, pro-inflammatory cytokines such as TNF-α and IL-6 play a role in the pathogenesis of psoriasis and are implicated in sleep regulation [63]. These cytokines can be secreted by peripheral immune cells and by astrocytes and microglia in the central nervous system in response to poor sleep quantity and quality during the previous night, and produce the effect of daytime sleepiness, commonly seen in OSA. On the other hand, in chronic insomnia, there is a shift in the timing of TNF and IL-6 secretion, eliciting a modification of the hypothalamus–pituitary–adrenal axis and the hypersecretion of cortisol, that plays a role in increased wakefulness. Therefore, it is well defined that sleep deprivation can affect the immunological integrity and nocturnal secretion of cytokines, increasing the risk of PsO. Hirotsu et al. [29] examined the influence of sleep loss in an animal model of PsO by measuring cytokine and stress-related hormone levels. Male adult Balb/C mice with or without PsO were subjected to 48 h of selective paradoxical sleep deprivation. Sleep deprivation enhanced the activities of serine proteases—kallikrein-5 and kallikrein-7—which led to desquamation in the skin of psoriatic groups. In addition, mice with PsO had significant increases in specific pro-inflammatory cytokines (IL-1β, IL-6 and IL-12) and decreases in the anti-inflammatory cytokine (IL-10) after sleep deprivation, which were normalized after 48 h of sleep rebound. Another cytokine in common between sleep disorders and PsO is vascular endothelial growth factor (VEGF). Its levels are elevated in patients with severe OSA and stimulate angiogenesis and inflammation in psoriatic skin [64,65]. Moreover, studies observed that both PsO and sleep deprivation are associated with reduced levels of adiponectin, an anti-inflammatory adipokine, or increased ghrelin and decreased leptin levels, an imbalance leading to a raised feeling of hunger and appetite that increases caloric intake and subsequently the risk of overweight or obesity conditions [66]. Obesity remains one of main comorbidities associated with both PsO and sleep disorders, suggesting a potential common pathomechanism.

3.2. Effects of Conventional and Biological Psoriasis Therapy on Sleep Disorders

Interestingly, conventional therapies may also influence sleep quality. In fact, cyclosporin improves sleep with a fast efficacy in skin lesions [67], and methotrexate is slower but also acts on joint pain in PsA patients [14]; conversely, acitretin may de-regulate the circadian rhythm, causing insomnia [68].

To date, the impact of biological treatment (or immunotherapy) on sleep outcomes in PsO patients is poorly considered in clinical research [55]. Nevertheless, Thaçi and colleagues [31] examined PsO patients with PASI greater than 10 that deserved biological therapy (etanercept) and observed them for 24 weeks, also checking sleep parameters. They concluded that by antagonizing TNF-alpha, sleep quality drastically improved in these patients, and the current results may sustain the hypothesis that systemic inflammation may trigger/elicit or even maintain sleep disorders.

Vgontzas et al. [36] further confirmed a role of TNF-alpha in the pathogenesis of OSA by treating these patients with etanercept, obtaining greater results than with continuous positive airway pressure. Interestingly, Strober et al. tested another TNF-alpha inhibitor, namely, adamilumab, for 16 weeks, finding that beside DLQI and PASI, it also improved sleep quality in PsO patients. Remarkably, no animal models are available, and the cause–effect relation between TNF-alpha inhibition and sleep remains to be further demonstrated, since DLQI and PASI improvements may condition sleep quality. Beside direct and indirect PsO-related influence on sleep quality, these data suggest the pathogenic role of inflammatory molecules in the link between PsO and sleep disorders [17,18,21,69–72]. Interestingly, no data are present in the literature about sleep quality and inhibitors of the IL-17/IL-23 pathway. Sleep quality in PsA patients is associated with the extinguishment of joint pain, CRP and disease duration, as well as, in PsO patients, cutaneous severity, duration and patient age [14]. Thus, anti-psoriatic drug efficacy is the main sleep modulator.

4. Western Diet and Psoriasis

Recently, the western diet has started to be regarded as a prominent modulator of PsO severity and even as a risk factor for its development [73–75]. This dietary pattern, which has spread with the industrial revolution and the Modern Age, is characterized by being rich in saturated fats, trans fatty acids (FAs) and n-6 FAs, refined carbohydrates and salt, and reduced intake of n-3 FAs and monounsaturated fatty acids (MUFAs), as well as antioxidants, due to the high intake of red meat, dairies and sugars, and low intake of vegetables and fruits [76,77]. Solid evidence links this diet to the development of metabolic diseases including obesity and type 2 diabetes, as well as atherosclerosis, neurodegeneration and cancer, through mechanisms involving the instigation of chronic inflammation, oxidative stress and alterations in gut microbiota (dysbiosis) [78]. Patients with PsO presented unbalanced dietary habits resembling the western diet, as testified by their dysmetabolism clinically manifesting in obesity, metabolic syndrome and dysplipidemia [79]. This dietary habit was directly associated with the increased cardiometabolic risk profile, inflammatory markers and clinical severity of PsO [79]. On the contrary, PsO patients showed lower adherence to the Mediterranean diet, a popular and effective anti-oxidant diet, which was inversely correlated with inflammatory markers and PsO severity [80]. Since none of the single food components show to exert specific effects on the pathogenesis of PsO, it is reasonable to sustain that the biological anti-psoriatic effect is exerted by the food pattern, in other words, by the diet.

However, single foods in a diet should be carefully chosen in terms of nutrient richness and quality; for example, dietary lipids are essential to maintain cutaneous homeostasis and modulate skin immune and endocrine systems [80,81]. The composition of fatty acids (FAs) in dietary lipids significantly differs among dietary patterns. In the western diet, there is a very high intake of calories derived from fried products, butter and processed meat to the disadvantage of fish, nuts, fruits and vegetables. As such, saturated fatty acid (SFA) intake is elevated through the consumption of meat, butter and palm oil, while the intake of n-3 PUFAs, such as α-linolenic acid (18:3), eicosapentaenoic acid (EPA; 20:5) and docosahexaenoic acid (DHA; 22:6), which can be found in fish and nuts, is low. This fatty acid profile contributes to the negative health outcomes associated with the western diet by increasing the risk of dyslipidemia, obesity, diabetes or cardiovascular diseases, as well as total mortality [82].

Dietary SFAs represent a major risk factor for PsO exacerbation, even independently of obesity. In PsO patients, the serum levels of free FAs were associated with disease severity [83]. Furthermore, by introducing SFAs in the diet elicited a psoriatic flare in mice, suggesting the prominent pro-inflammatory role exerted by the western diet [83].

Interestingly, n-3 PUFAs play a key anti-inflammatory role in rodents (i.e., mice and rats), as well as in humans. Interestingly, PASI inversely correlates with the serum level of n-3 PUFA, and the SFA/unsaturated FA ratio increases with the duration of the disease [84].

4.1. Effects of Western Diet and Psoriasis on Microbiota

The perturbation of gut and/or skin microbiota may trigger systemic inflammation or even a flare of a pre-existent inflammatory condition (i.e., PsO) through pathobiont colonies increase [18,71,72,85,86]. During cutaneous inflammation, antimicrobial peptide release becomes less effective, and the interaction between bacteria and immune system, more frequent, thus acting as inflammatory triggers (i.e., *Staphylococcus aureus*) [87]. PsO patients exhibited a depletion of *Corynebacterium* spp., *Lactobacillus* spp., *Burkholderis* spp. and *Propionibacterium acnes* in cutaneous microbiota, as well as *Faecalibacterium prausnitzii* and *Akkermansia muciniphila* in gut microbiota. Conversely, Firmicutes and Actinobacteria spp. proliferated in the gut mucosa, reducing the pool of microbes capable of producing short chain fatty acids (SCFAs) [88–90]. Thus, microbiota represent the living, dynamic filter of dietetic nutrients, and its unbalance may have pro-inflammatory effects. This idea received a proof of concept with the clinical study by Deng et al., in which they described the beneficial role of specific probiotics (*Bifidobacterium infantis 35,624* and *Lactobacillus pentosus GMNL-77*) in decreasing imiquimod(IMQ)-induced psoriasiform eczema and its systemic inflammation (i.e., TNF-α and IL-6) [91–93].

Several studies suggested the influence of environmental factors such as dietary composition and, in particular, the western diet on microbial community and function [94–96]. The western diet was associated with intestinal barrier disruption and gut dysbiosis with an altered profile of bacterially produced metabolites, resulting in metabolic endotoxemia, immune system deregulation and systemic inflammation [78,97].

Since the IL-17/Il-23 pathway represents a bridge between innate and adaptive immunity against microbes, its modulation by nutrients is of particular interest. With specific regards to psoriasis, in a recent study in an IL-23–mediated model of PsO and PsA, Shi et al. revealed that a short-term western diet intake exacerbated both intestinal dysbiosis as well as psoriasis-like skin and joint inflammation [18]. Assuming that diet and inflammation may influence gut dysbiosis, they proved that, by switching from the western to a normal diet or even treating with broad spectrum antibiotics, IL-23-induced skin and joint inflammation mitigated. Furthermore, fecal microbiota transplantation from western-diet-fed donors into mice pretreated with broad-spectrum antibiotics revealed that gut microbiota triggered $\gamma\delta$ T-cell infiltration into the dermis. Thus, microbiota modulation may be the key to also improve drug response in psoriatic patients.

The high-fat diet (HFD), as well as the western diet, was associated with lower microbial production of SCFAs, including butyrate, propionate and acetate, which are fermentation products of dietary fibers produced in the colon and are able to exert systemic effects, including at the skin level. Interestingly, SCFAs can be also produced by commensal bacteria in the skin [97]. A protective role of SCFAs against PsO was shown by several lines of evidence. SCFAs are able to (1) promote T-reg differentiation, activation and function; (2) inhibit the intestinal dendritic-cell production of IL-23 while inducing the expression of anti-inflammatory genes; (3) reduce skin inflammation in a mouse model of PsO, as well as downregulate IL-17 expression, and induce IL-10 and Foxp3 expression in animal and human psoriatic skin lesions [98].

4.2. How Lipids Influence Immune System Responses

4.2.1. Inflammasomes

Several studies demonstrated that macronutrients typical of the western diet, including SFAs, are able to induce and/or amplify pro-inflammatory responses involved in PsO development and progression [99–101]. Interestingly, IMQ-treated mice subjected to a HFD presented more severe clinical and histological (micro-abscesses and scaling) PsO than the ones following a regular diet. The HFD was responsible for the pathological activation of the nucleotide-binding domain, leucine-rich repeats containing family, pyrin domain-containing-3 (NLRP3) inflammasome. Inflammasomes are cytoplasmic multiprotein complexes for intracellular signaling that activate IL-1β via caspase-1, in response to different triggers (i.e., trauma or infections). Inflammasome activation was linked to the pathogenesis of metabolic and cardiovascular disease and is genetically associated with PsO [102]. Recent findings reported evidence of NLRP3 inflammasome activation in peripheral blood cells in PsO patients, in parallel with increased caspase-1 reactivity and serum levels of inflammasome-generated IL-1β and IL-18 [103]. TNF-α inhibitors are capable to turn off the inflammasome, as testified by the decreased plasmatic levels of IL-1β [103], thus suggesting a role of TNF-α-mediated NLRP3 inflammasome activation in patients with PsO and its contribution to systemic inflammation. In IMQ mice, the HFD, but not the regular diet, increased the expression of activated caspase-1 and IL-1β in the skin. The HFD is also a strong activator of the IL-17 pathway, as testified by higher levels of IL-17A in both the dermis and serum of IMQ-treated mice. Concordantly, SFAs were reported to activate inflammasomes and subsequent IL-1β release in macrophages [104], thus possibly mediating the exacerbation of psoriatic dermatitis. In addition, SFAs are powerful activators of myeloid residential cells (i.e., plasmacytoid dendritic cells) capable of further activating keratinocytes by boosting the psoriasis-related inflammation signal [83]. This suggest that dietary lipids are involved in the development and progression of psoriasis via systemic inflammation and inflammasome activation.

To prove the concept, Christ and colleagues fed LDLR$^{-}/^{-}$ mice with the western diet and then shifted to a chow diet (pelleted obesogenic diet), finding that the western diet increased systemic inflammation and promoted a pro-inflammatory imprinting on granulocyte monocyte precursor cells (GMPs); the effects were mitigated when mice abandoned the western diet in favor of the chow diet [105]. In addition, the western diet caused dysregulation in bile acids synthesis and release, further amplifying lipid dysmetabolism [105].

4.2.2. Adipokines, Cytokines and Chemokines

Adipose tissue (AT) dysfunction is currently regarded as a trigger for several inflammatory conditions, both local and systemic, such as PsO, and can be further amplified in obese patients. AT is still regarded as an endocrine organ secreting FAs and a panel of cytokines/chemokines and adipokines, such as leptin, resistin and adiponectin. In conditions of obesity or under chronic inflammation, dysfunctional adipocytes change their secretory profile toward a more pro-inflammatory state that induces the cutaneous infiltration and pro-inflammatory activation of immune cells (e.g. macrophages, neutrophils, lymphocytes). The resulting low-grade inflammation state in obesity is a central pathogenetic moment is PsO development [106]. Several studies [6,107,108] observed that the serum level of leptin is particularly elevated in obese individuals and associated with leptin resistance [109].

Leptin displays a pleiotropic function, spacing from appetite modulation to acting as a pro-inflammatory mediator capable of sustaining Th-1 differentiation and T-reg inhibition [110]. Therefore, PsO patients subjected to high-fat diets have a higher risk of obesity and to display higher levels of leptin [111].

Conversely, Lihn et al. revealed that adiponectin exerted an anti-inflammatory effect opposed to leptin capable of inactivating the psoriatic microenvironment, restoring cutaneous homeostasis [112,113]. Since the blood adiponectin levels resulted low in both obese people and PsO patients, adiponectin regulation is a potential mechanism mediating the relationship between PsO and AT [113].

Remarkably, antimicrobial peptides (AMPs) may also enter in the adipokines, since during adipogenesis, as well as injuries, they are produced and released [114]. Among AMPs, there is also cathelicidin antimicrobial peptide LL-37, capable of binding double-strained DNA filaments both exogenously and endogenously, creating the dangerous contact between self-DNA and immune system, thus starting PsO [115]. Cytokines IL-17A and IL-22 seem to play a crucial role in the development of PsO. Kanemaru et al. examined the mRNA expression of these cytokines in an obese mouse model with IMQ-induced psoriasiform dermatitis, finding five times more mRNA related to these cytokines than in the controls, suggesting a prominent AT-triggered activation of the IL-17 pathway. Additionally, it is interesting to note that food intake restriction partially decreased cytokine production in obese mice. Furthermore, a high-fat dietary pattern and IMQ treatment together stimulated the production of Reg3γ (regenerating islet-derived 3γ), an antimicrobial protein critical in psoriatic epidermal hyperplasia, since it is able to activate phosphatidylinositol 3-kinase (PI3K) in keratinocytes [116]. In particular, AT also releases palmitic acid, which is capable of influencing in vivo and in vitro Th-17 differentiation, infiltration and migration to lymph nodes [117]. Interestingly, a HFD further enforced palmitic acid release and this fact may explain the harmful effect of a HFD on PsO patients [118].

4.2.3. n-6 PUFA-Derived Prostanoids and Leukotrienes

Prostanoids and leukotrienes (LTs) are two well-known pro-inflammatory n-6 long-chain PUFAs (20:4) derived, produced and released during traumas or in response to inflammatory cytokines. They also contribute to create and maintain the inflammatory microenvironment in PsO, but their potential pathogenetic role is far from being fully elucidated [119]. Ueharaguchi and colleagues performed a targeted lipidomics on IMQ-treated mouse skin, finding that thromboxane (Tx) A2 synthase was consistently upregulated; so, lipid mediator TxA2 increased in parallel with dermatitis [120]. Moreover, they used TXA-receptor-deficient mice and treated them with IMQ, observing a significant decrease in cutaneous infiltrating $\gamma\delta$ Th17. The treatment of wild-type mice with an inhibitor for TXA2 synthase was also effective in reducing IL-17 production and disease severity. Prostaglandin (PG) E2 is another prostanoid that seems to be able to increase psoriatic dermatitis via the regulation of the IL-23/IL-17 pathway. Since Schirmer et al. demonstrated in vitro that dendritic cells may increase IL-23 production upon the fibroblastic release of PGE2 [121], Lee et al. showed in vivo and in vitro that Th-17 cells used PGE2 as an autocrine pro-proliferative mediator to maintain the pro-inflammatory microenvironment typical of the psoriasis plaque [122]. Since nonsteroidal anti-inflammatory drugs inhibit COX activity, increasing the production of LTs and inducing the progression of PsO, LTs are supposed to be additional disease-promoting factors in PsO. In particular, Sumida et al. described the importance of LTB4-BLT1 (LTB4 high affinity-receptor) signaling in a IMQ-induced PsO model. In fact, the inhibition of LTB4 synthesis or the genetic deficiency of BLT1 attenuated neutrophil infiltration in the skin and improved the symptoms of psoriasis [123]. Despite dedicated clinical trials with leukotrien inhibitors did not sort any significant anti-psoriatic effects, in vitro leukotriens instructed and increased neutrophils infiltration via LTB4–BLT1 signaling [124].

4.2.4. EPAs and DHAs (n-3 PUFAs)

Surprisingly, n-3 long-chain PUFAs may also exert a significant anti-inflammatory role in the pathogenesis of PsO, and obviously, they are regarded as anti-psoriatic drugs in line with conventional and biological therapies. For this reason, the anti-psoriatic effects of n-3 PUFAs are being studied, and the specific mechanisms by which they are involved are now to be well defined. n-3 PUFAs display a biological anti-inflammatory action directly by competing for the arachidonic acid present in the membrane and indirectly with its metabolites, which contribute globally to inactivate the inflammatory microenvironment typical of psoriasis [125]. Interestingly, Qin et al. utilized fat-1 transgenic mice that had the characteristic to convert n-6 PUFAs into n-3 PUFAs and compared them with the wild type after IMQ cream treatment. Fat-1 mice displayed less Th-17 and a lower rate of Th17/T-reg in the spleen, demonstrating the crucial role of n-3 PUFAs in modulating the Il-17 pathway [126]. Similarly, the relative systemic cytokine spill over was also limited. Allen et al. suggested that EPA contributes to increase IL-6 receptors on the CD4+ T-cell surface via STAT3 [127]. Kong et al. in their study revealed that DHA-treated dendritic cells showed a reduced ability to induce Th17 differentiation and proliferation [128]. Moreover, recent evidence found that resolvin E1 (RvE1), a DHA metabolite, inhibited Leukotriene B4 receptor 1 (BLT1) and modulated the dendritic-cell capability to present antigens and $\gamma\delta$ T-cell pro-inflammatory effects [129–131]. Likewise, resolvin D1, another DHA metabolite, reduced IMQ-induced PsO by acting on Nuclear factor kappa-light-chain-enhancer of activated B cells (NF-κB), a prominent inflammatory pathway involved in PsO initiation and maintenance [132].

5. Sleep and Dietary Patterns: An Integrated View

A bidirectional relationship does exist between diet and sleep/sleep disorders [133]. In particular, sleep disorders or mistimed sleep are linked to increased body mass index (BMI), overfeeding and irregular eating patterns, as well as unhealthy eating behaviors [134–137]. Indeed, both single nutrients and food patterns influence the gut-associated lymphoid tissue (GALT) and melatoninergic ones. These changes in hormone levels may increase appetite and thus influence dietary intake, predisposing to weight gain and obesity [138–140]. Moreover, the fatigue and excessive daytime sleepiness associated with poor quality sleep restrict the possibility of physical activity and also lead to a compensatory rise in caloric intake [141]. Interestingly, patients with sleep disorders displayed central obesity, but the insomnia/OSA group showed higher adherence to the western diet [142–145]. On the other hand, unhealthy diets such as the western diet negatively influence sleep parameters, thus leading to sleep disturbances or disorders [146,147], through potential mechanisms including body weight gain and overweight/obesity, related metabolic and vascular diseases, pro-inflammatory and pro-oxidant action, decreased synthesis of melatonin, induction of gut dysbiosis and altered pattern of bacterially produced metabolites, which can include sleep regulators such as SCFAs, GABAs, glutamate and serotonin [148]. Interestingly, the perturbation of the circadian rhythm, which modulates both sleep and appetite, with intermittent circadian fasting (Ramadan fasting) displayed surprising anti-inflammatory effects in both dermatological [149–151] and rheumatic diseases [152], included PsO [153].

Contrarily, a healthy diet (i.e., Mediterranean one) was associated with a correct sleep behavior (quality and quantity), through mechanisms going beyond the recognized benefit of weight reduction [154]. Campanini and colleagues affirmed that the adherence to the Mediterranean diet is protective in terms of both preventing and improving sleep disorders [136,137].

Nowadays, literature records seem to claim a link between sleep disorders and inflammation, so sleep may benefit anti-inflammatory therapies in patients with IMIDs. At the same time, pharmacological interventions should be provided in synergy with a healthy lifestyle (i.e., Mediterranean diet) to maximize their efficacy. In fact, the Mediterranean diet has a recognized anti-inflammatory effect related to food compounds (i.e., MUFAs and n-3 PUFAs, resveratrol and other polyphenols) and food quality/quantity. The Mediterranean diet (i.e., lemons and tomatoes) contains melatonin, which further implements the endogenously produced one and contributes to maintaining and regulating sleep. Furthermore, melatonin is currently also supplemented for sleep disorders but has never been tested as a complementary therapy in PsO. Thus, a health dietary pattern can be helpful in the management of sleep disorders, but more clinical studies are necessary to validate this hypothesis [155,156]. Sleep disorders and the western diet are bidirectionally linked and synergically promote systemic inflammation by de-programming human lipid metabolism, potentially triggering also PsO [12]. Further human studies are warranted to specifically evaluate the presence of sleep disturbances and the diet in PsO patients, and their relative and cumulative contribution to PsO development and clinical course.

6. Conclusions

Sleep disorders and an unhealthy lifestyle, such as the western dietary pattern, negatively impinge on skin homeostasis and regulate or exacerbate the development, clinical course and outcomes of PsO [157,158]. Indeed, both factors and PsO disease have in common an alteration of immune-mediated responses and chronic inflammatory conditions [159,160]. Remarkably, exposure to other factors, such as alcohol, infections, vaccines, pollution and smoking, may also influence the patient course and modulate sleep quality, so together with the diet, clinicians should advocate a change in the patient's lifestyle [161–166]. As such, it is important for dermatologists and practitioners to consider a screening for both sleep disturbances and lifestyle factors such as the diet, which are tightly intertwined, often disregarded and never considered simultaneously, in the management of PsO disease. These considerations could be useful for ameliorating the standard clinical care for PsO patients [167].

Author Contributions: Conceptualization, I.C. and G.D.; methodology, E.S., A.B. and K.K.; investigation, E.S., S.G. and G.D.; resources, S.G. and G.D.; data curation, M.D.F. and G.D.; writing—original draft preparation, I.C., E.S., A.B., M.D.F. and G.D.; writing—review and editing, I.C., E.S., A.B., A.F., K.K., M.D.F., S.G. and G.D.; visualization, A.B., K.K. and M.D.F.; supervision, E.S., S.G. and G.D.; project administration, I.C. and G.D.; funding acquisition, S.G. All authors have read and agreed to the published version of the manuscript.

Funding: This research study received no external funding.

Institutional Review Board Statement: Not applicable.

Informed Consent Statement: Not applicable.

Conflicts of Interest: The authors declare no conflict of interest.

References

1. Boehncke, W.H.; Schon, M.P. Psoriasis. *Lancet* **2015**, *386*, 983–994. [CrossRef]
2. Damiani, G.; Bragazzi, N.L.; Aksut, C.K.; Wu, D.; Alicandro, G.; McGonagle, D.; Guo, C.; Dellavalle, R.; Grada, A.; Wong, P.; et al. The Global, Regional, and National Burden of Psoriasis: Results and Insights From the Global Burden of Disease 2019 Study. *Front. Med.* **2021**, *8*, 743180. [CrossRef]
3. Wohlrab, J.; Fiedler, G.; Gerdes, S.; Nast, A.; Philipp, S.; Radtke, M.A.; Thaci, D.; Koenig, W.; Pfeiffer, A.F.H.; Härter, M.; et al. Recommendations for detection of individual risk for comorbidities in patients with psoriasis. *Arch. Dermatol. Res.* **2013**, *305*, 91–98. [CrossRef] [PubMed]
4. Boehncke, W.-H.; Brembilla, N.C. Pathogenesis-oriented therapy of psoriasis using biologics. *Expert Opin. Biol. Ther.* **2022**, 1–11. [CrossRef] [PubMed]
5. Conic, R.R.; Damiani, G.; Schrom, K.P.; Ramser, A.E.; Zheng, C.; Xu, R.; McCormick, T.S.; Cooper, K.D. Psoriasis and Psoriatic Arthritis Cardiovascular Disease Endotypes Identified by Red Blood Cell Distribution Width and Mean Platelet Volume. *J. Clin. Med.* **2020**, *9*, 186. [CrossRef] [PubMed]
6. Seth, D.; Ehlert, A.N.; Golden, J.B.; Damiani, G.; McCormick, T.S.; Cameron, M.J.; Cooper, K.D. Interaction of Resistin and Systolic Blood Pressure in Psoriasis Severity. *J. Investig. Dermatol.* **2020**, *140*, 1279–1282.e1. [CrossRef]
7. Malerba, M.; Damiani, G.; Radaeli, A.; Ragnoli, B.; Olivini, A.; Calzavara-Pinton, P.G. Narrowband ultraviolet B phototherapy in psoriasis reduces proinflammatory cytokine levels and improves vitiligo and neutrophilic asthma. *Br. J. Dermatol.* **2015**, *173*, 1544–1545. [CrossRef] [PubMed]
8. Damiani, G.; Pacifico, A.; Rizzi, M.; Santus, P.; Bridgewood, C.; Bragazzi, N.L.; Adawi, M.; Watad, A. Patients with psoriatic arthritis have higher levels of FeNO than those with only psoriasis, which may reflect a higher prevalence of a subclinical respiratory involvement. *Clin. Rheumatol.* **2020**, *39*, 2981–2988. [CrossRef] [PubMed]
9. Damiani, G.; Tacastacas, J.D.; Wuerz, T.; Miller, L.; Fastenau, P.; Bailey, C.; Chawa, M.S.; Argenas, A.; Fiore, M.; Cooper, K.D.; et al. Cognition/Psychological Burden and Resilience in Cutaneous T-Cell Lymphoma and Psoriasis Patients: Real-Life Data and Implications for the Treatment. *BioMed. Res. Int.* **2022**, *2022*, 8802469. [CrossRef] [PubMed]
10. Ożóg, M.K.; Grabarek, B.O.; Wierzbik-Strońska, M.; Świder, M. Neurological Complications of Biological Treatment of Psoriasis. *Life* **2022**, *12*, 118. [CrossRef]
11. Tang, X. The risk of organ-based comorbidities in psoriasis: A systematic review and meta-analysis. *An. Bras. Dermatol.* **2022**, *97*, 612–623. [CrossRef] [PubMed]
12. Nowowiejska, J.; Baran, A.; Flisiak, I. Mutual Relationship Between Sleep Disorders, Quality of Life and Psychosocial Aspects in Patients With Psoriasis. *Front. Psychiatry* **2021**, *12*, 674460. [CrossRef] [PubMed]
13. Madhulika, A.; Gupta, F.; Simpson, C.; Aditya, K. Gupta Psoriasis and sleep disorders: A systematic review. *Sleep Med. Rev.* **2016**, *29*, 63–75.
14. Krajewska-Włodarczyk, M.; Owczarczyk-Saczonek, A.; Placek, W. Sleep disorders in patients with psoriatic arthritis and pso-riasis. *Reumatologia* **2018**, *56*, 301–306. [CrossRef] [PubMed]
15. Kocaaga, A.; Kocaaga, M. Psoriasis: An Immunogenetic Perspective. *Glob. Med. Genet.* **2022**, *9*, 82–89. [CrossRef] [PubMed]
16. Roszkiewicz, M.; Dopytalska, K.; Szymańska, E.; Jakimiuk, A.; Walecka, I. Environmental risk factors and epigenetic alternations in psoriasis. *Ann. Agric. Environ. Med.* **2020**, *27*, 335–342. [CrossRef] [PubMed]
17. Shi, Z.; Wu, X.; Rocha, C.S.; Rolston, M.; Garcia-Melchor, E.; Huynh, M.; Nguyen, M.; Law, T.; Haas, K.N.; Yamada, D.; et al. Short-Term Western Diet Intake Promotes IL-23–Mediated Skin and Joint Inflammation Accompanied by Changes to the Gut Microbiota in Mice. *J. Investig. Dermatol.* **2021**, *141*, 1780–1791. [CrossRef] [PubMed]
18. Shi, Z.; Wu, X.; Yu, S.; Huynh, M.; Jena, P.K.; Nguyen, M.; Wan, Y.-J.Y.; Hwang, S.T. Short-Term Exposure to a Western Diet Induces Psoriasiform Dermatitis by Promoting Accumulation of IL-17A–Producing $\gamma\delta$ T Cells. *J. Investig. Dermatol.* **2020**, *140*, 1815–1823. [CrossRef]

19. Nowowiejska, J.; Baran, A.; Lewoc, M.; Grabowska, P.; Kaminski, T.W.; Flisiak, I. The Assessment of Risk and Predictors of Sleep Disorders in Patients with Psoriasis—A Questionnaire-Based Cross-Sectional Analysis. *J. Clin. Med.* **2021**, *10*, 664. [CrossRef]
20. Yeroushalmi, S.; Hakimi, M.; Chung, M.; Bartholomew, E.; Bhutani, T.; Liao, W. Psoriasis and Exercise: A Review. *Psoriasis Targets Ther.* **2022**, *12*, 189–197. [CrossRef]
21. Damiani, G.; Bragazzi, N.; McCormick, T.S.; Pigatto, P.D.M.; Leone, S.; Pacifico, A.; Tiodorovic, D.; Di Franco, S.; Alfieri, A.; Fiore, M. Gut microbiota and nutrient interactions with skin in psoriasis: A comprehensive review of animal and human studies. *World J. Clin. Cases* **2020**, *8*, 1002–1012. [CrossRef] [PubMed]
22. Drago, F.; Ciccarese, G.; Indemini, E.; Savarino, V.; Parodi, A. Psoriasis and small intestine bacterial overgrowth. *Int. J. Dermatol.* **2018**, *57*, 112–113. [CrossRef] [PubMed]
23. Hawro, T.; Hawro, M.; Zalewska-Janowska, A.; Weller, K.; Metz, M.; Maurer, M. Pruritus and sleep disturbances in patients with psoriasis. *Arch. Dermatol. Res.* **2020**, *312*, 103–111. [CrossRef] [PubMed]
24. Mostaghimi, L.; Hetzel, S. Insomnia and other sleep complaints in inflammatory versus noninflammatory skin disorders: An observational case-control study. *Int. J. Dermatol.* **2019**, *58*, 976–981. [CrossRef] [PubMed]
25. Kabeloglu Ilbay, V.; Tas, B.; Altuntas, M.; Atakli, H.D.; Soysal, A. Risk of Obstructive Sleep Apnea Syndrome in Psoriasis Patients. *Arch. Iran Med.* **2019**, *22*, 137–143.
26. Saçmacı, H.; Gürel, G. Sleep disorders in patients with psoriasis: A cross-sectional study using non-polysomnographical methods. *Sleep Breath.* **2019**, *23*, 893–898. [CrossRef]
27. Jensen, P.; Zachariae, C.; Skov, L.; Zachariae, R. Sleep disturbance in psoriasis: A case-controlled study. *Br. J. Dermatol.* **2018**, *179*, 1376–1384. [CrossRef]
28. Gowda, S.; Goldblum, O.M.; McCall, W.V.; Feldman, S.R. Factors affecting sleep quality in patients with psoriasis. *J. Am. Acad. Dermatol.* **2010**, *63*, 114–123. [CrossRef]
29. Hirotsu, C.; Rydlewski, M.; Araújo, M.S.; Tufik, S.; Andersen, M.L. Sleep Loss and Cytokines Levels in an Experimental Model of Psoriasis. *PLoS ONE* **2012**, *7*, e51183. [CrossRef]
30. Henry, A.L.; Chisholm, A.; Carter, L.A.; Bundy, C.; Griffiths, C.E.; Kyle, S.D. The relationship between sleep disturbance, symptoms and daytime functioning in psoriasis: A prospective study integrating actigraphy and experience sampling methodology. *Sleep Med.* **2020**, *72*, 144–149. [CrossRef]
31. Thaçi, D.; Galimberti, R.; Amaya-Guerra, M.; Rosenbach, T.; Robertson, D.; Pedersen, R.; Yang, S.; Kuligowski, M.; Boggs, R. Improvement in aspects of sleep with etanercept and optional adjunctive topical therapy in patients with moderate-to-severe psoriasis: Results from the PRISTINE trial. *J. Eur. Acad. Dermatol. Venereol.* **2014**, *28*, 900–906. [CrossRef] [PubMed]
32. Strober, B.E.; Sobell, J.M.; Duffin, K.C.; Bao, Y.; Guérin, A.; Yang, H.; Goldblum, O.; Okun, M.M.; Mulani, P.M. Sleep quality and other patient-reported outcomes improve after patients with psoriasis with suboptimal response to other systemic therapies are switched to adalimumab: Results from PROGRESS, an open-label Phase IIIB trial. *Br. J. Dermatol.* **2012**, *167*, 1374–1381. [CrossRef] [PubMed]
33. Egeberg, A.; Khalid, U.; Gislason, G.; Mallbris, L.; Skov, L.; Hansen, P.R. Psoriasis and Sleep Apnea: A Dan.ish Nationwide Cohort Study. *J. Clin. Sleep Med.* **2016**, *12*, 663–671. [CrossRef] [PubMed]
34. Yang, Y.-W.; Kang, J.-H.; Lin, H.-C. Increased risk of psoriasis following obstructive sleep apnea: A longitudinal population-based study. *Sleep Med.* **2012**, *13*, 285–289. [CrossRef]
35. Cohen, J.M.; Jackson, C.; Li, T.Y.; Wu, S.; Qureshi, A.A. Sleep disordered breathing and the risk of psoriasis among US women. *Arch. Dermatol. Res.* **2015**, *307*, 433–438. [CrossRef] [PubMed]
36. Vgontzas, A.N.; Zoumakis, E.; Lin, H.-M.; Bixler, E.O.; Trakada, G.; Chrousos, G.P. Marked Decrease in Sleepiness in Patients with Sleep Apnea by Etanercept, a Tumor Necrosis Factor-? Antagonist. *J. Clin. Endocrinol. Metab.* **2004**, *89*, 4409–4413. [CrossRef] [PubMed]
37. Chiu, H.-Y.; Hsieh, C.-F.; Chiang, Y.-T.; Tsai, Y.-W.; Huang, W.-F.; Li, C.-Y.; Wang, T.-S.; Tsai, T.-F. Concomitant Sleep Disorders Significantly Increase the Risk of Cardiovascular Disease in Patients with Psoriasis. *PLoS ONE* **2016**, *11*, e0146462. [CrossRef]
38. Karaca, S.; Fidan, F.; Erkan, F.; Nural, S.; Pinarcı, T.; Gunay, E.; Unlu, M. Might psoriasis be a risk factor for obstructive sleep apnea syndrome? *Sleep Breath* **2013**, *17*, 275–280. [CrossRef]
39. Shalom, G.; Dreiher, J.; Cohen, A. Psoriasis and obstructive sleep apnea. *Int. J. Dermatol.* **2016**, *55*, e579–e584. [CrossRef]
40. Haugeberg, G.; Hoff, M.; Kavanaugh, A.; Michelsen, B. Psoriatic arthritis: Exploring the occurrence of sleep disturbances, fatigue, and depression and their correlates. *Arthritis Res. Ther.* **2020**, *22*, 1–10. [CrossRef]
41. Luca, M.; Luca, A.; Musumeci, M.L.; Fiorentini, F.; Micali, G.; Calandra, C. Psychopathological Variables and Sleep Quality in Psoriatic Patients. *Int. J. Mol. Sci.* **2016**, *17*, 1184. [CrossRef] [PubMed]
42. Henry, A.L.; Kyle, S.D.; Chisholm, A.; Griffiths, C.E.M.; Bundy, C. A cross-sectional survey of the nature and correlates of sleep dis-turbance in people with psoriasis. *Br. J. Dermatol.* **2017**, *177*, 1052–1059. [CrossRef] [PubMed]
43. Kaaz, K.; Szepietowski, J.C.; Matusiak, Ł. Influence of Itch and Pain on Sleep Quality in Atopic Dermatitis and Psoriasis. *Acta Derm. Venereol.* **2019**, *99*, 175–180. [CrossRef] [PubMed]
44. Gupta, M.A.; Simpson, F.C.; Vujcic, B.; Gupta, A.K. Obstructive sleep apnea and dermatologic disorders. *Clin. Dermatol.* **2017**, *35*, 319–327. [CrossRef]
45. Walia, H.K.; Mehra, R. Overview of Common Sleep Disorders and Intersection with Dermatologic Conditions. *Int. J. Mol. Sci.* **2016**, *17*, 654. [CrossRef]

46. Papadavid, E.; Dalamaga, M.; Vlami, K.; Koumaki, D.; Gyftopoulos, S.; Christodoulatos, G.S.; Papiris, S.; Rigopoulos, D. Psoriasis is associated with risk of obstructive sleep apnea independently from metabolic parameters and other comorbidities: A large hospital-based case-control study. *Sleep Breath.* **2017**, *21*, 949–958. [CrossRef]
47. Wong, I.T.Y.; Chandran, V.; Li, S.; Gladman, D.D. Sleep Disturbance in Psoriatic Disease: Prevalence and Associated Factors. *J. Rheumatol.* **2017**, *44*, 1369–1374. [CrossRef]
48. Melikoglu, M. Sleep Quality and its Association with Disease Severity in Psoriasis. *Eurasian J. Med.* **2017**, *49*, 124–127. [CrossRef]
49. Gezer, O.; Batmaz, I.; Sariyildiz, M.A.; Sula, B.; Ucmak, D.; Bozkurt, M.; Nas, K. Sleep quality in patients with psoriatic arthritis. *Int. J. Rheum. Dis.* **2017**, *20*, 1212–1218. [CrossRef]
50. Tas, B.; Kabeloglu, V.; Soysal, A.; Atakli, D. Sleep Quality in Psoriasis Patients and Its Relations with Possible Affecting Factors. *Med. Bull. Sisli Hosp.* **2020**, *54*, 181–187. [CrossRef]
51. Hirotsu, C.; Nogueira, H.; Albuquerque, R.G.; Tomimori, J.; Tufik, S.; Andersen, M.L. The bidirectional interactions between psoriasis and obstructive sleep apnea. *Int. J. Dermatol.* **2015**, *54*, 1352–1358. [CrossRef] [PubMed]
52. Henry, A.; Bundy, C.; Kyle, S.; Griffiths, C.E.M.; Chisholm, A. Understanding the experience of sleep disturbance in psoriasis: A qualitative exploration using the Common-Sense Model of Self-Regulation. *Br. J. Dermatol.* **2019**, *180*, 1397–1404. [CrossRef] [PubMed]
53. Papadavid, E.; Vlami, K.; Dalamaga, M.; Giatrakou, S.; Theodoropoulos, K.; Gyftopoulos, S.; Stavrianeas, N.; Papiris, S.; Rigopoulos, D. Sleep apnea as a comorbidity in obese psoriasis patients: A cross-sectional study. Do psoriasis characteristics and metabolic parameters play a role? *J. Eur. Acad. Dermatol. Venereol.* **2013**, *27*, 820–826. [CrossRef] [PubMed]
54. Ger, T.-Y.; Fu, Y.; Chi, C.-C. Bidirectional Association Between Psoriasis and Obstructive Sleep Apnea: A Systematic Review and Meta-Analysis. *Sci. Rep.* **2020**, *10*, 5931. [CrossRef] [PubMed]
55. Maari, C.; Bolduc, C.; Nigen, S.; Marchessault, P.; Bissonnette, R. Effect of adalimumab on sleep parameters in patients with pso-riasis and obstructive sleep apnea: A randomized controlled trial. *J. Dermatol. Treat.* **2014**, *25*, 57–60. [CrossRef] [PubMed]
56. Selmi, C.; Montano, N.; Furlan, R.; Keen, C.L.; Gershwin, M.E. Inflammation and Oxidative Stress in Obstructive Sleep Apnea Syndrome. *Exp. Biol. Med.* **2007**, *232*, 1409–1413. [CrossRef] [PubMed]
57. E Lewis, P.; Emasealu, O.V.; Rohrbeck, P.; Hu, Z. Risk of type II diabetes and hypertension associated with chronic insomnia among active component, U.S. Armed Forces, 1998–2013. *MSMR* **2014**, *21*, 6–13.
58. Loke, Y.K.; Brown, J.W.; Kwok, C.S.; Niruban, A.; Myint, P.K. Association of obstructive sleep apnea with risk of serious cardio-vascular events: A systematic review and meta-analysis. *Circ. Cardiovasc. Qual. Outcomes* **2012**, *5*, 720–728. [CrossRef] [PubMed]
59. Gupta, M.A.; Simpson, F.C. Obstructive Sleep Apnea and Psychiatric Disorders: A Systematic Review. *J. Clin. Sleep Med.* **2015**, *11*, 165–175. [CrossRef] [PubMed]
60. Drager, L.F.; Togeiro, S.M.; Polotsky, V.Y.; Lorenzi-Filho, G. Obstructive sleep apnea: A cardiometabolic risk in obesity and the metabolic syndrome. *J. Am. Coll. Cardiol.* **2013**, *62*, 569–576. [CrossRef] [PubMed]
61. Jean-Louis, G.; Zizi, F.; Clark, L.T.; Brown, C.D.; McFarlane, S.I. Obstructive Sleep Apnea and Cardiovascular Disease: Role of the Metabolic Syndrome and Its Components. *J. Clin. Sleep Med.* **2008**, *4*, 261–272. [CrossRef] [PubMed]
62. Besedovsky, L.; Lange, T.; Haack, M. The Sleep-Immune Crosstalk in Health and Disease. *Physiol. Rev.* **2019**, *99*, 1325–1380. [CrossRef] [PubMed]
63. Garbarino, S.; Lanteri, P.; Bragazzi, N.L.; Magnavita, N.; Scoditti, E. Role of sleep deprivation in immune-related disease risk and outcomes. *Commun. Biol.* **2021**, *4*, 1304. [CrossRef] [PubMed]
64. Imagawa, S.; Yamaguchi, Y.; Higuchi, M.; Neichi, T.; Hasegawa, Y.; Mukai, H.Y.; Suzuki, N.; Yamamoto, M.; Nagasawa, T. Levels of vascular endothelial growth factor are elevated in patients with obstructive sleep apnea–hypopnea syndrome. *Blood* **2001**, *98*, 1255–1257. [CrossRef] [PubMed]
65. Rosenberger, C.; Solovan, C.; Rosenberger, A.D.; Jinping, L.; Treudler, R.; Frei, U.; Eckardt, K.-U.; Brown, L.F. Upregulation of Hypoxia-Inducible Factors in Normal and Psoriatic Skin. *J. Investig. Dermatol.* **2007**, *127*, 2445–2452. [CrossRef]
66. Baran, A.; Flisiak, I.; My´sliwiec, H.; Chodynicka, B. Role of adiponectin in psoriasis. *Dermatol. Rev.* **2010**, *97*, 413–416.
67. Yasuda, K.-I.; Ishiuji, Y.; Endo, T.; Tanito, K.; Kamide, R.; Nobeyama, Y.; Asahina, A. Cyclosporine Improves Sleep Quality in Patients with Atopic Dermatitis. *Dermatol. Ther.* **2020**, *10*, 1359–1369. [CrossRef] [PubMed]
68. Ransom, J.; Morgan, P.J.; McCaffery, P.J.; Stoney, P.N. The rhythm of retinoids in the brain. *J. Neurochem.* **2014**, *129*, 366–376. [CrossRef]
69. Smolensky, M.H.; Portaluppi, F.; Manfredini, R.; Hermida, R.C.; Tiseo, R.; Sackett-Lundeen, L.L.; Haus, E.L. Diurnal and twenty-four hour patterning of human diseases: Acute and chronic common and uncommon medical conditions. *Sleep Med. Rev.* **2015**, *21*, 12–22. [CrossRef] [PubMed]
70. Paolino, G.; Buratta, S.; Mercuri, S.R.; Pellegrino, R.M.; Urbanelli, L.; Emiliani, C.; Bertuccini, L.; Iosi, F.; Huber, V.; Brianti, P.; et al. Lipidic Profile Changes in Exosomes and Microvesicles Derived From Plasma of Monoclonal Antibody-Treated Psoriatic Patients. *Front Cell Dev Biol.* **2022**, *10*, 923769. [CrossRef] [PubMed]
71. Damiani, G.; Franchi, C.; Pigatto, P.; Altomare, A.; Pacifico, A.; Petrou, S.; Leone, S.; Pace, M.C.; Fiore, M. Outcomes assessment of hepatitis C virus-positive psoriatic patients treated using pegylated interferon in combination with ribavirin compared to new Direct-Acting Antiviral agents. *World J. Hepatol.* **2018**, *10*, 329–336. [CrossRef]

72. Gisondi, P.; Cazzaniga, S.; Chimenti, S.; Giannetti, A.; Maccarone, M.; Picardo, M.; Girolomoni, G.; Naldi, L.; Psocare Study Group. Metabolic abnormalities associated with initiation of systemic treatment for psoriasis: Evidence from the Italian Psocare Registry. *J. Eur. Acad. Dermatol. Venereol.* **2013**, *27*, e30–e41. [CrossRef] [PubMed]
73. Damiani, G.; Cazzaniga, S.; Conic, R.R.; Naldi, L. Psocare Registry Network. Pruritus Characteristics in a Large Italian Cohort of Psoriatic Patients. *J. Eur. Acad. Dermatol. Venereol.* **2019**, *33*, 1316–1324. [CrossRef] [PubMed]
74. Herbert, D.; Franz, S.; Popkova, Y.; Anderegg, U.; Schiller, J.; Schwede, K.; Lorz, A.; Simon, J.C.; Saalbach, A. High-Fat Diet Exacerbates Early Psoriatic Skin Inflammation Independent of Obesity: Saturated Fatty Acids as Key Players. *J. Investig. Dermatol.* **2018**, *138*, 1999–2009. [CrossRef] [PubMed]
75. Honda, T.; Kabashima, K. Current understanding of the role of dietary lipids in the pathophysiology of psoriasis. *J. Dermatol. Sci.* **2019**, *94*, 314–320. [CrossRef]
76. Carrera-Bastos, P.; Fontes-Villalba, M.; O'Keefe, J.H.; Lindeberg, S.; Cordain, L. The western diet and lifestyle and diseases of civilization. *Res. Rep. Clin. Cardiol.* **2011**, *2*, 15–35. [CrossRef]
77. Cordain, L.; Eaton, S.B.; Sebastian, A.; Mann, N.; Lindeberg, S.; Watkins, B.A.; O'Keefe, J.H.; Brand-Miller, J. Origins and evolution of the Western diet: Health implications for the 21st century. *Am. J. Clin. Nutr.* **2005**, *81*, 341–354. [CrossRef] [PubMed]
78. García-Montero, C.; Fraile-Martínez, O.; Gómez-Lahoz, A.M.; Pekarek, L.; Castellanos, A.J.; Noguerales-Fraguas, F.; Coca, S.; Guijarro, L.G.; García-Honduvilla, N.; Asúnsolo, A.; et al. Nutritional Components in Western Diet Versus Mediterranean Diet at the Gut Microbiota-Immune System Interplay. Implications for Health and Disease. *Nutrients* **2021**, *13*, 699. [CrossRef]
79. Barrea, L.; Macchia, P.E.; Tarantino, G.; Di Somma, C.; Pane, E.; Balato, N.; Napolitano, M.; Colao, A.; Savastano, S. Nutrition: A key environmental dietary factor in clinical severity and cardio-metabolic risk in psoriatic male patients evaluated by 7-day food-frequency questionnaire. *J. Transl. Med.* **2015**, *13*, 303. [CrossRef] [PubMed]
80. Barrea, L.; Balato, N.; Di Somma, C.; Macchia, P.E.; Napolitano, M.; Savanelli, M.C.; Esposito, K.; Colao, A.; Savastano, S. Nutrition and psoriasis: Is there any association between the severity of the disease and adherence to the Mediterranean diet? *J. Transl. Med.* **2015**, *13*, 18. [CrossRef] [PubMed]
81. Kocic, H.; Damiani, G.; Stamenkovic, B.; Tirant, M.; Jovic, A.; Tiodorovic, D.; Peris, K. Dietary compounds as potential modulators of microRNA expression in psoriasis. *Ther. Adv. Chronic Dis.* **2019**, *10*, 2040622319864805. [CrossRef] [PubMed]
82. Billingsley, H.E.; Carbone, S.; Lavie, C.J. Dietary Fats and Chronic Noncommunicable Diseases. *Nutrients* **2018**, *10*, 1385. [CrossRef]
83. Yazdanpanah, M.J.; Vahabi-Amlashi, S.; Nematy, M.; Shaelaei, N.; Mohajeri, S.A.R.; Tafazzoli, Z. Association of serum lipid profiles and dietary intakes of vitamin E and fiber with psoriasis severity. *Caspian J. Intern. Med.* **2021**, *12*, 606–612.
84. Myśliwiec, H.; Baran, A.; Harasim-Symbor, E.; Myśliwiec, P.; Milewska, A.J.; Chabowski, A.; Flisiak, I. Serum fatty acid profile in psoriasis and its comorbidity. *Arch. Dermatol. Res.* **2017**, *309*, 371–380. [CrossRef] [PubMed]
85. Zákostelská, Z.; Málková, J.; Klimešová, K.; Rossmann, P.; Hornová, M.; Novosádová, I.; Stehlíková, Z.; Kostovčík, M.; Hudcovic, T.; Štěpánková, R.; et al. Intestinal microbiota promotes psoriasis-like skin inflammation by enhancing Th17 response. *PLoS ONE* **2016**, *11*, e0159539. [CrossRef]
86. Stehlikova, Z.; Kostovcikova, K.; Kverka, M.; Rossmann, P.; Dvorak, J.; Novosadova, I.; Kostovcik, M.; Coufal, S.; Srutkova, D.; Prochazkova, P.; et al. Crucial Role of Microbiota in Experimental Psoriasis Revealed by a Gnotobiotic Mouse Model. *Front. Microbiol.* **2019**, *10*, 236. [CrossRef]
87. Kabashima, K.; Honda, T.; Ginhoux, F.; Egawa, G. The immunological anatomy of the skin. *Nat. Rev. Immunol.* **2017**, *19*, 19–30. [CrossRef]
88. Fyhrquist, N.; Muirhead, G.; Prast-Nielsen, S.; Jeanmougin, M.; Olah, P.; Skoog, T.; Jules-Clement, G.; Feld, M.; Barrientos-Somarribas, M.; Sinkko, H.; et al. Microbe-host interplay in atopic dermatitis and psoriasis. *Nat. Commun.* **2019**, *10*, 4703. [CrossRef] [PubMed]
89. Hidalgo-Cantabrana, C.; Gómez, J.; Delgado, S.; Requena-López, S.; Queiro-Silva, R.; Margolles, A.; Coto, E.; Sánchez, B.; Coto-Segura, P. Gut microbiota dysbiosis in a cohort of patients with psoriasis. *Br. J. Dermatol.* **2019**, *181*, 1287–1295. [CrossRef]
90. Chen, Y.; Ho, H.J.; Tseng, C.; Lai, Z.; Shieh, J.-J.; Wu, C.-Y. Intestinal microbiota profiling and predicted metabolic dysregulation in psoriasis patients. *Exp. Dermatol.* **2018**, *27*, 1336–1343. [CrossRef] [PubMed]
91. Deng, Y.; Fang, Z.; Cui, S.; Zhao, J.; Zhang, H.; Chen, W.; Lu, W. Evaluation of probiotics for inhibiting hyperproliferation and in-flammation relevant to psoriasis in vitro. *J. Funct. Foods* **2021**, *81*, 104433. [CrossRef]
92. Chen, Y.-H.; Wu, C.-S.; Chao, Y.-H.; Lin, C.-C.; Tsai, H.-Y.; Li, Y.-R.; Chen, Y.Z.; Tsai, W.-H.; Chen, Y.-K. Lactobacillus pentosus GMNL-77 inhibits skin lesions in imiquimod-induced psoriasis-like mice. *J. Food Drug Anal.* **2017**, *25*, 559–566. [CrossRef] [PubMed]
93. Groeger, D.; O'Mahony, L.; Murphy, E.F.; Bourke, J.F.; Dinan, T.G.; Kiely, B.; Shanahan, F.; Quigley, E.M. Bifidobacterium infantis35624 modulates host inflammatory processes beyond the gut. *Gut Microbes* **2013**, *4*, 325–339. [CrossRef] [PubMed]
94. Hsu, D.K.; Fung, M.A.; Chen, H.-L. Role of skin and gut microbiota in the pathogenesis of psoriasis, an inflammatory skin disease. *Med. Microecol.* **2020**, *4*, 100016. [CrossRef]
95. Wang, X.; Li, Y.; Wu, L.; Xiao, S.; Ji, Y.; Tan, Y.; Jiang, C.; Zhang, G. Dysregulation of the gut-brain-skin axis and key overlapping inflammatory and immune mechanisms of psoriasis and depression. *Biomed. Pharmacother.* **2021**, *137*, 111065. [CrossRef]
96. Martinez, K.B.; Leone, V.; Chang, E.B. Western diets, gut dysbiosis, and metabolic diseases: Are they linked? *Gut Microbes* **2017**, *8*, 130–142. [CrossRef]

97. Malesza, I.J.; Malesza, M.; Walkowiak, J.; Mussin, N.; Walkowiak, D.; Aringazina, R.; Bartkowiak-Wieczorek, J.; Mądry, E. High-Fat, Western-Style Diet, Systemic Inflammation, and Gut Microbiota: A Narrative Review. *Cells* **2021**, *10*, 3164. [CrossRef]
98. Kanda, N.; Hoashi, T.; Saeki, H. Nutrition and Psoriasis. *Int. J. Mol. Sci.* **2020**, *21*, 5405. [CrossRef]
99. Endo, Y.; Asou, H.K.; Matsugae, N.; Hirahara, K.; Shinoda, K.; Tumes, D.J.; Tokuyama, H.; Yokote, K.; Nakayama, T. Obesity Drives Th17 Cell Differentiation by Inducing the Lipid Metabolic Kinase, ACC1. *Cell Rep.* **2015**, *12*, 1042–1055. [CrossRef]
100. Stelzner, K.; Herbert, D.; Popkova, Y.; Lorz, A.; Schiller, J.; Gericke, M.; Klöting, N.; Blüher, M.; Franz, S.; Simon, J.C.; et al. Free fatty acids sensitize dendritic cells to amplify TH1/TH17-immune responses. *Eur. J. Immunol.* **2016**, *46*, 2043–2053. [CrossRef]
101. Shimoura, N.; Nagai, H.; Fujiwara, S.; Jimbo, H.; Nishigori, C. Exacerbation and Prolongation of Psoriasiform Inflammation in Diabetic Obese Mice: A Synergistic Role of CXCL5 and Endoplasmic Reticulum Stress. *J. Investig. Dermatol.* **2018**, *138*, 854–863. [CrossRef] [PubMed]
102. Olsen, M.B.; Gregersen, I.; Sandanger, Ø.; Yang, K.; Sokolova, M.; Halvorsen, B.E.; Gullestad, L.; Broch, K.; Aukrust, P.; Louwe, M.C. Targeting the Inflammasome in Cardiovascular Disease. *JACC Basic Transl. Sci.* **2021**, *7*, 84–98. [CrossRef] [PubMed]
103. Verma, D.; Fekri, S.Z.; Sigurdardottir, G.; Eding, C.B.; Sandin, C.; Enerbäck, C. Enhanced Inflammasome Activity in Patients with Psoriasis Promotes Systemic Inflammation. *J. Investig. Dermatol.* **2021**, *141*, 586–595.e5. [CrossRef] [PubMed]
104. Karasawa, T.; Kawashima, A.; Usui-Kawanishi, F.; Watanabe, S.; Kimura, H.; Kamata, R.; Shirasuna, K.; Koyama, Y.; Sato-Tomita, A.; Matsuzaka, T.; et al. Saturated Fatty Acids Undergo Intracellular Crystallization and Activate the NLRP3 Inflammasome in Macrophages. *Arter. Thromb. Vasc. Biol.* **2018**, *38*, 744–756. [CrossRef]
105. Christ, A.; Günther, P.; Lauterbach, M.A.; Duewell, P.; Biswas, D.; Pelka, K.; Scholz, C.J.; Oosting, M.; Haendler, K.; Baßler, K.; et al. Western Diet Triggers NLRP3-Dependent Innate Immune Reprogramming. *Cell* **2018**, *172*, 162–175.e14. [CrossRef]
106. Setty, A.R.; Curhan, G.; Choi, H.K. Obesity, waist circumference, weight change, and the risk of psoriasis in women: Nurses' Health Study II. *Arch. Intern. Med.* **2007**, *167*, 1670–1675. [CrossRef]
107. Higashi, Y.; Yamakuchi, M.; Fukushige, T.; Ibusuki, A.; Hashiguchi, T.; Kanekura, T. High-fat diet exacerbates imiquimod-induced psoriasis-like dermatitis in mice. *Exp. Dermatol.* **2018**, *27*, 178–184. [CrossRef]
108. Guo, Z.; Yang, Y.; Liao, Y.; Shi, Y.; Zhang, L.J. Emerging Roles of Adipose Tissue in the Pathogenesis of Psoriasis and Atopic Der-matitis in Obesity. *JID Innov.* **2021**, *2*, 100064. [CrossRef]
109. Kyriakou, A.; Patsatsi, A.; Sotiriadis, D.; Goulis, D.G. Serum Leptin, Resistin, and Adiponectin Concentrations in Psoriasis: A Meta-Analysis of Observational Studies. *Dermatology* **2017**, *233*, 378–389. [CrossRef]
110. Dopytalska, K.; Baranowska-Bik, A.; Roszkiewicz, M.; Bik, W.; Walecka, I. The role of leptin in selected skin diseases. *Lipids Health Dis.* **2020**, *19*, 215. [CrossRef]
111. Zhu, K.-J.; Zhang, C.; Li, M.; Zhu, C.-Y.; Shi, G.; Fan, Y.-M. Leptin levels in patients with psoriasis: A meta-analysis. *Clin. Exp. Dermatol.* **2013**, *38*, 478–483. [CrossRef] [PubMed]
112. Shibata, S.; Tada, Y.; Hau, C.S.; Mitsui, A.; Kamata, M.; Asano, Y.; Sugaya, M.; Kadono, T.; Masamoto, Y.; Kurokawa, M.; et al. Adiponectin regulates psoriasiform skin inflammation by suppressing IL-17 production from γδ-T cells. *Nat. Commun.* **2015**, *6*, 7687. [CrossRef] [PubMed]
113. Lihn, A.S.; Bruun, J.M.; He, G.; Pedersen, S.B.; Jensen, P.F.; Richelsen, B. Lower expression of adiponectin mRNA in visceral adipose tissue in lean and obese subjects. *Mol. Cell. Endocrinol.* **2004**, *219*, 9–15. [CrossRef] [PubMed]
114. Zhang, L.J.; Gallo, R.L. Antimicrobial peptides. *Curr. Biol.* **2016**, *26*, R14–R19. [CrossRef]
115. Ganguly, D.; Chamilos, G.; Lande, R.; Gregorio, J.; Meller, S.; Facchinetti, V.; Homey, B.; Barrat, F.J.; Zal, T.; Gilliet, M. Self-RNA–antimicrobial peptide complexes activate human dendritic cells through TLR7 and TLR8. *J. Exp. Med.* **2009**, *206*, 1983–1994. [CrossRef]
116. Kanemaru, K.; Matsuyuki, A.; Nakamura, Y.; Fukami, K. Obesity exacerbates imiquimod-induced psoriasis-like epidermal hy-perplasia and interleukin-17 and interleukin-22 production in mice. *Exp. Dermatol.* **2015**, *24*, 436–442. [CrossRef]
117. Vasseur, P.; Serres, L.; Jégou, J.-F.; Pohin, M.; Delwail, A.; Petit-Paris, I.; Levillain, P.; Favot, L.; Samson, M.; Yssel, H.; et al. High-Fat Diet–Induced IL-17A Exacerbates Psoriasiform Dermatitis in a Mouse Model of Steatohepatitis. *Am. J. Pathol.* **2016**, *186*, 2292–2301. [CrossRef]
118. Nakamizo, S.; Honda, T.; Adachi, A.; Nagatake, T.; Kunisawa, J.; Kitoh, A.; Otsuka, A.; Dainichi, T.; Nomura, T.; Ginhoux, F.; et al. High fat diet exacerbates murine psoriatic dermatitis by increasing the number of IL-17-producing γδ T cells. *Sci. Rep.* **2017**, *7*, 14076. [CrossRef]
119. Honda, T.; Kabashima, K. Prostanoids and leukotrienes in the pathophysiology of atopic dermatitis and psoriasis. *Int. Immunol.* **2019**, *31*, 589–595. [CrossRef]
120. Ueharaguchi, Y.; Honda, T.; Kusuba, N.; Hanakawa, S.; Adachi, A.; Sawada, Y.; Otsuka, A.; Kitoh, A.; Dainichi, T.; Egawa, G.; et al. Thromboxane A2 facilitates IL-17A pro-duction from Vγ4+ γδ T cells and promotes psoriatic dermatitis in mice. *J. Allergy Clin. Immunol.* **2018**, *142*, 680–683.e2. [CrossRef]
121. Schirmer, C.; Klein, C.; von Bergen, M.; Simon, J.C.; Saalbach, A. Human fibroblasts support the expansion of IL-17–producing T cells via up-regulation of IL-23 production by dendritic cells. *Blood* **2010**, *116*, 1715–1725. [CrossRef]
122. Lee, J.; Aoki, T.; Thumkeo, D.; Siriwach, R.; Yao, C. T cell-intrinsic prostaglandin E2-EP2/EP4 signaling is critical in pathogenic TH17 cell-driven inflammation. *J. Allergy Clin. Immunol.* **2019**, *143*, 631–643. [CrossRef]

123. Sumida, H.; Yanagida, K.; Kita, Y.; Abe, J.; Matsushima, K.; Nakamura, M.; Ishii, S.; Sato, S.; Shimizu, T. Interplay between CXCR2 and BLT1 Facilitates Neutrophil Infiltration and Resultant Keratinocyte Activation in a Murine Model of Imiquimod-Induced Psoriasis. *J. Immunol.* **2014**, *192*, 4361–4369. [CrossRef] [PubMed]
124. Black, A.K.; Camp, R.D.R.; Mallet, A.I.; Cunningham, F.M.; Hofbauer, M.; Greaves, M.W.; Derm, F.F. Pharmacologic and Clinical Effects of Lonapalene (RS 43179), a 5-Lipoxygenase Inhibitor, in Psoriasis. *J. Investig. Dermatol.* **1990**, *95*, 50–54. [CrossRef] [PubMed]
125. Mari, N.L.; Simão, A.N.C.; Dichi, I. n-3 polyunsaturated fatty acids supplementation in psoriasis: A review. *Nutrire* **2017**, *42*, 5. [CrossRef]
126. Qin, S.; Wen, J.; Bai, X.C.; Chen, T.Y.; Zheng, R.C.; Zhou, G.B.; Ma, J.; Feng, J.Y.; Zhong, B.L.; Li, Y.M. Endogenous n-3 polyunsaturated fatty acids protect against imiquimod-induced psoriasis-like inflammation via the IL-17/IL-23 axis. *Mol. Med. Rep.* **2014**, *9*, 2097–2104. [CrossRef]
127. Allen, M.J.; Fan, Y.Y.; Monk, J.M.; Hou, T.Y.; Barhoumi, R.; McMurray, D.N.; Chapkin, R.S. n-3 PUFAs reduce T-helper 17 cell differen-tiation by decreasing responsiveness to interleukin-6 in isolated mouse splenic $CD4^+$ T cells. *J. Nutr.* **2014**, *144*, 1306–1313. [CrossRef] [PubMed]
128. Kong, W.; Yen, J.H.; Ganea, D. Docosahexaenoic acid prevents dendritic cell maturation, inhibits antigen-specific Th1/Th17 differentiation and suppresses experimental autoimmune encephalomyelitis. *Brain Behav. Immun.* **2011**, *25*, 872–882. [CrossRef] [PubMed]
129. Seki, H.; Tani, Y.; Arita, M. Omega-3 PUFA derived anti-inflammatory lipid mediator resolvin E1. *Prostaglandins Other Lipid Mediat.* **2009**, *89*, 126–130. [CrossRef]
130. Sawada, Y.; Honda, T.; Hanakawa, S.; Nakamizo, S.; Murata, T.; Ueharaguchi-Tanada, Y.; Ono, S.; Amano, W.; Nakajima, S.; Egawa, G.; et al. Resolvin E1 inhibits dendritic cell migration in the skin and attenuates contact hypersensitivity responses. *J. Exp. Med.* **2015**, *212*, 1921–1930. [CrossRef] [PubMed]
131. Sawada, Y.; Honda, T.; Nakamizo, S.; Otsuka, A.; Ogawa, N.; Kobayashi, Y.; Nakamura, M.; Kabashima, K. Resolvin E1 attenuates murine psoriatic dermatitis. *Sci. Rep.* **2018**, *8*, 11873. [CrossRef]
132. Xu, J.; Duan, X.; Hu, F.; Poorun, D.; Liu, X.; Wang, X.; Zhang, S.; Gan, L.; He, M.; Zhu, K.; et al. Resolvin D1 attenuates imiquimod-induced mice psoriasiform dermatitis through MAPKs and NF-κB pathways. *J. Dermatol. Sci.* **2018**, *89*, 127–135. [CrossRef] [PubMed]
133. Zuraikat, F.M.; Wood, R.A.; Barragán, R.; St-Onge, M.-P. Sleep and Diet: Mounting Evidence of a Cyclical Relationship. *Annu. Rev. Nutr.* **2021**, *41*, 309–332. [CrossRef]
134. Taheri, S.; Lin, L.; Austin, D.; Young, T.; Mignot, E. Short sleep dura¬tion is associated with reduced leptin, elevated ghrelin, and increased body mass index. *PLoS Med* **2004**, *1*, e62. [CrossRef] [PubMed]
135. Lopes, T.V.; Borba, M.E.; Lopes, R.V.; Fisberg, R.M.; Paim, S.L.; Teodoro, V.V.; Zimberg, I.Z.; Araújo, L.B.; Shivappa, N.; Hébert, J.R.; et al. Association between inflammatory potential of the diet and sleep parameters in sleep apnea patients. *Nutrition* **2019**, *66*, 5–10. [CrossRef]
136. Campanini, M.Z.; Guallar-Castillón, P.; Rodríguez-Artalejo, F.; Lopez-Garcia, E. Mediterranean Diet and Changes in Sleep Du-ration and Indicators of Sleep Quality in Older Adults. *Sleep* **2017**, *40*, zsw083. [CrossRef] [PubMed]
137. Xiao, R.S.; Simas, T.A.M.; Pagoto, S.L.; Person, S.D.; Rosal, M.C.; Waring, M.E. Sleep Duration and Diet Quality Among Women Within 5 Years of Childbirth in the United States: A Cross-Sectional Study. *Matern. Child Health J.* **2016**, *20*, 1869–1877. [CrossRef]
138. Grandner, M.A.; Jackson, N.; Gerstner, J.R.; Knutson, K.L. Dietary nutrients associated with short and long sleep duration. Data from a nationally representative sample. *Appetite* **2013**, *64*, 71–80. [CrossRef]
139. Dashti, H.S.; Scheer, F.A.; Jacques, P.F.; Lamon-Fava, S.; Ordovás, J.M. Short sleep duration and dietary intake: Epidemiologic evi-dence, mechanisms, and health implications. *Adv. Nutr.* **2015**, *6*, 648–659. [CrossRef] [PubMed]
140. Grandner, M.A.; Kripke, D.F.; Naidoo, N.; Langer, R.D. Relationships among dietary nutrients and subjective sleep, objective sleep, and napping in women. *Sleep Med.* **2010**, *11*, 180–184. [CrossRef]
141. Spiegel, K.; Tasali, E.; Penev, P.; Van Cauter, E. Brief Communication: Sleep Curtailment in Healthy Young Men Is Associated with Decreased Leptin Levels, Elevated Ghrelin Levels, and Increased Hunger and Appetite. *Ann. Intern. Med.* **2004**, *141*, 846–850. [CrossRef] [PubMed]
142. Tan, X.; Alén, M.; Cheng, S.M.; Mikkola, T.M.; Tenhunen, J.; Lyytikäinen, A.; Wiklund, P.; Cong, F.; Saarinen, A.; Tarkka, I.; et al. Associations of disordered sleep with body fat distribu-tion, physical activity and diet among overweight middle-aged men. *J. Sleep Res.* **2015**, *24*, 414–424. [CrossRef]
143. Stelmach-Mardas, M.; Mardas, M.; Iqbal, K.; Tower, R.; Boeing, H.; Piorunek, T. Quality of life, depression and dietary intake in Obstructive Sleep Apnea patients. *Health Qual. Life Outcomes* **2016**, *14*, 111. [CrossRef] [PubMed]
144. Vasquez, M.M.; Goodwin, J.L.; Drescher, A.A.; Smith, T.W.; Quan, S.F. Associations of Dietary Intake and Physical Activity with Sleep Disordered Breathing in the Apnea Positive Pressure Long-term Efficacy Study (APPLES). *J. Clin. Sleep Med.* **2008**, *4*, 411–418. [CrossRef] [PubMed]
145. Cheng, F.W.; Li, Y.; Winkelman, J.W.; Hu, F.B.; Rimm, E.B.; Gao, X. Probable insomnia is associated with future total energy intake and diet quality in men. *Am. J. Clin. Nutr.* **2016**, *104*, 462–469. [CrossRef]

146. Katagiri, R.; Asakura, K.; Kobayashi, S.; Suga, H.; Sasaki, S. Low intake of vegetables, high intake of confectionary, and unhealthy eating habits are associated with poor sleep quality among middle-aged female Japanese workers. *J. Occup. Health* **2014**, *56*, 359–368. [CrossRef] [PubMed]
147. Grandner, M.A.; Jackson, N.; Gerstner, J.R.; Knutson, K.L. Sleep symptoms associated with intake of specific dietary nutrients. *J. Sleep Res.* **2014**, *23*, 22–34. [CrossRef]
148. Godos, J.; Grosso, G.; Castellano, S.; Galvano, F.; Caraci, F.; Ferri, R. Association between diet and sleep quality: A systematic review. *Sleep Med. Rev.* **2021**, *57*, 101430. [CrossRef]
149. Bragazzi, N.L.; Sellami, M.; Salem, I.; Conic, R.Z.; Kimak, M.; Pigatto, P.D.M.; Damiani, G. Fasting and Its Impact on Skin Anatomy, Physiology, and Physiopathology: A Comprehensive Review of the Literature. *Nutrients* **2019**, *11*, 249. [CrossRef]
150. Bragazzi, N.L.; Trabelsi, K.; Garbarino, S.; Ammar, A.; Chtourou, H.; Pacifico, A.; Malagoli, P.; Kocic, H.; Conic, R.R.Z.; Young Dermatologists Italian Network; et al. Can intermittent, time-restricted circadian fasting modulate cutaneous severity of dermatological disorders? Insights from a multicenter, observational, prospective study. *Dermatol Ther.* **2021**, *34*, e14912. [CrossRef] [PubMed]
151. Damiani, G.; Mahroum, N.; Pigatto, P.D.M.; Pacifico, A.; Malagoli, P.; Tiodorovic, D.; Conic, R.R.; Amital, H.; Bragazzi, N.L.; Watad, A.; et al. The Safety and Impact of a Model of Intermittent, Time-Restricted Circadian Fasting ("Ramadan Fasting") on Hidradenitis Suppurativa: Insights from a Multicenter, Observational, Cross-Over, Pilot, Exploratory Study. *Nutrients* **2019**, *11*, 1781. [CrossRef]
152. Adawi, M.; Damiani, G.; Bragazzi, N.L.; Bridgewood, C.; Pacifico, A.; Conic, R.R.Z.; Morrone, A.; Malagoli, P.; Pigatto, P.D.M.; Amital, H.; et al. The Impact of Intermittent Fasting (Ramadan Fasting) on Psoriatic Arthritis Disease Activity, Enthesitis, and Dactylitis: A Multicentre Study. *Nutrients* **2019**, *11*, 601. [CrossRef]
153. Damiani, G.; Watad, A.; Bridgewood, C.; Pigatto, P.D.M.; Pacifico, A.; Malagoli, P.; Bragazzi, N.L.; Adawi, M. The Impact of Ramadan Fasting on the Reduction of PASI Score, in Moderate-To-Severe Psoriatic Patients: A Real-Life Multicenter Study. *Nutrients* **2019**, *11*, 277. [CrossRef]
154. Scoditti, E.; Tumolo, M.R.; Garbarino, S. Mediterranean Diet on Sleep: A Health Alliance. *Nutrients* **2022**, *14*, 2998. [CrossRef]
155. Castro-Diehl, C.; Wood, A.C.; Redline, S.; Reid, M.; Johnson, D.A.; Maras, J.E.; Jacobs, D.R., Jr.; Shea, S.; Crawford, A.; St-Onge, M.P. Medi-terranean diet pattern and sleep duration and insomnia symptoms in the Multi-Ethnic Study of Atherosclerosis. *Sleep* **2018**, *41*, zsy158. [CrossRef]
156. St-Onge, M.-P.; Roberts, A.; Shechter, A.; Choudhury, A.R. Fiber and Saturated Fat Are Associated with Sleep Arousals and Slow Wave Sleep. *J. Clin. Sleep Med.* **2016**, *12*, 19–24. [CrossRef]
157. Mercuri, S.R.; Gregorio, G.; Brianti, P. Quality of life of psoriasis patients measured by the PSOdisk: A new visual method for assessing the impact of the disease. *G. Ital. Dermatol. Venereol.* **2017**, *152*, 424–431. [CrossRef]
158. Naldi, L.; Mercuri, S.R. Epidemiology of comorbidities in psoriasis. *Dermatol Ther.* **2010**, *23*, 114–118. [CrossRef]
159. Paolino, G.; Di Nicola, M.R.; Currado, M.; Brianti, P.; Mercuri, S.R. Serum tryptase levels in patients with psoriasis: A pilot study for a possible useful biomarker. *Clin. Exp. Dermatol.* **2022**, *47*, 178–179. [CrossRef]
160. Paolino, G.; Di Nicola, M.R.; Brianti, P.; Bianchi, V.G.; Mercuri, S.R. New onset atopic dermatitis and psoriasis in the same patients under biologic treatments: The role of systemic treatments as a possible trigger. *Dermatol. Ther.* **2022**, e15814. [CrossRef]
161. Damiani, G.; Allocco, F.; Young Dermatologists Italian Network; Malagoli, P. COVID-19 vaccination and patients with psoriasis under biologics: Real-life evidence on safety and effectiveness from Italian vaccinated healthcare workers. *Clin. Exp. Dermatol.* **2021**, *46*, 1106–1108. [CrossRef]
162. Zhou, Q.; Vadakekolathu, J.; Watad, A.; Sharif, K.; Russell, T.; Rowe, H.; Khan, A.; Millner, P.A.; Loughenbury, P.; Rao, A.; et al. SARS-CoV-2 Infection Induces Psoriatic Arthritis Flares and Enthesis Resident Plasmacytoid Dendritic Cell Type-1 Interferon Inhibition by JAK Antagonism Offer Novel Spondyloarthritis Pathogenesis Insights. *Front Immunol.* **2021**, *12*, 635018. [CrossRef] [PubMed]
163. Gisondi, P.; Cazzaniga, S.; Chimenti, S.; Maccarone, M.; Picardo, M.; Girolomoni, G.; Naldi, L.; Psocare Study Group. Latent tuberculosis infection in patients with chronic plaque psoriasis: Evidence from the Italian Psocare Registry. *Br. J. Dermatol.* **2015**, *172*, 1613–1620. [CrossRef] [PubMed]
164. Cazzaniga, S.; Lo Scocco, G.; Schincaglia, E.; Mercuri, S.R.; Chimenti, S.; Saraceno, R.; Naldi, L. Randomized, within-patient, clinical trial comparing fluorine-synthetic fiber socks with standard cotton socks in improving plantar pustulosis. *Dermatology* **2014**, *228*, 166–171. [CrossRef] [PubMed]
165. Cazzaniga, S.; Lo Scocco, G.; Schincaglia, E.; Mercuri, S.R.; Chimenti, S.; Saraceno, R.; Naldi, L. Double-blind, within-patient, randomized, clinical trial comparing fluorine-synthetic fiber socks with standard cotton socks in improving plantar psoriasis. *J. Dermatol. Treat.* **2014**, *25*, 26–29. [CrossRef]
166. Naldi, L.; Mercuri, S.R. Smoking and psoriasis: From epidemiology to pathomechanisms. *J. Investig. Dermatol.* **2009**, *129*, 2741–2743. [CrossRef]
167. Bragazzi, N.L.; Riccò, M.; Pacifico, A.; Malagoli, P.; Kridin, K.; Pigatto, P.; Damiani, G. COVID-19 knowledge prevents biologics discontinuation: Data from an Italian multicenter survey during RED-ZONE declaration. *Dermatol. Ther.* **2020**, *33*, e13508. [CrossRef]

MDPI
St. Alban-Anlage 66
4052 Basel
Switzerland
Tel. +41 61 683 77 34
Fax +41 61 302 89 18
www.mdpi.com

Nutrients Editorial Office
E-mail: nutrients@mdpi.com
www.mdpi.com/journal/nutrients

www.ingramcontent.com/pod-product-compliance
Lightning Source LLC
Chambersburg PA
CBHW061446210926
51850CB00008B/1624
* 9 7 8 3 0 3 6 5 6 7 0 8 2 *